Biotechnology

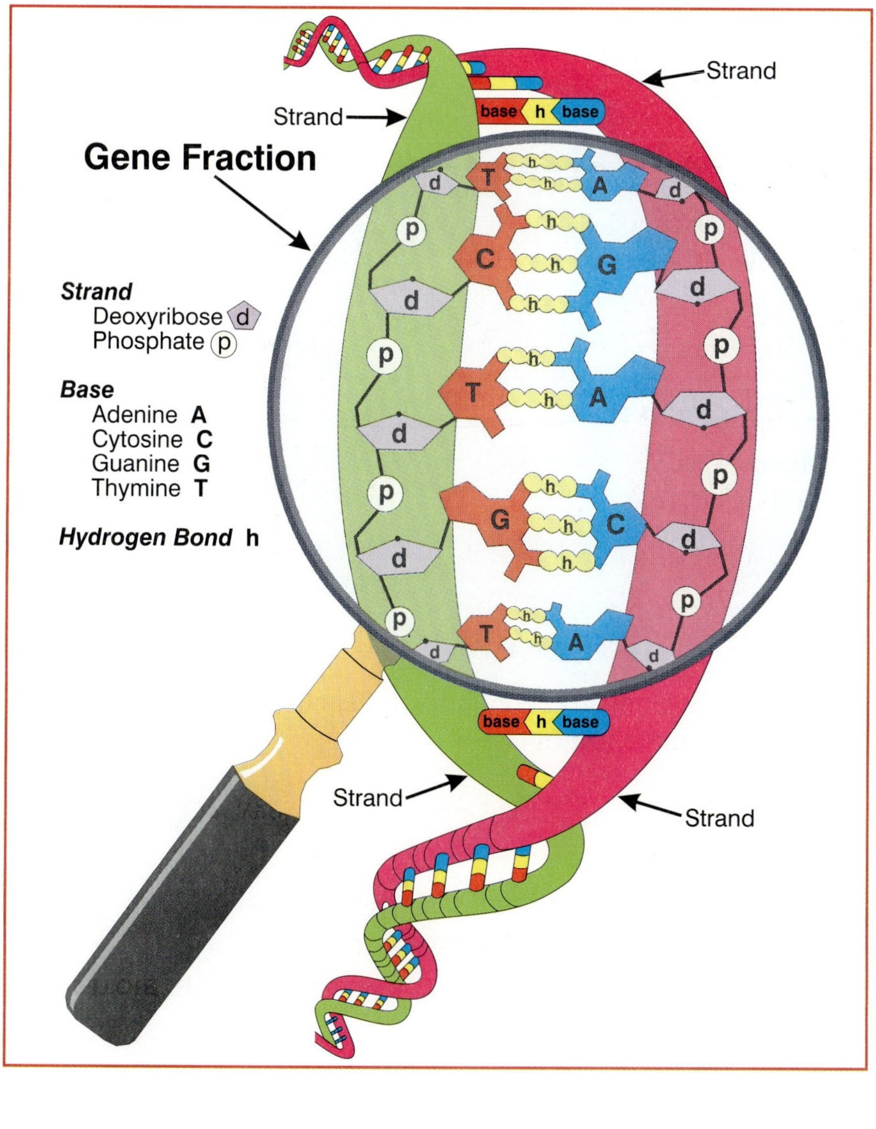

Stephen J. Lee

Chief for Organic Chemistry
U.S. Army Research Office
Research Triangle Park, North Carolina

Christy Mecey-Smith

Biotechnology Educator
Michigan State University

Elizabeth M. Morgan

AgriScience Teacher
Presque Isle, Maine

Ray E. Chelewski

AgriScience Teacher
Presque Isle, Maine

Randi Hunewill

Science and Agriculture Teacher
Smith, Nevada

Jasper S. Lee

Adjunct Professor
North Carolina State University

Biotechnology

IST

AgriScience and Technology Series
Jasper S. Lee, Series Editor

IPP

Interstate Publishers, Inc.
Danville, Illinois

BIOTECHNOLOGY

Library of Congress Catalog Card No. 99-76571

ISBN 0-8134-3161-1

3 4 5 6 7 8 9 10 04

Order from

INTERSTATE PUBLISHERS, INC.

P.O. Box 50
Danville, IL 61834-0050

Phone: (800) 843-4774
Fax: (217) 446-9706
Email: info-ipp@IPPINC.com

World Wide Web: http://www.IPPINC.com

PREFACE

Biotechnology! We hear about it quite often. Is it really that new? Some of it has been around for hundreds of years. We have grown accustomed to the "old" biotechnology and use it every day in our lives. The "new" biotechnology is embracing genetic methods that are on the cutting edge of science. These new methods are creating tremendous excitement.

Today's students and teachers need an educational tool to help them in the teaching and learning process. That is what this book—*Biotechnology*—is all about. Here is a little background to help you better understand the book.

Biotechnology is for the first course in high school biotechnology. It would be used in grades 10, 11, or 12 and, in some cases, grade 9. Students who have completed a biology or science-based agriculture course would find it most beneficial. The reading level and supporting illustrations focus on the high school grades. Learning will be enhanced with hands-on activities. These are likely to be demonstrations or research-based experiments. It is suggested that the *Activity Manual for Biotechnology* be used in conjunction with this textbook.

As with other textbooks in Interstate Publishers', Inc., IST Series, a student- and teacher-friendly format is used. The layout is based on national studies with students and teachers. Norm-reference test data also support the use of books following this general approach. They allow both students and teachers to be actively involved in an efficient learning process. The major features include:

- Full-color design
- Hundreds of photographs and drawings

- Objectives-based sequence and content
- High student appeal through technical models
- Carefully written and reviewed content
- Advocates responsible approaches in sustaining resources

Biotechnology is ideal for students and teachers as they begin the study of biotechnology. It uses an approach that will reinforce knowledge and motivate students to achieve. Look through the book. Begin with the Table of Contents. Move into the chapters. What you see will get your attention. It will surely promote excitement about learning among students.

Jasper S. Lee
Series Editor

ACKNOWLEDGMENTS

Many people throughout the United States helped the authors in many ways in preparing *Biotechnology*. Their assistance is gratefully acknowledged. Some are named below. Others are named in courtesy lines with the photographs they provided.

The individuals who reviewed the outline and manuscript were most helpful in assuring an accurate and appropriate book. They were:

- Jerry Wendt, Biotechnology and Agriscience Teacher, Stoughton, Wisconsin
- Annette VanVeghel, Bioscience Teacher, Stoughton, Wisconsin
- Chris Dickson, Agricultural Educator, Bakersfield, California
- Leslie Hudson, Agricultural Educator, Bastrop, Texas
- Diana Turner, Science Teacher, Starkville, Mississippi
- Linda Maston McMurry, Science Educator, Elgin, Texas

Other individuals who assisted in many ways are as follows:

- David Frazier, Agriscience Teacher, Snyder, Texas
- Anne Stewart Clark, Agriculture Teacher, Flemingsburg, Kentucky
- Lynn Wagner, Career Center Administrator, Newton, Mississippi
- Steven W. Walenta, College of Veterinary Medicine, Texas A&M University
- Jerry Gibson, Western Region Agricultural Education Coordinator, North Carolina

- Jeff Velton, USDA-ARS, Plant Stress and Water Conservation Lab, Lubbock, Texas
- Don Cawthon, Professor, Texas A&M University-Commerce
- Wilbur H. Palmer, Director of Vocational Education, Hudson, New Hampshire
- Ron Biondo, Field Advisor, FCAE, Countryside, Illinois
- Dan Upchurch, Plant Stress and Water Conservation Lab, Lubbock, Texas
- Carol Webb, Mineral County Schools, Keyser, West Virginia
- Jim Steeby, Aquaculture Specialist, Mississippi Cooperative Extension Service
- Keturah Harley, Information Office, Agricultural Research Service-USDA, Beltsville, Maryland
- Steve Lazelle, Aquaculture Educator, South Bend, Washington
- Tracy Westrom, Student, Piedmont College, Georgia
- Susan Combs, Commissioner, Texas Department of Agriculture
- Marcia Paterson, Administrator, Milton Hershey School, Pennsylvania
- Luther Talbert, Plant Scientist, Montana State University
- Clay Walker, Forestry Suppliers, Inc., Mississippi
- Lyle Westrom, University of Minnesota-Crookston, Minnesota
- John Shea, Horticultural Range Manager, Van Wingerden International, Inc., North Carolina
- Joe Keaschall, Maize Research Director, Pioneer Hi-Bred International, Inc., Iowa
- Ron Colbert, Danville Community Blood Bank, Illinois

The students and teachers at the following schools are thanked for their roles in performing experiments and serving as technical models for this book:

- Fleming County High School, Flemingsburg, Kentucky
- Park High School, Livingston, Montana
- Smith Valley High School, Smith, Nevada
- Carl Hayden Center for Agribusiness, Laveen, Arizona
- Alvirne High School, Hudson, New Hampshire
- South Bend High School, South Bend, Washington
- Newton High School, Newton, Mississippi

- Snyder High School, Snyder, Texas
- Paxton-Buckley-Loda High School, Paxton, Illinois
- Chino Valley High School, Chino Valley, Arizona
- Franklin County High School, Carnesville, Georgia

Other associations and agribusinesses acknowledged include:

- Elite Genetics, Waukon, Iowa
- American Soybean Association, St. Louis, Missouri
- Harris Farms, Coalinga, California
- Agriculture Division, Auburn University, Alabama
- Chemistry Department, Emory University, Atlanta, Georgia
- Chemistry Department, Georgia Tech, Atlanta, Georgia
- Texas Department of Agriculture, Austin, Texas
- Agricultural Research Service, USDA, Washington
- Fish and Wildlife Service, Washington
- The North Carolina Arboretum, Asheville, North Carolina

Each author expresses appreciation to those who assisted them in their work.

The staff of Interstate Publishers, Inc., is thanked for dedicated and enthusiastic efforts in producing this book. President Vernie Thomas is acknowledged for his untiring efforts in behalf of quality educational materials. Thank you to Kim Romine for the attractive design; Jane Weller and Ronald McDaniel for editorial assistance; Mary Carter for computer-generated graphics; and Britt Robinson and Dan Pentony for their efforts in planning marketing initiatives for the book.

CONTENTS

Preface v

Acknowledgements vi

Unit One— Biotechnology: Meaning and History

1 Introduction to Biotechnology 1

Biotechnology 3

Major Areas 4

Examples of Biotechnology 6

Meeting Human Needs 10

Synthetic Biology 13

2 History of Biotechnology 7

Stages of Biotechnology Development 19

People Who Made a Difference 27

Research and Development 31

3 Benefits of Biotechnology 35

Biotechnology and Quality of Life 37

World Food and Fiber Production 40

Medical Biotechnology 46

Materials and Environment 50

4 **Safety With Biotechnology** **55**

Safety and Hazards *57*

Potential Hazards and Being Safe *59*

Organism Safety *70*

Unit Two—Biotechnology: Science in Action

5 **Research: The Scientific Method in Action** **75**

Research *77*

The Scientific Method *79*

Identifying a Problem *80*

Investigate the Problem *81*

Form a Hypothesis *84*

Design Experimentation *86*

Conduct Experimentation and Gather Data *89*

Analyze the Data and Form Conclusions *90*

Report the Results *93*

6 **Preparing Reports** **97**

Reports: Sharing What You've Learned *99*

Research Log: An Important Tool *102*

Assembling the Report *103*

Presentations *114*

Visual Reports: Displays *117*

Avoiding the Pitfalls *120*

7 **Science Fundamentals** **125**

The Branches of Science *127*

Matter and Energy *130*

Classification of Living Organisms *133*

Life Processes *136*

Structures of Living Organisms *145*

DNA, Proteins, Enzymes, and Lipids *148*

8 Aseptic Technique 155

Aseptic Environments: "People vs. Germs" 157
Microbes 161
Aseptic Environments 165
Examples of Aseptic Conditions 171

Unit Three—Biotechnology: Organisms and Genetics

9 Basic Cell Structure 175

The Cell: Building Block of Life 177
Common Cell Structures 178
Cellular Processes 183
How Do We Know? 185
Cellular Reporduction 189

10 Genetics. 197

Mendel Unravels Mysteries 199
Predicting Genetic Outcomes 204
Molecular Genetics Unravels More Mysteries 208
Tools of DNA Science 211

11 Genetic Engineering and Transgenic Organisms. 221

Making Nature Better 223
Practical Applications of Genetic Engineering 224
Processes in Creating Transgenic Organisms 227
Genetic Engineering of Plants 232
Genetic Engineering of Animals 236

12 Microbial Biotechnology 239

Microorganisms and Biotechnology 241
Usefulness in Biotech 243
Types of Microbes 244
Microbe Identification and Morphology 253
Microbes in Biotechnology 258

13 **Plant Biotechnology** 263

Creating New Plants With Traditional Breeding 265
Creating New Plants With Genetic Engineering 268
Plant Tissue Culture 273
Why Plants Are Genetically Engineered 277

14 **Animal Biotechnology** 283

Roles of Animals in Biotechnology 285
Animal Reproduction 288
Animal Reproductive Technologies 292
Increasing Production Efficiency: Growth Hormones 299
Animal Genetic Engineering 300

15 **Human Biotechnology** 307

The Human Organism 309
Human Genomics 314
Methods in the HSGP 317
Helping Humans Live Better 320
Diagnostics 323

16 **Food Biotechnology** 327

Food Biotechnology 329
Food Science 329
Food Supply 333
Food Microbiology 337
Catalysis 341

17 **Environmental Biotechnology** 347

Environmental Quality 349
Bioremediation 353
Solid Wastes 356
Hazardous Wastes 359
Wastewater 360
Composting 364

Unit Four—Biotechnology: Future Impacts

18 **Using Biotech: Social and Environmental Impacts 369**

Biotechnology As Innovation 371
Considering Technology 373
Social Impact 377
The Environment 379

19 **Issues and Ethics. 387**

Issues and Biotechnology 389
Ethics and Biotechnology 391
Specific Ethical Issues 393

20 **Regulations in Biotechnology 399**

Nature and Purposes of Regulations 401
Agencies and Responsibilities 402
Organizations 407
Ethics and Regulations 408

Appendixes 413

Glossary 415

Bibliography 427

Index 431

1

INTRODUCTION TO BIOTECHNOLOGY

Biotechnology helps people! It helps us meet our basic needs. How well these needs are met determines our quality of life.

Basic needs include food, clothing, shelter, health, and safety. Each of these is improved by using science. Science helps in producing plants, animals, and other organisms. These are also used in having a good environment that promotes our well-being.

Many of the things that people enjoy today are a result of biotechnology. Excitement is high about recent and potential benefits. Some areas of biotechnology are new; others have been around for a long time. The future contributions of biotechnology are endless—limited only by the capacity of scientists to investigate and develop safe, new methods.

First, what is biotechnology? How is it making contributions to human well-being? These and other questions are covered in this chapter.

1-1. Students often apply biotechnology. (These Georgia students are applying biotechnology in a horticulture class.)

OBJECTIVES

This chapter explains the meaning of biotechnology, introduces important science areas, and relates biotechnology to quality of life. It has the following objectives:

1. Explain the meaning of biotechnology.
2. List and distinguish between the two major areas of biotechnology.
3. Identify examples of biotechnology.
4. Explain the importance of biotechnology in food and fiber production.
5. Relate synthetic biology to biotechnology.

TERMS

animal biotechnology
biopharmaceutical
biopharming
biotechnology
cloning
environmental biotechnology
food biotechnology
genetic engineering
human medicine
medicine
molecular biotechnology
multidisciplinary
organismic biotechnology
plant biotechnology
synthetic biology
therapeutant
transgenic
vesicle
veterinary medicine

BIOTECHNOLOGY

Biotechnology is using scientific processes to get new organisms or new products from organisms. It may involve changing an organism in some way to get a desired trait. At other times, biotechnology is used to promote the production of desired products by organisms.

Biotechnology is a large area. It includes many approaches and methods in science and technology. It is nearly impossible to define biotechnology in one sentence. Scientists often have different views on the meaning of biotechnology.

Government agencies and groups of scientists have tried to define biotechnology. The Office of Technology Assessment, which is a part of the U.S. Congress, defines biotechnology as "any technique that uses living organisms or substances from those organisms to make or modify a product, to improve plants or animals, or to develop microorganisms for specific uses."

1-2. A scientist is reading the DNA sequence of a soybean plant. (Courtesy, Agricultural Research Service, USDA)

THE AGRISCIENCE VIEW

Agriscientists have a broad view of biotechnology. In the agricultural industry, biotechnology is all of the applied science-based operations in producing food, fiber, shelter, and related products. New plants, animals, and other organisms and their products are included.

Many of the practices in growing crops and livestock and their products, such as milk, use biotechnology. New ornamental plants in horticulture and practices in landscaping are based on biotechnology. Biotechnology is also useful with wildlife, aquaculture, natural resources, and the environment.

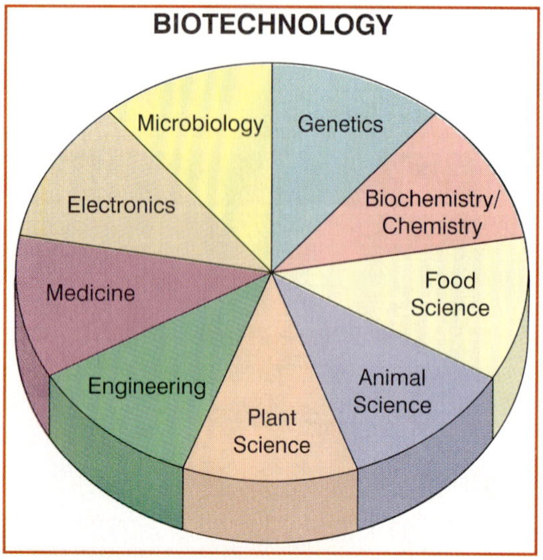

1-3. **Many areas of study are important in biotechnology.**

MANY DISCIPLINES

Biotechnology is said to be ***multidisciplinary***. This means that it involves many disciplines, or branches of learning. All areas of the life sciences, such as microbiology and genetics, are included. Computer science and related computer applications are used in biotechnology. Engineering and the physical sciences are often involved.

Doing biotechnology work means that scientists from different disciplines must work together. They plan, do, and assess their work as a team. Good communication is essential.

MAJOR AREAS

Biotechnology is carried out in two major levels: organismic and molecular. Each has made important contributions to human well-being.

ORGANISMIC BIOTECHNOLOGY

Organismic biotechnology is working with complete, intact organisms or their cells. The organisms are not genetically changed with artificial means. Practices are used to help the organisms live better or be more productive.

A major goal in organismic biotechnology is to improve organisms and the conditions in which they grow. Natural genetic variations are studied and used. Traditional breeding programs in agriculture use practices that are in organismic biotechnology. Many of the crops and animals produced today are from carefully designed breeding efforts.

One example of organismic biotechnology is cloning. ***Cloning*** is the process of producing new organisms from cells or tissues of existing organisms. Cloning does not alter genetic makeup. It uses existing genetic characteristics to achieve desired traits in new organisms. Wide publicity was made in

1-4. **Traditional approaches in animal breeding have been used to make improvements in beef cattle.**

1997 of the cloned sheep, Dolly. Some people became alarmed. Other people realized that forms of cloning had been used for years to improve plants, such as grafting peach trees.

MOLECULAR BIOTECHNOLOGY

Molecular biotechnology is changing the genetic make-up of organisms. This is achieved by altering the structure and parts of cells, particularly the genetic material. Complex processes are involved.

1-5. **These scientists are studying genetically altered guayule plantlets. (Guayule *[Parthenium argentatum]* is a rubber-producing plant that is native to southwestern Texas. Rubber from this plant does not cause allergies. Some people are allergic to the rubber from rubber trees.) (Courtesy, Agricultural Research Service, USDA)**

Genetic engineering, molecular mapping, and similar procedures are used in molecular biotechnology. ***Genetic engineering*** is changing the genetic information in a cell. A specific trait in the cell of one organism may be isolated, cut, and moved into the cell of another organism. The resulting new organism is said to have been "genetically engineered."

Organisms produced by genetic engineering are known as transgenic organisms. ***Transgenic*** means that the genetic material in an organism has been altered. The alteration is the result of methods other than those of breeding. The changes that occur are said to have been artificially made.

EXAMPLES OF BIOTECHNOLOGY

Many examples of biotechnology can be included. Some have been important for a long time. Others are new and emerging. A good way to look at examples of biotechnology is in broad groups—medicine, the environment, agriculture, forestry, and the processing of food and beverages.

MEDICINE

Medicine includes the art and science of promoting health among living animal organisms. ***Human medicine*** is the area that deals with human health. Human medicine includes preventing disease and healing individuals who have contracted a disease. Some of the new developments delve into the hereditary material of humans, known as gene therapy.

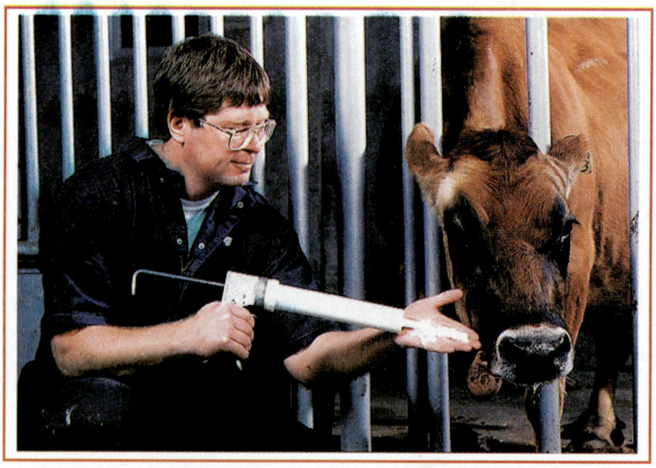

1-6. To prevent milk fever, a veterinarian is preparing to orally administer a calcium gel to a dairy cow. (The medicine alters biological processes that cause an energy deficiency immediately after giving birth.) (Courtesy, Agricultural Research Service, USDA)

Another area of medicine is veterinary medicine. ***Veterinary medicine*** is a branch of medicine that deals with the health of animals other than humans. It is applied to farm animals, such as cattle and horses, as well as companion animals, such as dogs and cats. The health of aquatic animals, such as trout and catfish, is included in veterinary medicine. Producers of animal products are well aware of the importance of veterinary medicine. New developments in veterinary medicine are often useful in human medicine.

Products to treat disease are developed through biotechnology. A ***therapeutant*** is a product used to maintain health or treat disease. Many products are used in both human medicine and veterinary medicine. New products are being developed. Some new kinds of therapeutants are known as biopharmaceuticals. A ***biopharmaceutical*** is a drug or vaccine that is developed through biotechnology. These products may be called "designer drugs."

Biopharming is the production of pharmaceuticals in cultured organisms. This includes microorganisms as well as some common farm animals. Biopharming is said to be a combination of agriculture and the pharmaceutical industry. For example, certain blood-derived products needed in human medicine can be produced in the milk of goats!

THE ENVIRONMENT

The environment is the world around us–air, water, soil, and other conditions that influence our surroundings. How people relate to the environment has a lot of influence on its quality. For example, releasing a harmful substance into the water of a stream degrades the water. It may make the water unsafe and lower the quality of the environment.

1-7. Composting makes use of natural microbe decomposition processes to convert wastes into useful products, such as mulch.

Career Profile

ENVIRONMENTAL BIOTECHNOLOGIST

An environmental biotechnologist studies and plans ways of using biotechnology to insure a good environment. The work may involve ways of cleaning up after accidents or disposing of waste materials. The work may be done in a laboratory or outside studying problems. The ability to make accurate observations is essential.

Environmental biotechnologists need a college degree in environmental science, biology, or a closely related area. Many have master's or doctoral degrees. Practical experience is useful in getting a job.

Most jobs are with government agencies or large companies involved with environmental biotechnology. This photograph shows water measurements being made on an Oregon research plot that uses organic decomposition processes with crop residues. (Courtesy, Agricultural Research Service, USDA)

Biotechnology has many uses in assuring a good environment. An area that relates biotechnology and the environment is known as environmental biotechnology. ***Environmental biotechnology*** is any biotechnological process that may promote a good environment. The processes may be used to prevent pollution or to clean up pollution should it get into the environment.

Many environmental problems are naturally solved by microorganisms. Bacteria, fungi, and other microorganisms may break down harmful forms of materials into forms that are less harmful or are not harmful at all.

AGRICULTURE AND FORESTRY

Many advances in biotechnology have been in agriculture and forestry. This includes horticulture as well as wildlife and renewable natural resources. Two large areas within agriculture and forestry are plant biotechnology and animal biotechnology.

Plant biotechnology is the use of biotechnology to improve plants and the products that are produced from plants. Many approaches are being used. Some plants are being "engineered" to resist insects and disease. Other plants are being "engineered" to produce products with desired characteristics. In still other cases, new methods are being used to reproduce plants to assure desired traits in the new plants. The "new" plants focus on helping to meet the demand for food and clothing needed by an increasing human population.

Animal biotechnology is the use of biotechnology with animals. The goal is to improve animals or the products that they produce. Birds, cattle, pigs, and insects are examples of animals where biotechnology has been used. In some cases, the animals are being used to produce products that promote human health. Much of the work has been to improve the productivity of animals and promote their well-being. Animals that live on the land as well as in water have been studied with animal biotechnology.

FOOD AND BEVERAGES

Biotechnology is important in food and beverage products. Of course, some of these applications begin on the farm with plants and animals. Other uses occur in how food is processed and made available in desired forms. Throughout, emphasis must be on providing a wholesome and nutritious food product at an affordable cost.

Food biotechnology is the use of technology in producing and processing food. The nature of food biotechnology varies from emerging genetic engineering to practices that have been used for hundreds of years.

1-8. A vat of fermenting milk being made into cheese.

Biotech Connection

ASSESS PERSONAL INTERESTS EARLY

Students often wonder if they would like a biotechnology career. You can begin assessing your interests now. Here are a few things to do.

- Conduct science or agriscience projects related to biotechnology.
- Job shadow a biotechnologist (accompany the person for a day or so on the job).
- Have supervised experience with a biotechnologist.
- Take a biotechnology, environmental science, or natural resources class.
- Visit a college that offers biotechnology-related areas and investigate the nature of the studies.
- Use biotechnology sites on the Internet to gather information.
- Talk with a career counselor or interview a person in biotechnology.

Once you have collected information, make a decision based on your personal preferences. Avoid emotion and make choices based on what you know. (This photo shows a counselor discussing careers with a student.)

People have different opinions about the use of biotechnology with food. Some uses are felt to be acceptable, such as the use of yeast in baking bread. Other uses may be questioned, such as the use of a genetically altered crop to produce food.

How do you feel about biotechnology in the food that you eat? Some people have strong feelings about it. Most people are more concerned with nutritious food that promotes their well-being. (The advantages and disadvantages of biotechnology in food production are covered later in the book.)

MEETING HUMAN NEEDS

Biotechnology helps meet human needs. Three of these needs are food, clothing, and shelter. The products of plants and animals are used in manufacturing food, clothing, and materials for homes. Biotechnology is also used to make products that are more desirable, such as the conversion of milk into cheese.

EFFICIENCY

Most people have limited money with which to buy things. This means that the costs of making the products must be kept as low as possible.

Biotechnology results in greater efficiency in producing for the needs of people. One simple example is the inoculation of seed of some crops with bacteria so that fertilizer is naturally formed in the soil. Producers will not need to invest money in buying and applying artificial fertilizer.

Biotechnology has resulted in trees that grow faster and produce wood materials that are more desirable. This helps meet the demands of consumers for housing, paper, furniture, and other wood-based products.

GREATER PRODUCTION

Biotechnology makes it possible for crops and animals to have higher yields. This means that fewer plants or animals make larger amounts of desired products.

In recent years, animals have been given hormones to help them be more productive. An example is the use of a bST with dairy cattle to increase milk production. More milk is available from fewer cows!

Biotechnology also improves the productivity of crops, such as corn and wheat. Genetically improved varieties can be planted. These may resist weed or insect damage.

1-9. Genetically altered crops may be resistant to certain weed-killing chemicals known as herbicides.

HEALTH-PROMOTING FOODS

Foods with unique traits may be useful. Some may contain therapeutants to prevent or treat disease. A therapeutant is a substance that promotes health or cures disease. Other foods may be designed with nutrient enrichment, which will satisfy the special nutritional needs of people.

Humans have varying nutrient needs in different stages of life. An example is a child. If wheat with a higher calcium content could be produced, the needs of the child for calcium might be met by bread.

SAFETY

People want to be sure that the food they have is safe. They want to enjoy food without worrying about substances that could damage health. New methods in food production and processing help assure safety.

People also want the food to provide needed nutrients. In some cases, they want food products with enhanced nutrients, such as needed vitamins or minerals. At the same time, people do not want the enhanced foods to have side effects that result in health problems.

EASE OF PREPARATION

Many people have busy lives. They have limited time for preparing food. They want foods that can be prepared quickly yet be nutritious. Biotechnology can help assure tasty, tender food products that meet consumer demand.

1-10. A supermarket display of genetically altered tomatoes. (Courtesy, James Leising, Oklahoma State University)

One example is the MacGregor's Flavr-Savr Tomato. Reaching the market in the early 1990s, this tomato has a longer shelf life. This means that it can be stored longer than other tomatoes. Keeping tomatoes from spoiling is good for the tomato producer as well as assures a tomato that can be prepared easily. Consumers do not want tomatoes that have rotten spots or are soft.

SYNTHETIC BIOLOGY

Synthetic biology is closely related to biotechnology. Some people feel that it should be included. Others feel that it is a separate field. Regardless, synthetic biology is still in the infant stage.

WHAT IS IT?

Synthetic biology is creating lifelike characteristics through the use of chemicals. Some people may refer to this as creating living organisms in a lab. With synthetic biology, the traits of living organisms can be created. This includes events in the life span of an organism, such as birth, growth, and death. Processes needed for an organism to live can also be created with synthetic biology.

Synthetic biology has been based on creating structures similar to those found in living organisms. Careful study of naturally occurring living organisms was needed. This study focused on the structure of organisms. The first step was to study the structure of cells. Cells are the building blocks of organisms. Scientists felt that to create synthetic life required the development of synthetic cells. This lead to the laboratory creation of vesicles. A ***vesicle*** is a tiny, rounded structure with cell-like traits. A membrane was needed to surround the synthetic cell. Tiny structures similar to soap bubbles were created. These were often quite small and visible only with a powerful microscope. Once the cell membrane has been successfully developed, scientists will then initiate development of the materials within the cell.

WHY IS IT IMPORTANT?

Synthetic biology takes scientists one step closer to creating life in a laboratory from chemicals. The work is not easy and requires much time and resources. It has a long way to go to be successful.

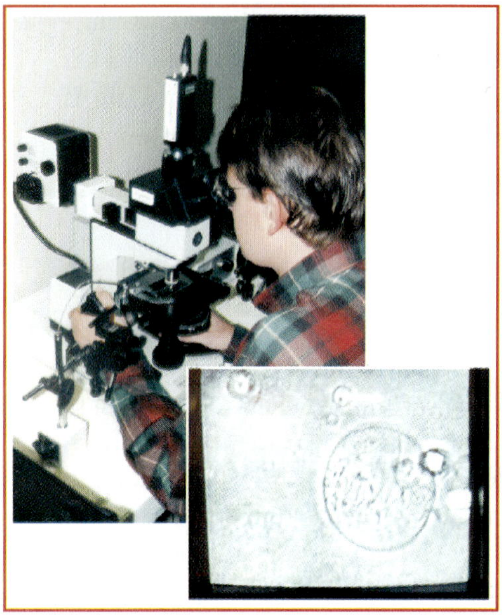

1-11. Powerful microscopes and computers are used to view vesicles. (This shows a microscope being used to view vesicles and a greatly enlarged vesicle.) (Courtesy, Emory University, Atlanta)

The research has focused on the benefits of creating structures for human medicine and not on the creation of synthetic living organisms. Through the methods being used, cells and tissues may be developed to treat human injury and prevent loss of life to disease.

Some people are concerned that such research should not be done. They feel that some "monster" organism might be created that would get loose in the world. Informed people know this not going to happen. What do you think? Isn't a living organism a collection of chemical processes? The processes carried out in cells, tissues, and organs involve chemical elements. A few top researchers, such as those at Emory University, are continuing to investigate synthetic biology and its useful applications in human well-being.

REVIEWING

MAIN IDEAS

Biotechnology helps people in many ways. It makes it easier to supply the basic needs—food, clothing, shelter, health, and safety—of a growing population.

Biotechnology is using scientific processes to obtain new and/or improved products. Some of the processes are simple and may occur naturally in nature. Other processes require the skill and knowledge of a highly trained scientist. Many different science disciplines, as well as computer and communication technologies, are involved.

Biotechnology is carried out at two major levels: organismic and molecular. Organismic has been used for hundreds of years to produce products that are much a part of human life. Molecular biotechnology is newer and results in organisms with altered genetic material.

Some areas of biotechnology are new; others are emerging as routine in use. Important areas related to the needs of people are medicine, the environment, agriculture, forestry, and the production of food and beverages. Biotechnology contributes to the efficiency and enhancement of many products.

Synthetic biology is an area closely related to biotechnology. It involves creating lifelike characteristics in a laboratory setting. Chemical substances are used to create the lifelike characteristics.

QUESTIONS

Answer the following questions using correct spelling and complete sentences.

1. What is biotechnology? Why is biotechnology difficult to define?
2. How do agriscientists view biotechnology?
3. What is meant by biotechnology being "multidisciplinary?"
4. What are the two major areas of biotechnology? Distinguish between the two.
5. How is biotechnology important in medicine?
6. What is the role of biotechnology in the environment?
7. What are the two important areas in agriculture and forestry biotechnology? Briefly explain each.
8. What is food biotechnology?
9. What are five areas in which biotechnology contributes to meeting human needs?
10. What is synthetic biology? Why is it important?

EVALUATING

Match the term with the correct definition. Write the letter of the term in the blank that is provided.

a. synthetic biology
b. food biotechnology
c. environmental biotechnology
d. therapeutant
e. biopharming
f. organismic biotechnology
g. molecular biotechnology
h. biotechnology
i. multidisciplinary
j. genetic engineering

_____ 1. An area of biotechnology that deals with changing the genetic makeup of an organism.

_____ 2. Changing the genetic information in the cell of an organism.

_____ 3. The use of biotechnology in producing and processing food.

_____ 4. Creating lifelike characteristics from chemicals.

_____ 5. A term that describes the many areas of life science and technology involved in biotechnology.

_____ 6. The use of scientific processes to get new organisms or products from organisms.

_____ 7. Producing pharmaceuticals in situations somewhat like agriculture.

_____ 8. A product that is used to maintain health and treat disease.

_____ 9. Biotechnology that involves working with complete, intact organisms or their cells.

_____ 10. Processes that involve using biotechnology to promote a good environment.

EXPLORING

1. Prepare a report that describes the use of oil-eating bacteria to clean up after oil spills. Identify an example of a spill and describe how the bacteria were used.

2. Form small groups to survey students in the school about biotechnology. Prepare a short survey form that asks students questions that assess their thoughts. (Note: A short-answer format may be preferred, such as "yes" or "no," "agree" or "disagree," and a rating scale of 1 to 5, where 1=no benefit and 5=highest benefit, with numbers in between representing different levels of benefit.)

3. Use the Internet to prepare an oral presentation on an area of biotechnology, such as medicine, agriculture, and the environment. Begin your search with one of the following web sites:

The Biotechnology Information Resource — www.nal.usda.gov/bic/
Lawrence Livermore National Laboratory —
www-bio.llnl.gov/bbrp/bbrp.homepage.html
Biotechnology Center — www.life.uiuc.edu/biotech/

2

HISTORY OF BIOTECHNOLOGY

Many people think that biotechnology is new. It is not new. In one form or another, biotechnology has been around for hundreds of years. The nature and uses of biotechnology have changed over the years.

Older uses are still important in some areas; these include making cheese, baking bread, distilling vinegar, and pickling foods. No one gives much thought to when these uses started or how they began. Older methods have served as the foundation on which modern biotechnology has been built.

The amount of science needed to use the new methods in biotechnology has greatly increased in recent years. Computers, communication, and other technology made these new methods in biotechnology possible. Today, biotechnology includes, among others, such things as DNA structure, gene expression, and recombinant DNA methods.

2-1. **Animal domestication was an early step in biotechnology. (Today, many people enjoy companion animals, such as dogs.)**

OBJECTIVES

This chapter provides a brief introduction to the history and development of biotechnology. It has the following objectives:

1. Trace the historical development of biotechnology in three stages.
2. List and describe the work of important individuals in the development of biotechnology.
3. Describe the importance of research and development in biotechnology.

TERMS

ancient biotechnology
antibiotic
classical biotechnology
development
domestication
fermentation
fermenter
field plot
food preservation
genetics
heredity
modern biotechnology
prototype
recombinant DNA process
research
variability

STAGES OF BIOTECHNOLOGY DEVELOPMENT

The history of biotechnology is identified in three stages: ancient, classical, and modern. Susan Barnum clearly justifies these three stages in her book, *Introduction to Biotechnology*.

Ancient biotechnology begins with early history of human civilizations. It includes important developments in agriculture and food production. In some cases, records do not exist. Archeologists dig into the remains of civilizations to study their artifacts in an effort to gain information about how these people went about life. Art in the form of carvings and sketches is often a source of information.

Classical biotechnology follows ancient biotechnology. It makes widespread use of the methods from ancient biotechnology, especially fermentation. The methods were adapted to industrial production, which made it possible to produce large quantities of food products and other materials in short amounts of time. This made it possible for manufacturers to meet the demands of an increasing population. Many of the methods that emerged through classical biotechnology are widely used today.

Modern biotechnology is the manipulation of the genetic material within organisms. It is based on genetics and the use of microscopy, biochemical methods, and related sciences and technologies. Modern biotechnology is often known as genetic engineering. Its roots involved the investigation of genes.

2-2. Modern biotechnology involves the use of sophisticated instruments, such as NMR instrumentation. (NMR is "nuclear magnetic resonance." It is used to identify an unknown compound.) (Courtesy, Emory University, Atlanta)

ANCIENT BIOTECHNOLOGY

No one knows exactly when biotechnology began. Its beginnings largely focused on having food and other human needs. Useful plants were brought from the wild. They were planted near caves where people stayed for shelter. The people found it easier to get food if the sources were close by. Searching deep into forests was not needed unless the supply ran short.

As food was available, the ability to store and preserve it emerged. This protected food from spoilage. Early food preservation most likely came from unplanned events, such as a fire or freeze.

Domestication

What would life be like if you had to eat what you found in the forest? Imagine life 15,000 years ago. Wild fruit, berries, nuts, mushrooms, leaves, seed, and other plant foods might be used. Large animals were hard to capture. In most cases, people only had meat when they found a dead animal. Some people came up with ways of capturing fish and small animals. Food supplies were often seasonal. What was available in the spring was likely not available in the fall. Winter food supplies might get quite low.

Scientists often list the domestication of plants and animals as the beginning of biotechnology. ***Domestication*** is the adaption of organisms so they can be cultured. It most likely began 11,000 to 12,000 years ago in the Middle East.

2-3. Beef cattle have long been domesticated and are raised throughout North America. (This shows a young Kentucky woman who is proud of her herd of Limousine cattle.)

2-4. **Wild foods often consisted of berries and nuts, such as those shown here (clockwise from top right): persimmons, mulberries, Indian strawberry, and chestnut.**

Early domestication involved collecting seed from useful plants and growing crude crops from the seed. This involved knowing that seed had to be mature and properly planted in the soil to grow. These prehistoric farmers also learned of the need for water, light, and other conditions for plant growth. The earliest plant domestication likely involved grains and other seeds used for food.

Raising animals in captivity began about the same time in history. People found that it was far easier to have an animal close by than to hunt and capture a wild one. They learned that animals need food and water. They also learned about simple animal breeding and how to raise young animals. Cattle, goats, and sheep were the first domesticated food animals.

About 10,000 years ago, people in some places had learned enough about plants and animals to grow their food. This was the beginning of farming.

Food

Domestication resulted in food supplies being greater at certain times of the year. To have food year-round, the products were gathered and stored in some way. People noticed that some foods rotted. Others changed form and continued to be good to eat. Foods stored in a cool cave did not spoil as quickly as foods stored in warmer places. Foods that were heated by fire also did not spoil as quickly. Immersing foods in sour liquids helped prevent food decay.

Food preservation gradually came about. ***Food preservation*** is using processes that prevent or slow spoilage. This may involve heating or cooling foods to keep microorganisms from growing. Some foods were stored in bags of leather or jars of clay. Fermentation occurs if certain microorganisms are present. This creates an acid condition in which spoilage does not readily occur.

2-5. Foods must be kept under proper conditions to prevent spoilage, such as the decay of these peaches.

Cheese was one of the first food products made through biotechnology. It began some 4,000 years ago by nomadic tribes in Asia. Strains of bacteria were added to milk. This caused acid to form, resulting in sour milk. An enzyme was added. The enzyme, named rennet, was taken from the linings of the stomachs of calves. Today, genetically engineered rennet produced by bacteria is used in some cheese making.

Yeast has long been used in food preservation and preparation. Bread baking involves using yeast. The yeast produces a gas in the uncooked mix causing the bread to "rise." Breads cooked without yeast are unleavened and do not rise. Fermented products, including vinegar, require the use of yeast in at least one stage of production. Yeasts are species of fungi. Some are useful; others that cause disease are not.

Vinegar is an ancient product that is used to preserve food. The juices and extracts from fruits and grains can be fermented. ***Fermentation*** is a process in which yeast enzymes chemically change compounds into alcohol. In mak-

Biotech Connection

REVERTING BACK TO THE WILD

Once domesticated, some animals will revert to the wild form if the opportunity exists. Dogs, horses, and other species occasionally revert back. These animals are known as feral, meaning that they have returned to the wild state. They usually do so because they have been abandoned by their owners.

Horses are often allowed to exist as feral animals. The eastern shore area of Virginia has feral horses that roam the coastline. Most are found on protected lands, such as wildlife refuges.

This photograph shows feral ponies in Chincoteague National Wildlife Refuge. This is on an island located just off the shore of Virginia.

ing vinegar, the first product of fermentation is alcohol. Alcohol is converted to acetic acid by additional microbe activity. This acid gives vinegar a sour taste. Vinegar prevents the growth of some bacteria and, therefore, keeps foods from spoiling. It is used in pickling. Pickled foods include cucumbers, peppers, beets, and selected meat products. Biblical references to wine indicate the use of fermentation some 3,000 years ago.

CLASSICAL BIOTECHNOLOGY

Classical biotechnology came about as people learned more about the methods of ancient biotechnology. They were able to use these methods in large-scale factory production. Most of these developed along with the Industrial Revolution in the United States.

Fermentation Control

A major step in classical biotechnology was the control of the fermentation process. In ancient times, fermentation likely occurred by accident. This was changing by the early 1700s. Significant strides occurred in the 1800s and early 1900s.

Fermenters were used to advance the fermentation process. A ***fermenter*** is a specially designed chamber that promotes the fermentation process. This technology allowed better control over the fermentation, especially with vinegar. New products resulted, such as glycerol, acetone, and citric acid. The development of yeasts that were predictable and readily available led to the modern baking industry.

ANTIBIOTICS

The use of fermentation hastened the development of antibiotics. An ***antibiotic*** is a drug used to combat bacterial infections. Penicillin, a very effective antibiotic, was developed in the late 1920s and introduced in the 1940s. It was the first drug produced by microbes in the *Penicillium* genus. These microbes were molds (kinds of fungi).

Career Profile

POULTRY PRODUCER

Poultry producers care for chickens, ducks, turkeys, and other fowl. The work may be with computers, inside poultry houses, or outside. Producers must observe birds for disease and other harmful conditions. A good environment must be provided. Most producers grow birds under contract with an integrator. An integrator is a company that directs all aspects of production and markets the finished product. Growers follow the specifications of integrators.

The educational level of poultry producers varies. Some have college degrees. Others have a high school education. Taking agriculture classes in high school provides basic preparation. Practical experience with poultry is beneficial. Continuing education is needed to keep up with new developments.

Poultry producers are found wherever poultry are produced. In some cases, an integrator must be present for a market for the birds to exist. This shows a poultry producer using a computer system to monitor and control the conditions in several poultry houses from one location.

Many kinds of antibiotics are available today. Some limitation has been placed on their use. The limitations keep disease-producing microbes from developing immunity to the antibiotics. A good rule is to use antibiotics only when needed. Overuse may mean that an antibiotic will not be effective when you have a real need later. Some disease organisms are now resistant to antibiotics.

Antibiotics are used in both human and veterinary medicine. Immunity of microbes to antibiotics can occur with both uses.

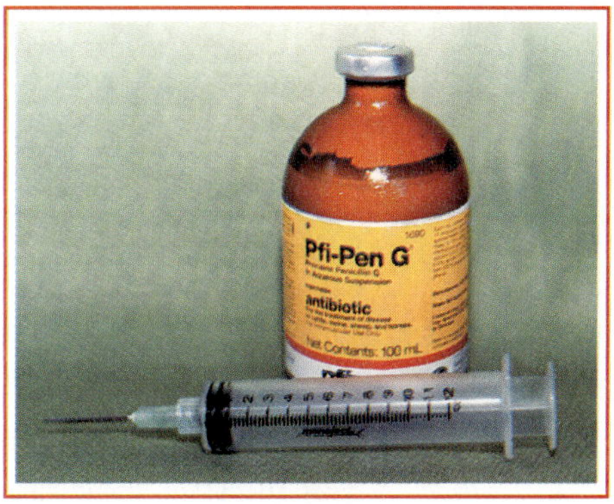

2-6. **A veterinary medicine antibiotic given with a syringe and hypodermic needle.**

MODERN BIOTECHNOLOGY

Modern biotechnology deals with manipulating the genetic information in an organism. Various methods are used in the process. These require in-depth knowledge of science. Microscopy and advanced computer technology are also used. Modern biotechnology is based on genetics research from the mid-1800s.

Genetics is the study of the heredity of living organisms. Most of the work has focused on animal and plant genetics. Genes are the determiners of heredity. They carry the genetic code for an organism. Understanding the genetic structure is essential for genetic engineering.

How traits are passed from parents to offspring is ***heredity***. Members of the same species pass the characteristics of that species to offspring. Within each species, differences exist. These differences are known as ***variability***. Scientists use the principles of heredity and variability in modern biotechnology.

2-7. **Scientists are studying the transfer of a gene from eastern gama grass into a corn plant to create genetically altered corn. (Courtesy, Agricultural Research Service, USDA)**

Table 2-1. A Few Major Events in Modern Biotechnology

Date	Event	Developer
1953	DNA double helix discovered	James Watson and Francis Crick (England)
1968	A process developed for identifying chromosomes	T. O. Caspersson and Lore Zech (Switzerland)
1973	First successful combining of DNA from unrelated organisms (First Recombinant DNA)	Stanley Cohen and Herbert Boyer (USA)
1978	First test-tube baby (Louise Brown) born in England	
1984	Sheep and goat embryo cells fused to create a sheep-goat	Steen Willadsen (England)
1986	Light-producing gene of firefly transferred to tobacco creating tobacco plants with glowing leaves	
1987	Calf cloned from an embryo	Randall Prather and Neal First (USA)
1991	Sheep created that makes drug in milk to treat cystic fibrosis (Drug was alpha-1 antitrupsin)	Ian Wilmut (Scotland)
1994	First transgenic food approved (Flavr Savr tomato)	Calgene Corp. (USA)
1996	Goat injected with a human gene produces an anticancer drug	Genzyme Transgenics Corporation (USA)
1997	Adult sheep cloned ("Dolly")	Ian Wilmut (Scotland)
1998	Unfertilized cow eggs used to reprogram genes from another species	Neal First (USA)

The use of modern biotechnology to produce new life forms emerged in the mid-1900s. The techniques are often known as genetic engineering. Major advances occurred in the late 1900s. These were possible because of ***recombinant DNA processes*** in which scientists moved genetic material from one chromosome to another. The materials involved are quite small. Even with the new technology, the processes are challenging and often controversial. For various reasons, some people have negative views of biotechnology.

PEOPLE WHO MADE A DIFFERENCE

A number of people have made contributions to biotechnology. In some cases, the individuals were ahead of their time. Their work was not recognized until years after it was done. The individuals were often no longer living.

The work of a few individuals is described here.

ZACHARIAS JANSSEN

Zacharias Janssen discovered the principle of the compound microscope in 1590. He was a Dutch eye spectacle maker and amateur scientist. His work led to further developments of the microscope by Anton van Leeuwenhoek.

ANTON VAN LEEUWENHOEK

Anton van Leeuwenhoek used the earlier work of Janssen and developed a single lens microscope in the 1670s. With the microscope, he was the first person to observe tiny organisms and record his observations. His work led to modern microscopes with power to make even the smallest objects visible. The development of the electron microscope in 1931 by a group of German scientists furthered the ability to see tiny materials.

2-8. The compound microscope was one of the first technologies that made biotechnology possible.

GREGOR JOHANN MENDEL

Gregor Mendel formulated the basic laws of heredity during the mid-1800s. As an Austrian botanist and monk, he conducted experiments with common peas. He studied the inheritance of seven pairs of traits, including short or tall plants and rounded or wrinkled seeds. Mendel bred

2-9. **Gregor Mendel is known for his study of heredity. (Courtesy, Stock Montage, Inc.)**

and crossbred thousands of plants in his trials. He concluded that offspring receive hereditary material from each parent. Some traits were said to be dominant; others were recessive. Mendel's work served as the basis for future efforts in identifying and naming genes and other genetic material. His work was important in the emergence of modern biotechnology. Mendel's findings were published in 1866 but were largely ignored for 34 years.

JOHAN FRIEDRICH MIESCHER

John Friedrich Miescher, a Swiss biologist, isolated the nuclei of white blood cells in 1869. This work led to the identification of nucleic acid by another scientist, Walter Flemming.

WALTER SUTTON

Walter Sutton determined in 1903 that chromosomes carried units of heredity that had been identified by Mendel. These units were named "genes" in 1909 by Wilhelm Johannsen, a Danish botanist.

THOMAS HUNT MORGAN

Thomas Hunt Morgan studied the genetics of fruit flies. In the early 1900s, he experimented with eye color and gender of flies at Columbia University. He concluded that some traits of fruit flies were related to the gender of the flies. Morgan bred white-eyed male fruit flies to red-eyed female fruit flies. All first-generation offspring had red eyes. In the second generation, some of the flies had white eyes and in every case those with white eyes were males. His work contributed to knowledge of X (female) and Y (male) chromosomes. Morgan received the Nobel Peace Prize in 1933 for his research in gene theory.

ERNST RUSKA

Ernst Ruska built the first electron microscope in 1932. Ruska was a German electrical engineer. His first microscope was not sophisticated but did offer 400x magnification. Improvements made it possible for future scientists to study cell biochemical processes.

ALEXANDER FLEMING

Alexander Fleming discovered penicillin in 1928. It was the first antibiotic drug used in treating human disease. He observed the growth of molds (*Penicillium* genus) in a dish that also contained bacteria. The bacteria that were close to molds were dead. Extracting and purifying the molds into forms that could be used took a decade of research. Penicillin was first used to treat human disease in 1941. It is credited with saving many lives during World War II when wounded soldiers developed infections.

ROSALIND ELSIE FRANKLIN

Rosalind Elsie Franklin carried out research in France and England in the mid-1900s that led to the discovery of the structure of DNA. Her early research was to produce an atomic bomb but her interests were in the mystery of living organisms. Her work at King's College in England included setting up an x-ray diffraction laboratory. Franklin's photographs of DNA showed that it could have a double helix structure. Some questions surround the theft of her early work in 1952, including x-ray photographs.

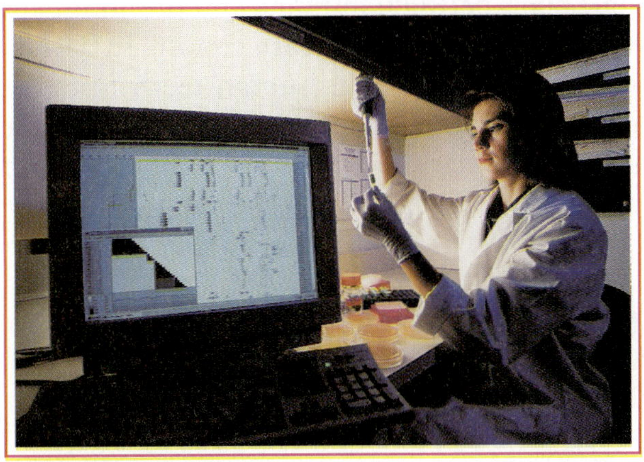

2-10. Today's study of DNA is possible because of the work of earlier scientists. This scientist is withdrawing DNA from soil bacteria in a study of the wheat disease known as take-all. (Courtesy, Agricultural Research Service, USDA)

JAMES WATSON AND FRANCIS CRICK

James Watson and Francis Crick collaborated to produce the first model of DNA structure in 1953. X-rays were studied to determine the nature of

DNA as a double helix structure. They described DNA dimensions and spacing of base-pairs. Their work had a major impact on genetic engineering as carried out today. Watson was born in the United States and Crick in England. A brilliant scientist, Watson met Crick and later moved to Cambridge University in England to collaborate with him.

NORMAN E. BORLAUG

Norman E. Borlaug is recognized for developing wheat varieties capable of producing high yields of grain. The 1960s research focused on semi-dwarf wheat that could be produced in climates where existing wheat varieties would not thrive. His work was carried out in Mexico. Many nations now grow wheat varieties based on Borlaug's research. Borlaug was awarded the Nobel Peace Prize in 1971. This research is credited with helping relieve widespread hunger in some nations.

MARY-CLAIRE KING

Mary-Claire King is known for research into the nature of DNA during the late 1900s. The Illinois native determined that 99 percent of human DNA is identical to that of a chimpanzee. In 1975, she found that similar gene pools between humans and the chimpanzee made it possible to research hereditary causes of breast cancer. In 1992, King perfected techniques for comparing DNA from human teeth to find matches with unidentified human remains. Law enforcement uses her methods in solving crimes. King took an active role in the huge Human Genome Project to map and sequence all DNA in the human genome.

IAN WILMUT

Ian Wilmut achieved several important developments. One receiving worldwide attention was when he announced the cloning of a sheep named Dolly on February 20, 1997. He worked with a team of Scottish researchers. Public opinions varied widely on the news of cloning from an adult sheep. (Previous animal cloning had been carried out using animals in the early stages of embryo development.) The announcement led to considerable debate over the possibility of cloning humans. Many political leaders spoke in opposition to human cloning.

2-11. **A student in an Arizona high school lab is cloning a Venus flytrap plant.**

RESEARCH AND DEVELOPMENT

Systematic methods are used to produce new information in biotechnology. These methods involve research and development, which is also known as R&D.

RESEARCH

Research is the use of systematic methods to answer questions. The questions are usually based on a problem that needs to be solved. Problems may be basic or applied. Basic problems require generating new information to gain understanding. Applied problems involve the use of knowledge already acquired.

Research supplies facts that can be used to improve a process or the product that results. The settings for research vary. Some research is carried out in elaborate laboratories where precision and careful observation are critical. Other research is carried out in fields, forests, factories, and families. Agricultural research often makes use of field plots. A ***field plot*** is a small area of land that is used to test questions or hypotheses. The belief is that the same results would be obtained if the research was carried out on a larger scale. Research with field plots is often tested several times, which is known as replication.

2-12. Research with cotton plants is being carried out using cylinders of sand and irrigation.

Research is done by agencies, universities, private companies, and individuals. Biotechnology research in agriculture is carried out by agricultural experiment stations and large private agri-manufacturing corporations, such as Monsanto, DuPont, and Novartis. These are but a few examples. Many others could be listed.

DEVELOPMENT

Development is the creation of new products or methods based on the findings of research. New developments in biotechnology are carefully studied before they are put into large-scale use. New things do not just happen; someone has to plan and carefully carry out the research to support possible new developments.

New products must be tested before they are approved. Government agencies, such as the Food and Drug Administration, are involved. A first step is to produce a prototype. A ***prototype*** is a research model that is carefully tested. It becomes the pattern for the production of similar products in the future. After a prototype has been fully tested, full-scale production may begin.

REVIEWING

MAIN IDEAS

Biotechnology has been a part of human life for thousands of years. Some early uses have been improved and expanded, such as cheese making.

The history and development of biotechnology is in three stages: ancient, classical, and modern. Ancient biotechnology begins with domestication of plants and animals and food preservation. Fermentation is an ancient process that has been improved and is widely used today.

Classical biotechnology expanded the use of ancient methods. It set the stage for modern biotechnology. Developments allowed the use of biotechnology in large-scale manufacturing.

Modern biotechnology is concerned with the manipulation of genetic material. Principles of heredity and genetics form the core of processes used with recombinant DNA and related technologies.

Most of the developments in biotechnology can be attributed to the efforts in biology, technology, and related areas. Microscopy, heredity, and techniques for manipulating the smallest of materials were important. Anton van Leeuwenhoek, Gregor Mendel, Rosalind Franklin, James Watson and Francis Crick, and Mary-Claire King are a few people who made big contributions to biotechnology.

All efforts are strengthened by research and development. Research is used to systematically answer questions. From the research, new products may be developed. Testing and prototype development occur before a new product is released.

QUESTIONS

Answer the following questions using correct spelling and complete sentences.

1. What are the three stages of development in biotechnology? Distinguish between each.
2. What is domestication? Why was it important?
3. How did domestication affect food supplies?
4. What is fermentation? Why is it important?
5. What is an antibiotic?
6. What is genetics?
7. What major contributions were made by the following individuals: Gregor Johann Mendel, Thomas Hunt Morgan, Alexander Fleming, James Watson and Francis Crick, Norman E. Borlaug, and Mary-Claire King?
8. What major announcement was made by Ian Wilmut? How did people respond to his announcement?
9. What is research?
10. What is development? How are prototypes used in the development process?

EVALUATING

Match the term with the correct definition. Write the letter of the term in the blank that is provided.

a. research
b. domestication
c. food preservation
d. fermentation
e. genetics
f. recombinant DNA
g. prototype
h. variability
i. heredity
j. antibiotic

_____ 1. Moving genetic material from one chromosome to another.
_____ 2. The study of the heredity of living organisms.
_____ 3. How traits are passed from parents to offspring.
_____ 4. Naturally occurring differences within a species.
_____ 5. Using systematic methods to answer questions.
_____ 6. A research model that is carefully tested.
_____ 7. Adapting organisms that grow wild so they can be cultured.
_____ 8. Using processes to prevent the spoilage of food.
_____ 9. The process in which yeast enzymes chemically transform compounds into alcohol.
_____ 10. A drug used to combat bacterial infections.

EXPLORING

1. Select one of the people listed in the chapter as having made important contributions to biotechnology. Research the individual's background, education, and other contributions. Prepare an oral report for the class on your findings.

2. Identify an animal or plant that has been domesticated. Prepare a report that describes the importance of the organism. Include the care requirements for the organism to gain efficient growth.

3. Select one of your favorite foods or an ingredient of a favorite dish. Identify how it is preserved to prevent spoilage. Assess if it spoils quickly or slowly under natural conditions. Prepare a written report on your findings.

4. Make yogurt or cheese. Use the procedures in Chapter 25 of *Biological Science Applications in Agriculture* by Edward W. Osborne. This book is available from Interstate Publishers, Inc.

3

BENEFITS OF BIOTECHNOLOGY

Six billion people live on the planet Earth. Each of these has the same basic needs. How the needs are met varies widely. Some people have plenty. They have food, housing, and clothing. They live in a safe environment. Other people lack what they need to have a productive life.

Biotechnology can help solve many problems. Food, health, housing, and other areas are involved. For example, pest-resistant crops can be used to improve yields and help prevent hunger. Another example is tissue engineering to create new tissues, such as skin, cartilage, and blood vessels. And a third example is using microbial biotechnology to produce new disease-preventing vaccines.

The benefits of biotechnology are numerous. Much of our quality of life is based on using biotechnology in some way. The needs of people worldwide are more adequately met.

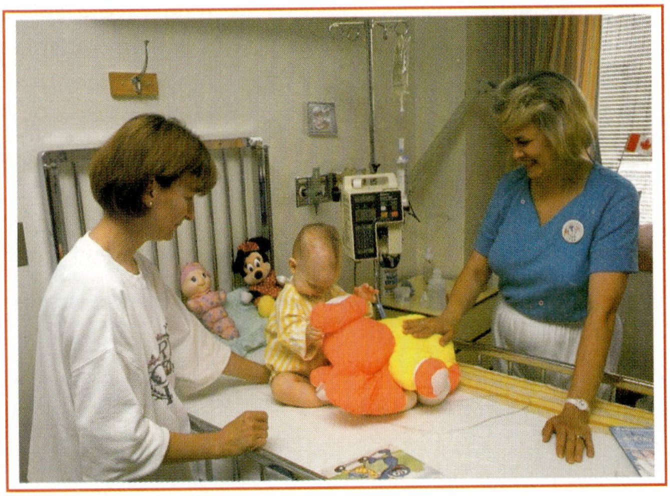

3-1. Life expectancy and quality of life of an infant are related to advances in biotechnology.

OBJECTIVES

This chapter provides information on how biotechnology contributes to quality of life. Future contributions are also suggested. It has the following objectives:

1. Describe the role of biotechnology in contributing to the quality of life.
2. Explain how biotechnology contributes to the supply of food and fiber.
3. Explain the benefits of biotechnology in health.
4. Describe the role of biotechnology in materials and the environment.

TERMS

antibody
bovine somatotropin
crop
designer drug
disease
environment
fiber
food
gene therapy
health
healthcare
life expectancy
livestock
material
nutraceutical
organ transplant
pharmaceutical
polymer
poultry
quality of life
tissue engineering
vaccine

BIOTECHNOLOGY AND QUALITY OF LIFE

Quality of life is having a good environment in which to live. Some places are said to have a better quality of life than others. Why? It is mostly related to housing, healthcare, education, jobs, social and recreational opportunities, and the environment. These and other conditions make up the quality of life.

LIFE CONDITIONS

Quality of life promotes the well-being of people. It is an environment that meets the basic needs of people.

People must have food and water. These provide needed nourishment. The food must contribute to good health. With children, food must contribute to growth and development. Without good food and water, people do not have a good quality of life.

Shelter in the form of clothing and housing contributes to a good quality of life. The clothing should be clean and in good repair. Further, it should meet the needs of the people. Housing should protect from the weather and provide a place for a family to go about life. It should promote their well-being.

3-2. People want to have a quality of life that includes recreation and time to relax.

Career Profile

BIOCHEMIST

A biochemist may do research on the chemistry of living organisms. This combines chemistry and biology in an effort to understand living things. A biochemist will often study a small series of chemical events among the thousands of chemical reactions that occur in a living organism. The biochemistry of many different living organisms is studied including animals, plants, fungi, and bacteria. Some biochemists use biotechnology in their work, such as shown in this photo.

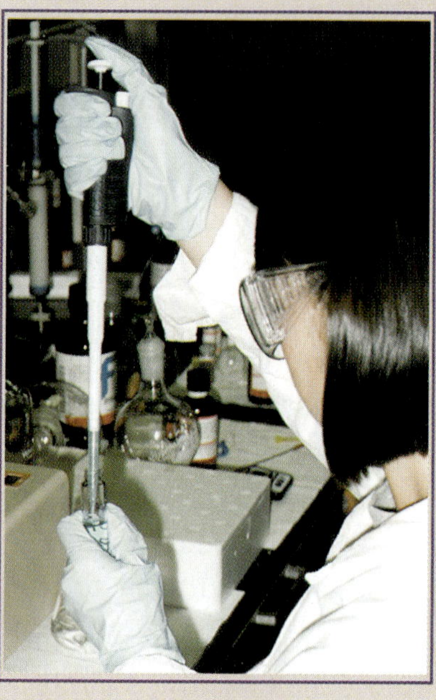

Biochemists have college degrees in chemistry, biology, biochemistry, molecular biology, or related areas. Advanced positions may require master's or doctor's degrees. Many have education beyond the doctoral level. Preparation to be a biochemist can begin with high school classes in biology, chemistry, physics, biotechnology, and related areas.

Jobs are with universities, research agencies, and businesses that use biochemistry. (Courtesy, Emory University, Atlanta)

Social and recreational opportunities have much to do with how people view quality of life. The absence of these results in a low quality of life. Opportunities to interact with other people promote human well-being.

HEALTHCARE

An important part of quality of life is access to good healthcare. ***Healthcare*** includes the services, information, equipment, and other conditions that promote good health. Physicians and nurses, hospitals, and pharmacies are associated with healthcare. ***Health*** is the state of the human body as related to disease and vigor. Health may be impaired when disease occurs. A ***disease*** is an abnormal condition of the body. Some diseases affect only certain organs. Quality of life includes healthcare practices that prevent disease as well as treat and heal diseases when they occur.

3-3. Good healthcare is a part of having a good quality of life.

LIFE EXPECTANCY

Life expectancy is the number of years that an individual may be expected to live. It is from the time of birth until death. Our life expectancy has increased. When people live longer, they add to the population numbers.

Life expectancy in the United States is 76 years. Females, with 80 years, have a longer life expectancy than males. Overall life expectancy in Canada is 79 years. The average life expectancy is higher in Sweden and Switzerland but lower in the developing areas of Africa, Asia, and Central America.

The use of biotechnology has contributed to life expectancy. Scientists are seeking new products and to improve existing products to further extend life expectancy. These should contribute to worldwide quality of life.

3-4. Quality of life enhances life expectancy.

WORLD FOOD AND FIBER PRODUCTION

Everyone needs food! A good supply of food promotes a better quality of life. ***Food*** is the solid and liquid material humans consume for nutrients. Nutrients are needed to carry out life processes. Adequate amounts of wholesome foods must be available. Most all foods are from plant and animal sources.

In addition to food, humans use fiber crops. ***Fiber*** is slender threadlike material used to make clothing, paper, and shelter. Fiber comes from plants and animals, though plants are most important. Some fiber is artificial, such as nylon.

Biotechnology in agriculture has a positive impact on world food production through increased yields and quality. Crops and livestock are modified and enhanced using biotechnology.

Traditional ways of growing crops and animals are too limited to meet food needs. Human population is increasing at a rapid rate. Food productivity must keep up. Biotechnology is needed to assure adequate food. Without new methods, food supplies may be short and more people would die of starvation. New methods can also help protect Earth's fragile environment.

3-5. Traditional ways of gathering food, such as coconuts, have given way to modern production systems.

PLANT BIOTECHNOLOGY

Many species of plants are used for food and shelter. Some are carefully tended to assure yield and quality. A ***crop*** is a plant that is grown and harvested for food and other products. Important crops include wheat, corn, rice, cotton, canola, soybeans, and many kinds of vegetables. Vast amounts of land are planted with these crops. Approved practices are used to tend the crops and assure productivity from the land on a long-term basis.

Table 3-1. Examples of Important Crops

Rice — An important cereal grain food crop worldwide; a good source of carbohydrates; China and India are the leading rice-growing countries; Arkansas, California, Texas, and Louisiana are the leading states in rice production.

Wheat — The most important food crop worldwide; planted on more land than any other crop; used in bread, pasta, and breakfast cereal; China, Russia, and the United States are leading wheat-growing countries; Kansas, North Dakota, and Oklahoma are the leading states in wheat production.

Corn — The second most important food crop behind wheat; used for grain and oil; it is important in human food products and animal feeding; leading corn-producing countries are the United States, China, Argentina, Brazil, and France; Iowa, Illinois, Nebraska, Minnesota, and Indiana are the leading corn-producing states.

Oats — A cereal grain used for human and animal food; oats have higher food value than other cereal grains; used for oatmeal, cookies, and breakfast cereals; Russia, the United States, and Canada are leading producers of oats; Iowa and Minnesota lead other states in oat production.

Soybeans — Soybean seeds are used for oil, with the meal byproduct from the seed used for animal feed; some soybeans are eaten as vegetables; the largest source of vegetable oil in the United States; leading soybean-producing countries are the United States, Argentina, Brazil, and China; Iowa and Illinois are the leading soybean-producing states.

Canola — A rapidly increasing crop used for cooking oil; major producers are China, India, Northern Europe, and the United States; leading the production in the United States are Montana, North Dakota, and others in the upper plains states.

Cotton — The most important fiber used in making clothing; seeds are used to make oil and animal feed; the leading countries in cotton production are China and the United States; the leading cotton-producing states are Texas and California.

Vegetables — Many plant species are used as vegetables; examples include tomatoes, lettuce, onions, peas, cauliflower, asparagus, carrots, potatoes, and lima beans; China, India, and the United States are leading vegetable-producing countries; California, Texas, Idaho, Florida, and Wisconsin lead in vegetable production.

Fruits and Nuts — Many plant species are grown for fruit and nuts, including oranges, apples, and pears for fruit and walnuts, peanuts, and pecans for nuts; leading fruit-producing states are California, Florida, and Washington; leading nut-producing states are California, Georgia, and Texas.

3-6. Potatoes are staples in the diets of many people.

Plant biotechnology is the application of biotechnology with plants. It is used to have plants with desired qualities. Plants can be engineered for insect, virus, bacteria, fungal and herbicide resistance. The crops are modified using biotechnology so the pests, including insects and disease, cannot damage the plant and reduce food production.

Several plants are now available with insect-resistant traits. The potato is an example. In 1995, the NewLeaf Russet Burbank Potato was approved. This is a potato with a gene inserted from the *Bacillus thuringiensis* bacterium. These bacteria naturally produce an insecticide that kills certain kinds of insects. Known as Bt, the trait causes the potato plant to resist insect pests. Spraying with pesticide is not needed.

Cotton and corn have also had the Bt gene inserted. These are known as Bt cotton and Bt corn. On farms, these transgenic crops have worked out reasonably well. Some problems came about when the resistance to insects appeared not be present. Some scientists feel that insect pests will develop resistance to the

3-7. Transgenic Bt corn plants produce a natural insecticide that prevents damage and promotes good yields. (Insect damage has destroyed the productivity of the corn plants on the left, while those on the right have produced large ears.)

Biotech Connection

ALTERNATIVE AUTOMOBILE FUEL

Biotechnology is being used to produce fuel to run your automobile. Scientists have modified bacteria so they use grains, crop residue, and other materials to produce ethanol. The cellulose in plant stems cannot be digested by humans and is often considered waste.

The new bacteria can convert cellulose to ethanol, which can be used as fuel. For example, leftover rice hulls and sugarcane waste can be used to produce ethanol. In the future, lawn clippings, corn stalks, and sawdust may be used to get you to school or work. This puts a waste to good use and conserves the use of scarce fossil fuels.

This photo shows an automobile that runs on a fuel of 85 percent ethanol and 15 percent gasoline.

Bt plants. Other scientists feel that the amount of Bt expressed in the plant may not be enough to control the insects. Some problems have arisen over the destruction of desirable insects, such as Monarch butterflies.

Biotechnology can also be used to improve the quality of a food product. Flavr Savr tomatoes developed by Calgene, Inc., are modified so they can be harvested at a time when they have better taste. Ordinarily, tomatoes must be picked before they are ripe so they will be in the store before they begin to over ripen. Flavr Savr tomatoes are picked when ripe and have a better taste, but will not over-ripen before they get to the market.

Research is developing crops that contain vaccines for diseases that affect humans. A ***nutraceutical*** is a food that has benefits other than nutrition. Nutraceuticals may contain disease-fighting drugs. Instead of having syringe-injected immunizations against disease, people would simply eat the vaccine-containing crop. No more shots!

ANIMAL BIOTECHNOLOGY

Animal biotechnology is using biotechnology methods to improve animals or their products. It has been used to improve livestock and poultry.

Livestock refers to species of domesticated animals found on farms and ranches. They are often used to produce food and other products. Common livestock include beef cattle, dairy cattle, hogs, and sheep. Horses, mules, and donkeys are kept for non-food uses in the United States but are used for food in some countries.

Table 3-2. Examples of Important Animals

Beef cattle — These animals are used primarily for meat; hides and other by-products are also used; cattle are typically raised on pasture grassland with animals finished on grain diets; a typical feed-lot-finished animal weighs 1,100 to 1,300 pounds and is at least two years old; beef cattle are raised in nearly all states with the leading beef-producing states being Texas, Nebraska, and Kansas.

Dairy cattle — These animals are used for milk production, with non-producers going for meat; cows are kept on pasture, in barns, or in dry lots; milk production is tied to the reproductive process; milking begins after giving birth to a calf (the calf nurses the cow only a day or so); most cows are milked 305 days a year, which gives 50 to 60 days before the next calf is born; average milk production per cow is 16,000 pounds a year or 6.3 gallons a day; California, Wisconsin, New York, and Pennsylvania lead in milk production.

Swine (hogs) — Swine are raised primarily for meat, though some are raised for medical purposes; hogs are fed grains to reach a desired finish weight of 220 to 230 pounds at an age of 6 to 8 months; leading swine-producing states are North Carolina, Illinois, Iowa, and Indiana.

Sheep — Sheep are raised for meat and wool; the primary source of meat is young animals, known as lambs; mature sheep are used for wool; the leading sheep states are Texas, Wyoming, and California.

Chickens — Chickens are used for meat and eggs; most are raised in specially designed houses; young chickens for meat are broilers; 6 weeks are needed to raise a baby chick to a weight of 5 pounds; chickens used for egg production are layers; the leading broiler-producing states are Georgia, Arkansas, Alabama, and Mississippi; the leading egg-producing states are California, Ohio, and Indiana.

Poultry is domesticated fowl produced for food and other products. Examples include chickens, ducks, and turkeys. These species produce meat, eggs, and feathers. The consumption of chickens has increased in recent years.

Biotechnology is used to insure the health and productivity of animals. An example of the use can be seen with dairy cows. Milk production is regulated by a hormone, known as ***bovine somatotropin*** (bST), produced in the pituitary gland. Scientists have produced bST with bacteria using copies of a cow's genes. When given to cows, milk production is increased about 20 percent. Some side effects have been noted. These are handled with good management of the cows.

3-8. Genetically improved chicks will reach market weight of 5 pounds in six weeks in this automated poultry house.

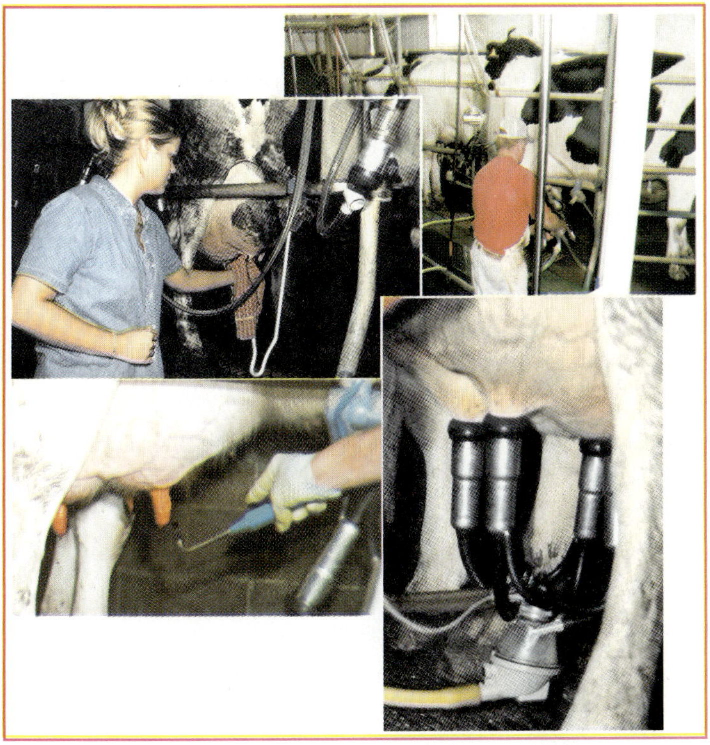

3-9. bST promotes wholesome milk production in modern, clean facilities on this Kentucky dairy farm. Shown here are: milking parlor (top right), cleaning teats before attaching milkers (top left), properly attached milkers in operation (bottom right), and spraying teats after milking with an iodine solution to assure teat health (bottom left).

FUTURE BIOTECHNOLOGY IN FOOD PRODUCTION

Biotechnology can be used in food production to improve the quality of life. The promise of biotechnology is to have higher yields of food and eliminate the problems of famine and disease. Pest- and disease-resistant crops can help eliminate starvation and malnourishment. Drought-resistant crops will enable adequate food production during periods of low precipitation.

Some people disagree with the use of biotechnology to produce food products. They feel that human health may be impaired by the food. Part of the problem may be that people need more education about biotechnology.

Table 3-3. Examples of Biotechnology Benefits

Improved supplies of safe food:

- Better crop yields
- Improved animal productivity
- Nutraceuticals that promote health
- Reduced use of pesticides
- Improved quality of products
- Healthier companion animals

Better shelter:

- More efficient production of wood products
- Improved cotton and other fiber crops

Improved health care:

- New drugs to help treat disease
- Gene therapy to overcome genetic diseases
- Tissue engineering to replace or grow needed tissues
- Disease-free environment

Improved materials:

- New products that conserve scarce resources
- Readily biodegradable materials
- Sustained use of resources

Protection of the environment:

- Microbes that speed degrading of materials
- Less pollution from pesticides, engine exhausts, etc.

MEDICAL BIOTECHNOLOGY

Biotechnology is used in human and animal medicine. Human medicine deals with areas related to the health of human beings. Animal medicine, also known as veterinary medicine, deals with the health of livestock, companion animals, and other domesticated and wildlife animals. The focus of this section is human medicine.

VACCINES

A ***vaccine*** is a substance introduced into the body to prevent disease. Vaccines enter the body as injections, orally, or in other ways. Most vaccines stimulate the production of antibodies by the human body. An ***antibody*** is a protein molecule produced as defense against disease organisms. The production of antibodies usually results in immunity to a disease. Vaccines are a very efficient and cost-effective way to prevent diseases before they occur.

Vaccines protect people from specific diseases. Different diseases require different vaccines. Most vaccines contain a "dead" or less-active (modified-live) copy of the virus or bacteria responsible for the disease.

Almost all children in the United States are immunized using vaccines against such diseases as polio, mumps, measles, tetanus, diphtheria, and rubella. Thanks to a worldwide vaccination campaign, the small pox disease has been eradicated from our planet. Polio has been isolated to a few nations. A few cases result each year from the vaccine.

Most domestic animals receive vaccinations to promote health. Our pets, such as dogs and cats, have routine vaccination schedules. Livestock are vaccinated for disease. Chickens are vaccinated as developing embryos on the 18th day of incubation (normal incubation is 21 days). The vaccine is delivered *in vivo* or through the egg shell.

Biotechnology provides new methods for making vaccines and leads to the development of new vaccines against different diseases. Recent research has lead to new vaccines against chicken pox and childhood ear infections.

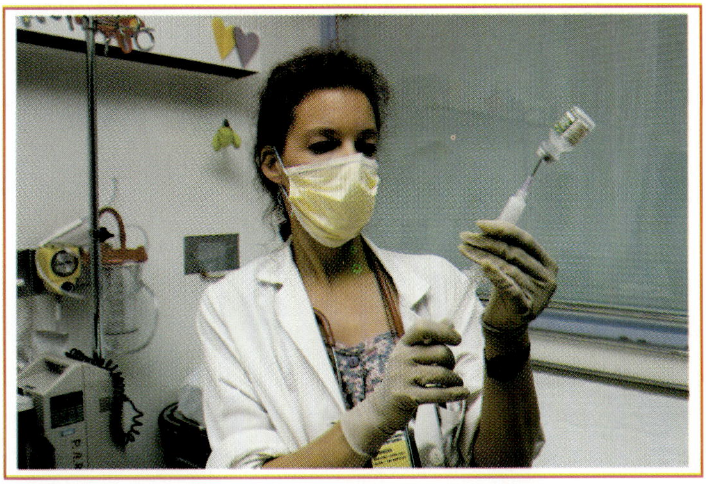

3-10. Many vaccines are given by injection.

Of course, most people are familiar with getting their vaccine injections or "shots." Biotechnology is also developing new ways to deliver vaccinations instead of using a syringe and needle. Scientists are developing fruits and vegetables that contain vaccines. A person simply eats the fruit containing the vaccine, and they become immunized with no "shots!"

PHARMACEUTICALS

A ***pharmaceutical*** (or drug) is a chemical substance used in the prevention, treatment, cure, or diagnosis of a disease. These are drugs that help the body of an organism fight a specific disease.

Originally, pharmaceuticals were isolated from plants or fungi, such as penicillin. Many of the pharmaceuticals that are currently available are designed or engineered drugs.

Biotechnology provides new drugs that work better than traditional drugs. An example of a biotechnology-engineered pharmaceutical is human insulin. Human insulin is currently used to treat diabetes. Originally, diabetics used insulin that was isolated from pigs. Now, bacteria are used to produce genuine human insulin to treat diabetes.

Biotechnology has an increasingly important role in producing drugs. Biotechnology techniques are now utilized to manufacture vitamin C and vitamin B-2. Production of these vitamins using biotechnology is cheaper and safer than making them with other methods.

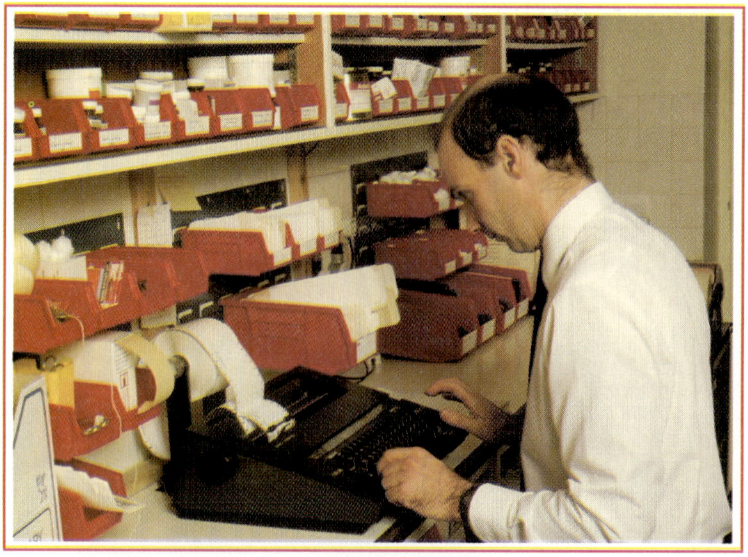

3-11. A hospital pharmacist prepares drugs for delivery to patients.

Computers help in drug development. Those developed with computer models are sometimes referred to as designer drugs. A ***designer drug*** is one that is developed as a three-dimensional model on a computer before it is tried out in a lab. It is easy to turn and manipulate models of protein on a computer screen. Laboratory testing follows computer models that appear promising.

Gene therapy is changing the genetic makeup of a human being. Different approaches are used with gene therapy. One involves removing blood cells from the human, genetically altering them, and placing them back in the body. Another approach involves using carriers or vectors. It is used to treat people affected with genetic diseases, such as cystic fibrosis and hemophilia, and cancer. Human gene therapy is not widely available, but the therapy could become a very efficient way to cure diseases.

TISSUE ENGINEERING

Every year people die because organs, such as the liver, heart, or pancreas, fail to function properly. These lives can be saved with use of mechanical devices (such as pacemakers and dialysis machines) or by organ transplants. Mechanical devices are limited to the replacement of only a few body functions and can be seen as a constraint, such as kidney dialysis. Organ replacement or transplanting may have more promise.

An ***organ transplant*** is replacing bad or damaged organs with organs from a donor. An organ transplant is not an ideal solution either, because there are fewer organs available than people who need them and the body does not always accept the new organ. To avoid rejection of the organ by the body, people need to take large amounts of anti-rejection medication.

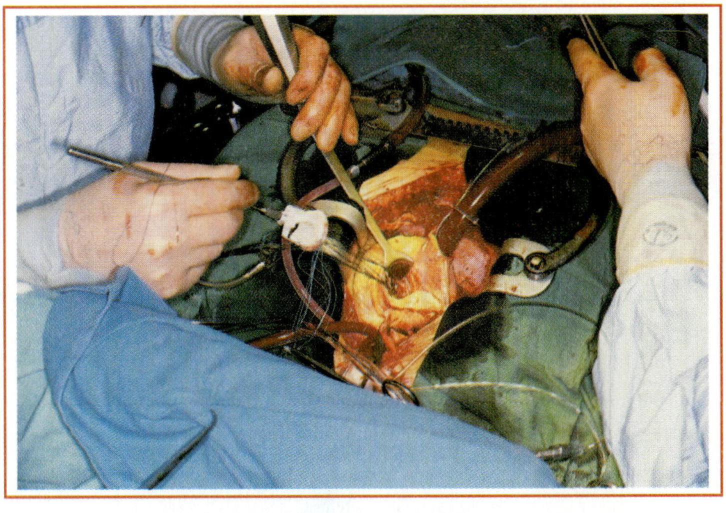

3-12. A replacement valve from a pig being sutured into place in a human.

Biotechnology can lead to promising alternatives. Good cells can be injected into the damaged organ to restore its activity. The animal-to-human transplant of organs is another possibility. The main problem of an animal-to-human transplant, other than ethical issues, is the rejection of the animal organ. Specific proteins are being engineered using biotechnology to prevent rejection of a transplanted organ. Scientists are finding similarities in the organs and tissues of pigs and humans. For example, a defective human heart valve may be replaced with one from a pig.

Severe human burns have often involved skin transplants or grafts. New methods may make it unnecessary to use transplants. Tissue engineering is involved. ***Tissue engineering*** is the process of causing new, healthy tissues to grow. New skin tissue may be grown where an injury occurred; or the new skin may be grown in a laboratory environment. Research is being conducted to grow blood vessels, tendons, and bone. In the future, organ donors may not be necessary. Physicians will be able to repair the damaged organ or grow a new organ in the laboratory.

Artificial organs and tissues are also used. These are manufactured of plastic and metal materials. Defective organs or tissues are removed and replaced with the artificial devices that help the body function. An example is knee replacement. This is done when the knee joint no longer functions properly. The body must accept the artificial materials and use them in normal life processes.

Medical biotechnology holds the key to the future of medicine. Each day, research enables us to live longer, healthier lives.

MATERIALS AND ENVIRONMENT

Humans make use of many material things. The clothes we wear, the houses in which we live, the automobiles we drive, and many other gadgets and conveniences are material things. Most of these are used and discarded. Biotechnology can help in reducing pollution of the environment.

MATERIALS

Materials play an important part in our lives. A ***material*** is a physical object or chemical that has been produced by humans. Examples of materials that have changed our lives are plastic, synthetic fabrics, and silicon chips in computers.

3-13. A plastic bag is used for selecting tomatoes in a French supermarket.

Polyester polymers are used to make many things. A ***polymer*** is the product of the chemical union of two molecules to the same compound to produce another compound. A polyester polymer is the kind of polymer used to make plastics and textile fibers. Examples of products include electrical tape, yard waste bags, diapers, overhead transparencies, and decals.

The silk that spiders produce is very strong. In fact, the spider silk is stronger than steel, but it is very difficult to isolate the silk from spiders. Currently, studies are being done to use bacteria to produce spider silk.

In the past, much manufacturing used scarce fossil fuels. These used up the supply and often released smoke and other pollutants into the environment. Biotechnology can help produce many materials, including polyesters, in a more cost-effective manner. Bacteria can be transformed so they will produce polymers using biotechnology. This will conserve scarce fuel resources.

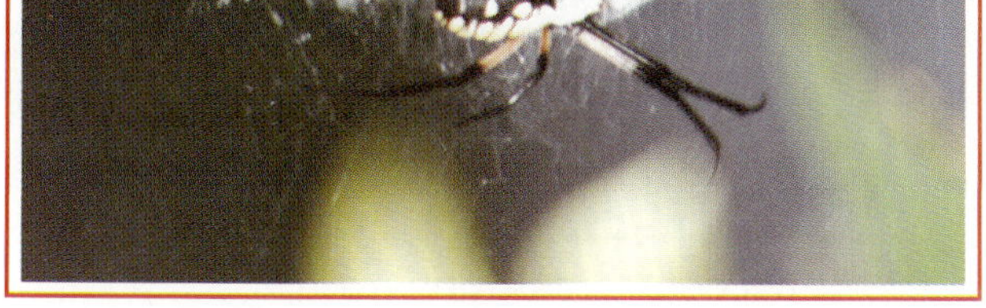

3-14. Just how strong is a spider web? Stronger than steel!

In the future, biotechnology promises to create new materials and new methods to produce materials. Scientists are using biotechnology to create stronger and more durable plastics. Common objects in your classroom, such as tables and pencils, may be products of biotechnology.

THE ENVIRONMENT

People want a good environment for living. Their well-being is promoted by a good environment. Our ***environment*** is our surroundings. Other organisms also have environments. These determine the quality of their lives just as it shapes the quality of the lives of human beings. Environment includes air, water, soil, wildlife, and other natural and artificial features of Earth.

Many kinds of gaseous, solid, and liquid materials are released. These often degrade the environment. Biotechnology methods can be used to help alleviate pollution problems.

Materials may pollute the environment. Discarded products must be disposed of. Materials that get into the environment accidentally need to be dealt with to prevent damaging the environment.

In nature, microbes and other processes break down materials. For example, microbes decay leaves and wood. The leaves and wood become organic matter. In the woods, this decay process is how soil is restored.

3-15. **Oil-eating bacteria are being sprinkled to remove an oil spot from a driveway.**

Bacteria have been modified to be useful in cleaning up materials. The first patented form of life from genetic engineering was an oil-eating microbe. Ananda Chakrabarty received the patent in 1980. A short time later, the U. S. Congress and the Environmental Protection Agency (EPA) got involved. Regulations were passed that restricted the use of such organisms outside of controlled laboratories. Some scientists feel that these set back the development of transgenic organisms. The actions were based on fear of releasing engineered life forms into the environment. Strains of non-transgenic bacteria are now used to break down oil.

REVIEWING

MAIN IDEAS

Biotechnology is being used to improve our quality of life. Advances in biotechnology provide us with better food, clothing, housing, medicine, and environment.

Biotechnology will have a large effect on the future. Scientists around the world are using biotechnology to improve our food supply, health, and quality of life. Biotechnology will be used to provide food for everyone. Scientists are working to create pest-, drought-, and disease-resistant crops.

Some people hope to modify our food so it contains more of the nutrients we need and tastes better. Instead of taking pills and injections for an illness we may simply eat a piece of fruit containing the pharmaceutical medicine. Medical biotechnology and gene therapy will hopefully provide the knowledge to stop cancer, cystic fibrosis, and many other diseases. Material scientists hope to create new, less expensive, and environmentally friendly methods for producing plastics. The future of biotechnology will change our lives.

Pest-resistant crops are being developed to produce more food. New materials and fuel sources help provide shelter and energy. New drugs and vaccines have been and are being developed to improve health and protect from disease. Biotechnology discoveries are improving our quality of life, which will help us live longer, happier lives.

QUESTIONS

Answer the following questions using correct spelling and complete sentences.

1. What is quality of life?
2. What life conditions contribute to quality of life?
3. How is plant biotechnology used to improve world food production?
4. What is an example of the use of a hormone with animals?
5. What is the general trend in life expectancy?
6. What are two examples of plant biotechnology?
7. What are three areas in medicine where biotechnology is used?
8. How do vaccines help the body resist disease?
9. What is material? How is biotechnology used in material production?
10. How does biotechnology relate to the environment?

EVALUATING

Match the term with the correct definition. Write the letter of the term in the blank that is provided.

a. pharmaceutical
b. designer drug
c. life expectancy
d. gene therapy
e. crop
f. livestock
g. vaccine
h. health
i. disease
j. nutraceutical

_____ 1. A food with disease-fighting benefits.

_____ 2. Plants grown for food and fiber.

_____ 3. Animals that are used to produce food and fiber.

_____ 4. State of the body as related to disease.

_____ 5. Used to prevent diseases before they occur.

_____ 6. Drugs used to help fight specific diseases.

_____ 7. Computer modeling is used before laboratory testing.

_____ 8. Holds promise to treat genetic diseases.

_____ 9. The number of years that a person may be expected to live from his or her date of birth.

_____ 10. An abnormal condition of the body that impairs normal functions.

EXPLORING

1. Investigate the use of transgenic crops. Contact a local seed dealer, extension agent, or other individual for information. Collect brochures and other materials on genetically modified plants used in your area. Visit a crop demonstration plot and interview the person in charge of the plot. Prepare a written report on your findings. Be sure to offer your assessment of the technology in the report.

2. Make a field trip to a local college or university to observe a biochemist or molecular biologist at work. Ask about the kind of biotechnology he or she is using. Provide an oral report in class on your findings.

3. Conduct a plant tissue test. Study the presence of nitrogen, phosphorus, and potassium in plant tissue. Use a kit, such as the LaMotte Plant Tissue Test Kit, for green plants. Test three freshly cut samples, such as corn, clover, or beans. Keep a log of your findings. Prepare a report on your findings. (For additional information, refer to Chapter 4 of *Biological Science Applications in Agriculture* by Edward W. Osborne and available from Interstate Publishers, Inc., Danville, Illinois.)

4

SAFETY WITH BIOTECHNOLOGY

Accidents sometimes happen. They are events that occur unintentionally. What if a person knew an accident was going to happen? The person would change what he or she was doing so the accident would not happen!

Some accidents can be avoided. Following the proper practices helps to prevent accidents. When hazards exist, we can take steps to assure that we are safe. Taking the appropriate steps requires knowledge of potential hazards. It also requires that we act properly around hazards.

"Think safety" is a slogan in many places where accidents might occur. Work sites, laboratories, homes, and other places have hazards. By thinking about and following through with safety, most hazards can be handled. This reduces personal injury and property damage.

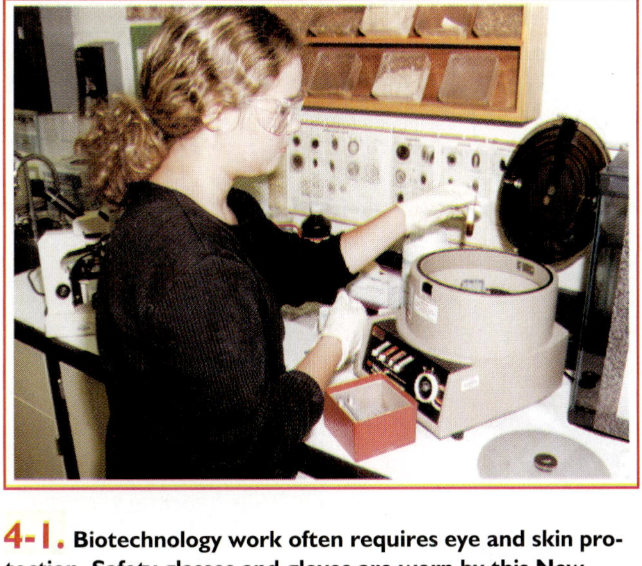

4-1. Biotechnology work often requires eye and skin protection. Safety glasses and gloves are worn by this New Hampshire student in preparing to centrifuge a bovine blood sample.

OBJECTIVES

This chapter covers important areas of safety in biotechnology. It has the following objectives:

1. Explain the meaning of safety as related to biotechnology.
2. Identify potential hazards associated with biotechnology and describe approaches in reducing risks.
3. Evaluate safety guidelines with biotechnology research.

TERMS

animal well-being
Asilomar Conferences
biohazard
containment
ectotherm
endotherm
euthanasia
hazard
hazardous waste
industrial biotechnology
material safety data sheet (MSDS)
negligence
personal protective equipment (PPE)
protocol
risk
safety

SAFETY AND HAZARDS

Safety is using practices that prevent injury or loss. These are precautions that people take so they are safe. No one wants to be injured personally nor create conditions that harm other people or the environment. Taking precautions reduces the chance of something causing an injury.

4-2. Signs in this North Carolina plant growth facility remind people of safety.

HAZARDS AND RISK

Hazards are present with most of the things that people do. Biotechnology work is no exception. A ***hazard*** is a danger. Some sort of risk is present. ***Risk*** involves the chance of a loss or injury. People can reduce risk and the chance that a hazard will cause an injury. This means that they take steps to assure a safe place for work.

Examples of hazards in biotechnology work are:

- liquids splashing into the eyes
- caustic liquids coming into contact with the skin
- biological organisms or their substances getting on the body causing allergies or injury, including the transmission of disease to humans
- electrical shock from improperly installed and grounded equipment
- injuries from animals, plants, and other organisms
- falls resulting from slick floors or tripping over objects
- loud noises injuring hearing
- cuts or other abrasions from broken test tubes, knives, or similar devices
- organisms being released into the environment creating a hazard

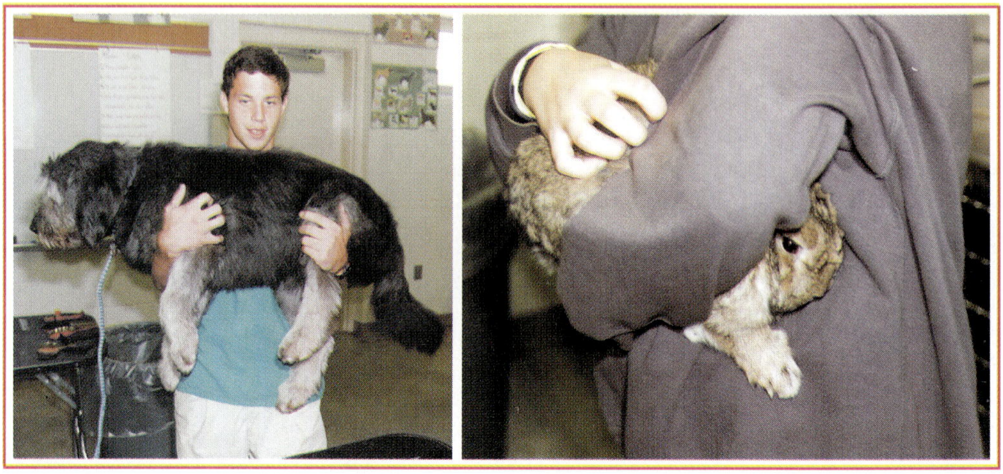

4-3. **Properly handling animals helps prevent injury. These photos show how to properly carry a large dog and a rabbit.**

TAKING RESPONSIBILITY

People can take responsibility for safety. They can use practices that protect themselves as well as other people, equipment, other organisms, and the environment.

Exercising Care

People must use care in biotechnology work. Risk is reduced by following safe practices. People who do not exercise care are said to be negligent. ***Negligence*** is failing to take a reasonable degree of care. It especially applies to situations that involve other people, such as students in the school lab or coworkers on the job.

Carelessness and neglect are closely related to negligence. Carelessness occurs when a person does not pay attention to his or her work. The individual fails to realize that the work may pose a risk to himself or herself, to others, or property. Always go about biotechnology work in a responsible manner. Keep facilities and equipment in good condition.

Acting in Good Faith

"Acting in good faith" is doing what is judged as best. People are wise and cautious. They take no unnecessary risks. They are careful in their work. They remind others to follow safe practices.

Biotech Connection

DRESSING TO PROTECT ORGANISMS

Wow! What are these biologists doing? They are wearing special clothing for a safety reason. It is not for their own safety but for the safety of the organism involved.

The biologists are trying to prevent the extinction of a species: the whooping crane. They are moving eggs of endangered whooping cranes into an incubator. Extra care is needed to assure that the eggs hatch. They do not want the young to contract disease.

The biologists are wearing special clothing to assure a clean environment. They remove their street clothes before going into the facility. Special lab frocks are worn that completely cover the body. Upon leaving the lab, the frocks are removed. The frocks undergo sterile laundering and storage before they are worn again. The biologists change back into their street clothes when leaving the facility. This keeps down the chance of a disease organism entering the hatching area. (Courtesy, Luther Goldman, U.S. Fish and Wildlife Service)

Courts of law frequently use "acting in good faith" when cases arise where a person had a loss. The liability of an individual is often reduced when it can be shown that the person was "acting in good faith."

POTENTIAL HAZARDS AND BEING SAFE

Areas of hazard and safety to consider in biotechnology are: personal protection, facility safety, environmental safety, and organism safety.

PERSONAL PROTECTION

People can take steps to help prevent injury to themselves. A person who does not do so is not very smart!

Personal protective equipment, also known as PPE, includes devices that people wear to protect themselves from injury. PPE is commonly available to protect the eyes, hearing, skin, hands, and respiratory system. Each individual must have and properly use the appropriate PPE.

4-4. PPE that may be needed in biotech work.

Eye PPE

Eye PPE is needed in situations where liquids or solid particles might splash or fly into the eyes. When bright lights are present, lenses are needed to prevent eye burns. Protecting the eyes from splashing liquid is likely the most pressing need in biotechnology work.

Eye PPE includes safety glasses, goggles, and face shields. In most biotechnology labs, goggles are recommended over safety glasses and face shields.

When properly designed, fitted, and used, goggles provide the best protection. They should fit snugly around the eyes. No loose fitting spaces should be present. A loose fit will allow liquids splashed on the forehead or face to run into the eyes. The strap should be adjusted to assure a snug fit. The best goggles have a double plastic lens and vents that prevent a condensation build-up. The lens should be kept clean. Each person should have their own goggles and not share them with others. Often, it is necessary to share. If shared, a sterilizing cabinet with ultraviolet light should be used as the storage cabinet. Goggles should be exposed to the light between uses.

Safety glasses offer a great deal of protection in most situations. The best safety glasses have browguards and sideshields. These features assure a good fit. The lens in safety glasses should be break-resistant plastic.

Face shields are worn when the entire facial area needs protection. Safety glasses or goggles are worn underneath face shields.

Both goggles and face shields can be worn over corrective vision glasses. In some cases, corrective vision glasses can be obtained that meet safety glasses requirements.

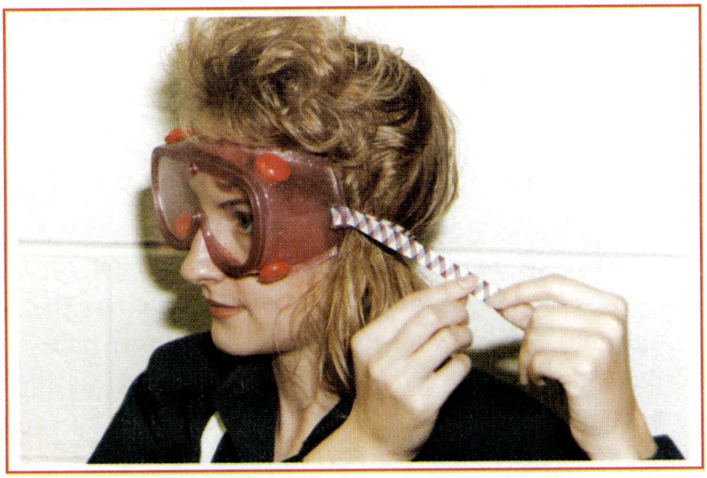

4-5. Adjusting goggles for a good fit.

Career Profile

RESEARCH VETERINARIAN

A research veterinarian studies diseases and other areas related to animal health and well-being. The work is often in labs but may be outside in research areas. Some use biotechnology in their work.

A research veterinarian needs a degree in veterinary medicine and a doctorate in a related area of science. Most have undergraduate degrees in an area of science or agriculture. High school classes in science and mathematics help build a good base for future education. Courses in agriculture and practical experience with animals are beneficial.

Jobs for research veterinarians are found where veterinary medical research is carried out. Colleges and universities, government agencies, and some businesses may have them. This photo shows a research veterinarian studying a hazardous poultry virus known as the Hong Kong Flu. (Courtesy, Agricultural Research Service, USDA)

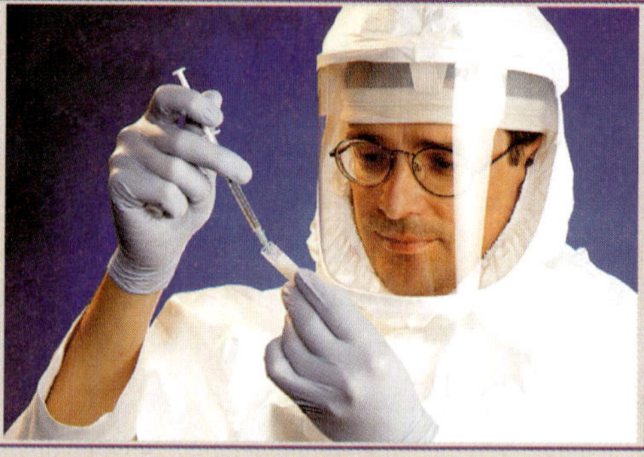

In some cases, laboratory equipment may have built-in shields to provide some protection. Always wear goggles in addition to assure good protection.

Most biotech labs recommend that contact lenses not be worn. This is mostly due to the risk involved when working with organic solvents. Such solvents can cause reactions with the contact lens material, which could injure the eyes. Of course, wearing the proper goggles reduces the risk of solvent getting into eyes of contact wearers.

Hearing PPE

Hearing PPE is needed if the work place has noise levels that can damage hearing. The noise may be continuous, such as the whir of motors, or occasional or repeated, such as loud bangs from hammers or equipment. If in doubt, a decibel meter can be used to determine if the noise is sufficient to damage hearing. Interestingly, hearing damage is often not immediate. Years may pass before hearing deteriorates due to excess noise.

Two kinds of hearing PPE are typically used: earplugs and earmuffs. An earplug is a soft pliable device shaped to fit snugly in the ear. Earplugs reduce the level of sound that reaches delicate eardrums. Both ears must be protected. Earmuffs for hearing protection fit over the ears and are held in place by a metal or plastic band. Earmuffs designed for outside wear in cold weather do not afford protection from noise. In some places of work, earmuffs are attached to hardhats. To be effective, the exterior of earmuffs should be pliable and readily adjust to the size and shape of the ear.

Facilities can be designed to abate sound. Carpet, draperies, ceiling tile, or specially designed sound abatement materials can be used on walls and ceilings or around equipment that produces loud noise. Carpet and draperies may not be recommended in labs. Most biotech labs do not have excessive noise.

Skin and Hand PPE

Clothing and gloves are often needed to keep caustic materials from coming into contact with the skin. Most often, people working in biotechnology wear gloves to protect the hands. The kind of glove worn depends on the hazard involved. Some hazards involve only liquids or solid materials while other hazards may involve heat or cold. Gloves may also be needed to protect the hands from cuts or other physical injury. Some gloves are long and may extend to near the elbow or above the elbow. Other gloves are short and reach only to the wrist.

Gloves to protect from liquid or other substances are usually made of rubber or plastic materials. These gloves prevent materials from coming into contact with the skin. Proper use and disposal of used gloves are essential.

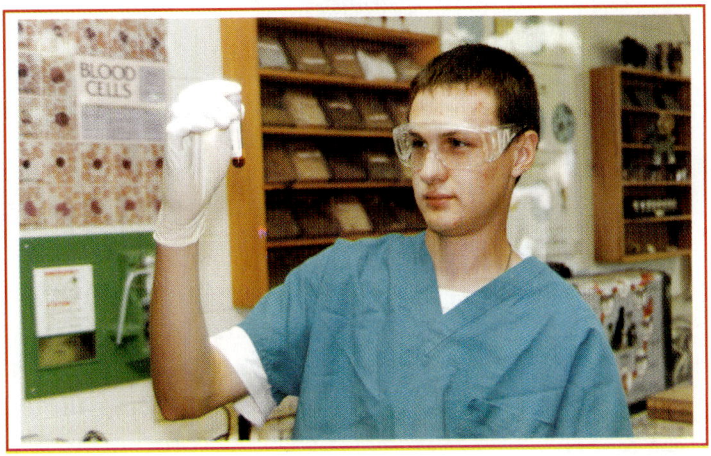

4-7. Protection depends on the hazards present. This student has eye protection, gloves, and a cotton gown over his street clothes.

Gloves also prevent the contamination of aseptic areas by contact with human skin. If the gloves are contaminated, they will spread contamination.

Aprons, body suits, and other PPE to protect the body may be needed. This depends on the nature of the work.

Respiratory PPE

Respiratory PPE is needed in areas where hazardous fumes, dust, or other airborne materials may enter the respiratory system. Particle masks offer some protection from dust in the air. A particle mask removes only the larger airborne materials.

A respirator may be needed with hazardous materials. Respirators filter the air and remove nearly all materials regardless of size. Respirators must fit properly. Filter cartridges should be regularly replaced to assure that the respirator is working properly.

Most biotechnology laboratories are equipped with fume hoods. Experiments that release fumes into the air should be carried out inside a fume hood. Proper ventilation is essential. The experiment should be set up so it is carried out in a good location. Always be sure a fume hood system is turned on before beginning an experiment.

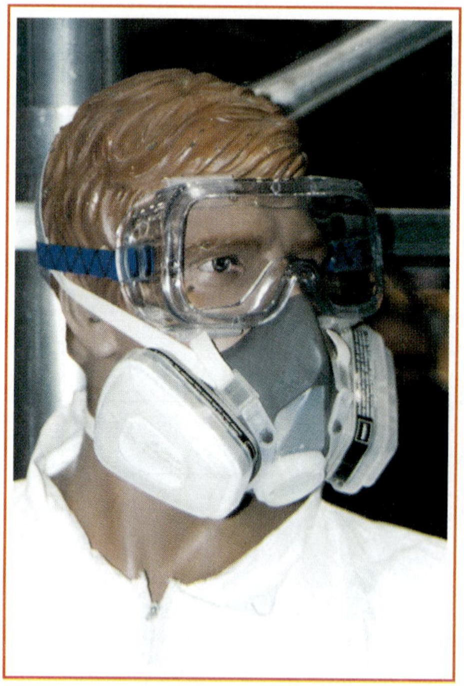

4-8. **A mannequin shows the use of a respirator and goggles.**

In some laboratories, supplied air systems are used. These are important where respirators and fume hoods do not provide clean air for breathing. Such systems may involve covering the entire body or only the head. The system is connected by a hose to a source of air. In special situations, individuals might wear systems that have air tanks attached and carried by the person.

PROTOCOL SAFETY

Protocol is the plan for carrying out a scientific research procedure. Protocols vary depending on the nature of the research being done. In most cases, a definite procedure is a part of the protocol.

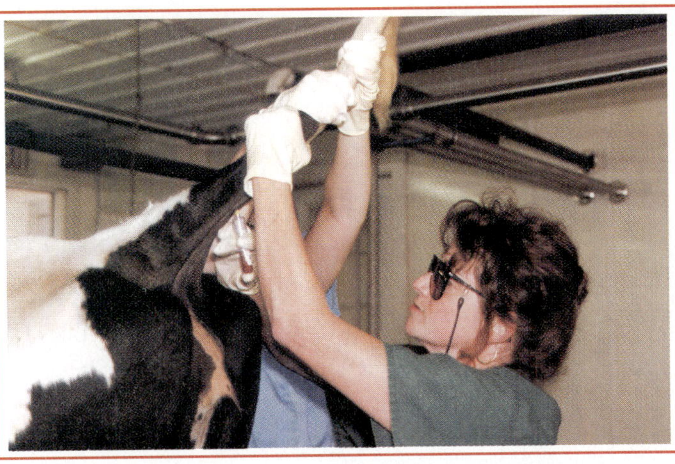

4-9. Protocol safety in collecting blood from the tail of a cow involves hand and eye safety.

This assures that appropriate scientific methods are used. Following the right protocol promotes all aspects of safety.

Good protocol is evident in how a person goes about the work. The person respects the safety of himself or herself as well as the organisms or tissues being used and the environment. Safe practices should be a part of all protocol. Unsafe practices may make the results invalid.

Protocol includes several areas. Equipment must be used properly. Work areas must be appropriately clean. Tests or treatments must follow the proper procedures. Contamination of the work must be prevented. Observations must be accurately made and recorded. A failure at any point jeopardizes the work. Failures also jeopardize safety.

An example of proper protocol is the use of agrobacterium. This involves using two species of soil bacteria with the scientific names of *Agrobacterium tumefaciens* and *A. rhizogenes*. These bacteria are used by some biotech-

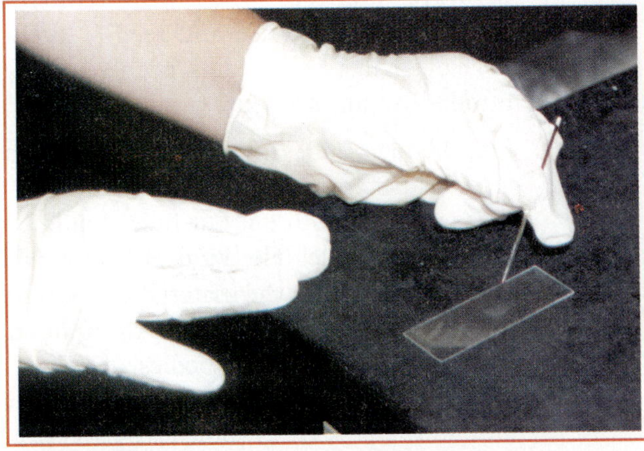

4-10. Protocol safety in preparing a slide with bovine blood.

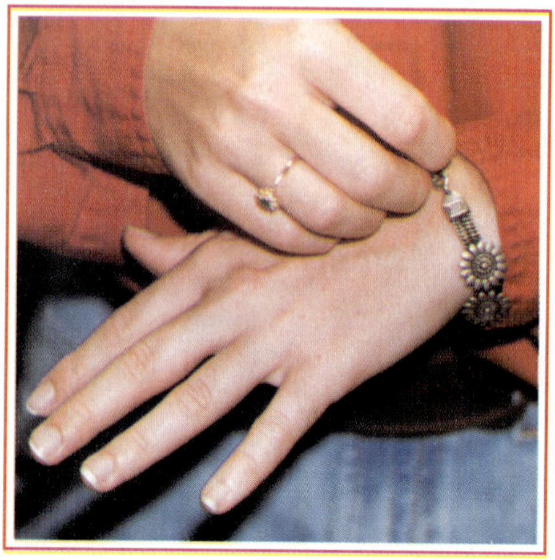

4-11. Always remove jewelry before working in the lab.

4-12. A combination emergency eye wash and drench shower.

nologies in making gene transfers. Following the proper protocol is needed to assure success and prevent unsafe conditions.

FACILITY SAFETY

The facility in which biotechnology work is carried out should promote safety. This includes a wide range of safety features. Here are a few examples:

- Emergency eye wash equipment—Emergency eye wash equipment is needed in any place where substances that cause eye injury may get into the eyes. A hand-held eyewash bottle is best if several are readily available around a lab. All people in a lab should know how to use them. Stationary emergency eye wash fountains are often found in labs. These are excellent if kept fresh by frequently running water through the system. A system that is not regularly flushed or used can have a build-up of bacteria that cause disease.
- Drench shower—A drench shower is needed in any facility where people might get spills or splashes on their clothing or bodies. The showers may be freestanding or wall-mounted. The on-off control should be easy to use and readily available even if a person has been partially blinded by an accident.
- Fire blanket and extinguishers—In labs where substances that can burn are handled, equipment to put out fires should be readily accessible. The kind of equipment needed depends on the material that might burn. Fire blankets are used to smother fires on the clothing or bodies of people. The blanket is thrown over or wrapped around the person. In some cases, the per-

son falls to the floor or ground and rolls in the fire blanket. Whatever is done must happen quickly to minimize the injury from burns. The kind of fire extinguisher depends on the fuel for the fire. Water and dry chemical extinguishers are commonly found in labs.

- Construction materials/furnishings—Facilities may need noise abatement materials on walls, ceilings, and other locations. Floors may need mats or carpets. Lab floors should be well drained to prevent slipping or falling in the event water has accumulated on them. For ease of cleaning, lab floors often have drains. The floor surface should have a gradual slope toward the drain. Fans and other ventilation equipment should be built-in.
- Layout—Facilities should be arranged to promote safety. This includes the location of doors, light switches, relationships between different work areas and equipment, and convenience of areas where supplies and other materials are stored. Air locks and devices that limit access to facilities are needed with medium- and high-risk activities.

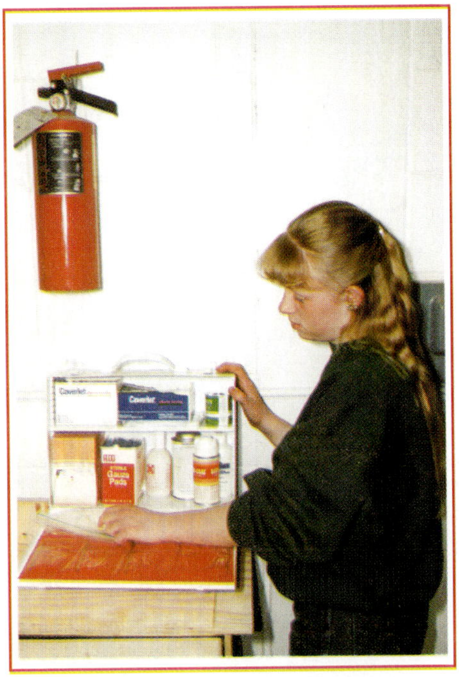

4-13. **First aid kit with a fire extinguisher nearby in a Montana school.**

ENVIRONMENTAL SAFETY

Environmental safety is protecting the environment. It is carrying out biotechnology work so organisms, substances, and other materials are not improperly released. Some people are concerned about the accidental release of transgenic organisms.

All laboratory work should be carried out using the proper protocol. Experiments should be properly designed. People should take pride in their work and do a good job. Sloppy work has no place in a lab.

Hazardous Wastes

Hazardous wastes need special care. A ***hazardous waste*** is a substance, tissue, or organism that is potentially dangerous to human health or the en-

4-14. **Hazardous materials are stored in a safety cabinet. Note the clipboard on top of the cabinet for recording materials use.**

vironment. These wastes may be solids, liquids, or vapors. If released or improperly disposed of, hazardous wastes pose danger in the environment. Pollution and dangerous residues may result.

Proper disposal is needed with wastes. Some wastes pose definite hazards. These wastes should be disposed of within approved guidelines. Tissues may need to be incinerated (burned). Empty containers may need to be returned to the source or properly cleaned. All biotechnologists should be well aware of and follow proper disposal procedures. Some materials may be stored and picked up by a disposal company. For example, a separate freezer may be used for storing animal tissues for disposal.

Some substances may not be washed down drains into sanitary sewer systems. These substances may damage plumbing or harm natural degrading processes in lagoons or other disposal systems.

Safety Information Sheets

Chemical materials should have material safety data sheets (MSDS). A ***material safety data sheet*** is a summary of information about the safe use of a chemical. Companies that sell chemical supplies furnish MSDS with new

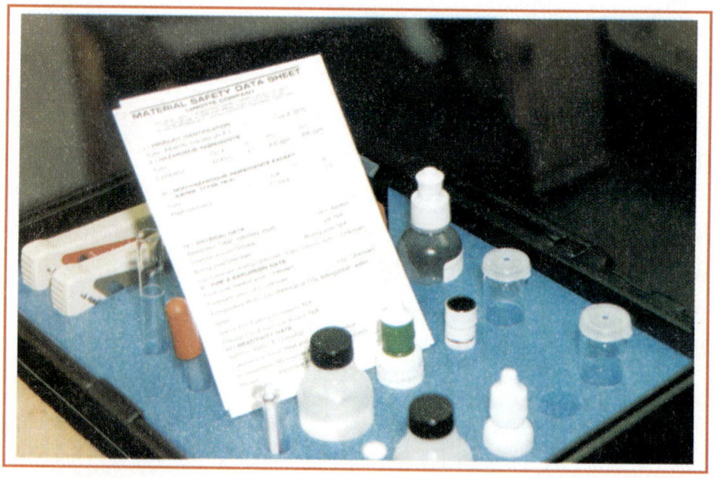

4-15. **Material safety data sheets accompany chemical products when sold.**

shipments. These sheets should be available in notebooks, posted on bulletin boards, or instantly available on a computer. All people using chemicals should refer to MSDS before using the substances. This provides the needed information to safely use the material.

ANIMAL WELL-BEING

Biotechnology work sometimes involves the use of living animals. ***Animal well-being*** is caring for animals so their needs are met and the animals do not suffer. This means that the animals should have proper housing, be fed properly, live in a clean area, and not have cruel treatment. Animals should be appropriately transported and handled. Agencies and businesses may have policies on the use of animals. If so, these should be followed.

It is sometimes necessary to cause the death of an animal. Animals that are sick or severely injured may live for a while. Those born with abnormalities or injured when mistakes are made in research should be properly disposed of. To some people, euthanasia is an acceptable practice to relieve animal suffering. ***Euthanasia*** is causing death without pain or stress. Farms that produce animals are developing euthanasia plans. Such plans cover safe and economical ways of disposing of animals without pain. Methods are used based on the size of the animal, the cost, and aesthetics. Once death has been induced, the carcass should be carefully disposed of by incineration or burying. In some cases, composting may be acceptable. Composting is now used on poultry farms for disposing of birds that die of natural causes.

Opinions differ on animal well-being. Some people feel that animals should never be used in research or experiments. Others feel that it is okay to

4-16. A clean pen provides for the well-being of this puppy.

do so if the animals are treated properly. People should get the facts before forming opinions. What is your opinion? You may want to refer to other sources for information. Always know and follow the procedures of your school or employer with animals.

ORGANISM SAFETY

Organism safety is a concern some people have with biotechnology. This refers primarily to biotechnology that involves disease organisms or those with genetic manipulation. A few people feel that "uncontrollable monster organisms" may be produced and accidentally released by laboratories. Such problems are not likely. Practices are used to assure that it does not happen.

The concern also refers to industrial-scale biotechnology. ***Industrial biotechnology*** is the mass production of substances or organisms using a factory approach. It is the efficient approach to provide large quantities of biotechnology products. This large-scale production is sometimes referred as factory production.

People doing biotechnology work should determine the risks with their work. They should take appropriate steps to reduce risk.

U.S. GUIDELINES

Several efforts have been made by scientists to develop guidelines for biotechnology. The ***Asilomar Conferences*** were among the first efforts. This involved two conferences held at the Asilomar Conference Center in California to focus on biohazards. A ***biohazard*** is a potential risk from working with or developing organisms. The risk may be to people and other organisms, the environment, and property.

The first Asilomar Conference, held in 1973, dealt with the study of viruses. The second Asilomar Conference was held in 1975 to deal with the use of recombinant DNA safety. The National Academy of Sciences, the National Institute of Health, and the National Science Foundation promoted the Conferences.

The guidelines have focused on laboratory safety and containment of organisms. ***Containment*** is taking steps to control the accidental release of potentially harmful organisms or their products. It includes protecting the scientists who are working with the organisms.

Three levels of containment are used:

- Low Containment—Low containment guidelines are with organisms that normally exchange genetic information (reproduce within species). These are lower-order species including plants, invertebrates, and ectothermic vertebrates. An ***ectotherm*** is an organism whose body temperature adjusts to its environment, such as snakes, fish, and lizards. (These are sometimes referred to as "cold-blooded animals.") Basic guidelines include:
  - using good techniques,
  - following normally expected safety practices, such as wearing lab coats, and
  - using mechanical pipettes.

  As with all labs, eating in the lab is prohibited.

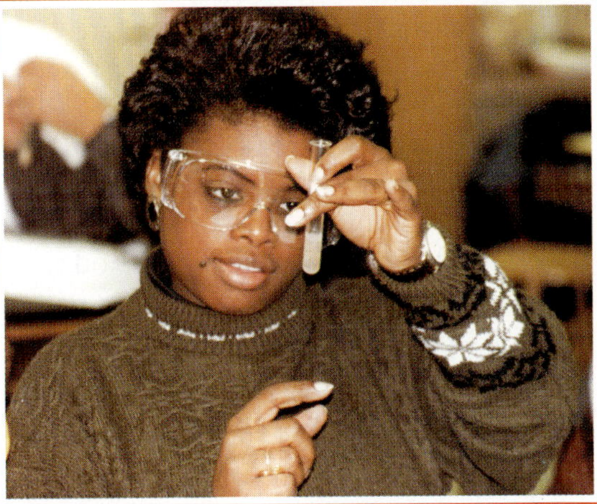

4-17. The small amount of DNA material in this tube might cause problems if not disposed of properly.

- Moderate Containment—Moderate containment is used with moderately risky organisms or their products. This includes those that deal with disease, such as cancer. Experiments with any endothermic organism are included. An ***endotherm*** is an organism that maintains a constant body temperature, such as dogs, mice, and chimpanzees. (They are sometimes known as "warm-blooded animals.") Practices include using gloves, biological exchange cabinets, limited lab access, and filters for air or other materials that may be released.

- High Containment—High containment guidelines cover organisms or DNA materials that might produce toxic or pharmaceutical products. Any study that might result in a catastrophe if organisms were accidentally released is covered. Facilities need to be isolated and include air locks for entry. People who enter will change clothes and shower before leaving. People must wear protective clothing. Environmental controls reduce the air pressure in the facility.

All levels of containment require the proper disposal of wastes. PPE should always be used. Determine the risk involved and follow appropriate guidelines.

OTHER GUIDELINES

Guidelines vary outside the United States. The World Health Organization does not have evidence that genetically altered organisms pose greater hazards than some other organisms.

Most guidelines address the likelihood that an organism will escape. Both physical and biological methods are used to contain organisms. This includes barriers that prevent escape as well as proper cleaning.

Guidelines in other nations are similar to those in the United States. For example, the United Kingdom classifies organisms as minimal risk, normal risk, high risk to individuals, and high risk to individuals and community.

OUTSIDE SOURCES

Facilities need to be secure to keep out unauthorized people. Intruders may damage equipment, release animals, or otherwise foul research and development work. Good locks, gates, security devices, and other means are needed to secure a facility.

4-18. Some facilities must be protected with security fences.

REVIEWING

MAIN IDEAS

Safety is using practices that prevent injury and loss. It is the responsibility of everyone in biotechnology work. Taking precautions reduces the risk of a hazard being a safety problem. Failing to take responsibility is negligence.

The first step in safety is to use personal protective equipment (PPE). This equipment protects the eyes, hearing, skin and hands, and respiratory system. PPE should be properly selected and used. Safety must be part of the protocol in biotechnology work.

The facilities for biotechnology work should have safety features. Facilities should have emergency eye washes, drench showers, fire blankets and extinguishers, proper noise abatement, allow for safe movement, and a convenient layout. Hazardous wastes should be properly disposed of. Material safety data sheets (MSDS) should be posted and/or conveniently filed. The practices followed with MSDS should be prudent and within guidelines.

Animal well-being is essential in biotechnology work using animals. The animals should be properly fed, housed, and otherwise cared for. Injured or diseased animals need help. Medical attention may be needed. Based on the severity of injury, euthanasia may be used.

Guidelines have been developed for biotechnology research. Those followed in the United States arose from the Asilomar Conferences. In general, these guidelines focus on the need for containment as low, moderate, or high.

QUESTIONS

Answer the following questions using correct spelling and complete sentences.

1. What is safety? Why is it important?
2. List at least three examples of hazards in biotechnology work.
3. What is "acting in good faith"? Why is it important?
4. What is personal protective equipment?
5. What personal protective equipment is used for eye safety? Which is best?
6. When is hearing PPE used?
7. When is skin and hand PPE used?
8. What facility considerations promote safety?
9. Why is animal well-being important? When is euthanasia used?
10. What three levels of containment are typically practiced in the United States? Distinguish between the levels.

EVALUATING

Match the term with the correct definition. Write the letter of the term in the blank that is provided.

a. safety
b. euthanasia
c. animal well-being
d. endotherm
e. hazard
f. protocol
g. hazardous waste
h. material safety data sheet (MSDS)
i. biohazard
j. Asilomar Conferences

____ 1. The risk potential of working with or developing organisms.

____ 2. Events where scientists gathered to develop guidelines for biotechnology work.

____ 3. Using practices that prevent injury or loss.

____ 4. A danger.

____ 5. Caring for animals so their needs are met.

____ 6. Causing death without pain or stress.

____ 7. A summary of information about the safe use of a chemical.

____ 8. An organism that maintains a constant body temperature.

____ 9. The plan for carrying out an experiment, including safe practices.

____ 10. A substance, organism, or tissue that poses potential danger to humans and the environment.

EXPLORING

1. Tour a biotechnology facility. Observe the safety practices that are followed. Have the safety officer of the facility address important safety issues. Prepare a report on your observations.

2. Study the safety practices followed in the school lab. Assess the adequacy of the practices and suggest needed changes to improve safety.

3. Try on various items of personal protective equipment. Properly adjust the equipment. Clean and store the equipment properly. Be sure you know how to use the equipment in laboratory situations. Demonstrate your skills to your teacher. Correct any errors in the use of the equipment.

5

RESEARCH: THE SCIENTIFIC METHOD IN ACTION

Have you ever had good luck? Nothing beats good luck. Just ask someone who has recently had good luck! In science, the word "serendipity" is used. Serendipity is making desirable discoveries by chance. What about reality? We cannot rely on our luck.

In reality, few scientific breakthroughs occur by chance. They begin because scientists are looking for something. When careful scientists make cautious observations in controlled environments, the unexpected may happen. Bang! We have something new.

Successful science carefully follows a process developed over time and accepted by scientists worldwide. The process is the scientific method. The process is used to solve problems carefully and methodically. It is the foundation for scientific discoveries. If you understand and follow the scientific process, your work will be accepted and understood by scientists everywhere.

5-1. The ripening of tomatoes is being studied using the scientific method.

OBJECTIVES

This chapter introduces the scientific method. It describes the role of the scientific method in research, particularly in biotechnology. A typical biotechnology activity is used to describe the scientific method. An opportunity is provided for you to apply the scientific method to activities of your own. It has the following objectives:

1. Describe the importance of research and name two major types.
2. Explain the scientific method (process) and list the steps in the process.
3. Describe how to identify a problem.
4. Explain why literature review is important and how it is done.
5. Define hypothesis and explain the kinds that are used.
6. Describe how to design an experiment.
7. Explain how to conduct an experiment and gather data.
8. Describe how data are analyzed and used in forming conclusions.
9. Explain the importance of a report and how a report is organized.

TERMS

applied research
basic research
conclusion
constant
control
data
descriptive statistics
directional hypothesis
experimentation
hypothesis
inferential statistics
literature review
mean
median
mode
multiple hypotheses
null hypothesis
objective
replication
research protocol
scientific method
statistics
subjective
technical journal
variable

RESEARCH

Research is using systematic methods to answer questions. It provides new information and helps improve our quality of life. Research is used to solve problems. How it is carried out is important.

5-2. Research in medicine is important for the welfare of human life.

Research must be done so the findings are credible. Credible means that people believe in the research. It is truthful and represents a careful approach to scientific inquiry. The person who carries out research (known as a researcher or scientist) must use good methods and make accurate observations. Bad research may result in the wrong decisions. Good research should help in making the right decisions.

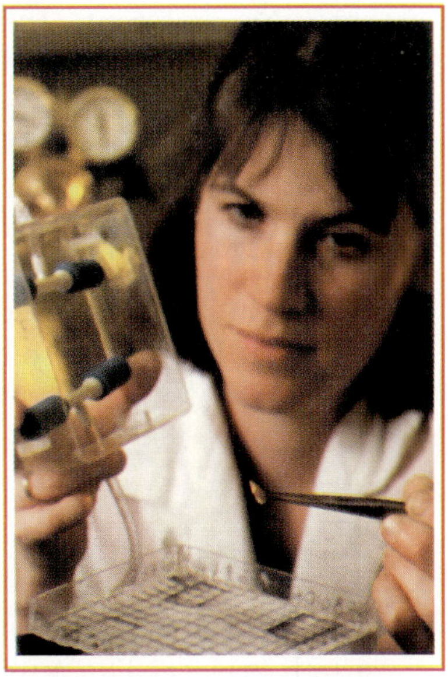

5-3. Basic research provides general information about a specific topic. (Courtesy, Agricultural Research Service, USDA)

KINDS OF RESEARCH

Two kinds of research are typically carried out: basic and applied.

When identifying a research problem, scientists may have one of two motivations. If they simply desire information, they conduct what is known as basic research. ***Basic research*** adds to

5-4. Applied research is carried out in field plots at this experiment station. By studying the cultural practices of crop production, scientists hope to improve oat yields.

the body of knowledge about a specific issue. That research may later be used to solve problems related to the issue. For example, studying how animals digest food is basic research. It provides knowledge about an important life process.

If, on the other hand, the scientists seek to solve a specific issue to meet a specific need, they are conducting applied research. ***Applied research*** is designed to directly solve an immediate problem. Most agricultural biotechnology research is applied research. It is intended to meet a specific need. For example, studying the kinds of feed that produce the best rate of gain in raising turkeys is applied research. It has practical application in turkey production.

HOW RESEARCH IS DONE

Research may be used in two broad ways: 1- to observe something or 2- to test what happens if something is done. Observing something is getting information about it. Testing something involves seeing the results of an action.

Observing Something

Have you ever filled out a survey form? The survey was likely part of a research project. The purpose was to get information to describe something. This is known as survey research. The survey form was designed to collect certain information. It is often used to assess human feelings or actions. This kind of research often provides useful information.

Testing Something

Testing an action involves experimentation. ***Experimentation*** is the process of conducting tests which, together, establish the validity of a hypothesis. Careful planning is a key to successful use of the scientific method.

In designing the experimentation, two goals exist. The first is to summarize activities to this point. The second is to describe the future activities, or experiments, and the strategies to be used. The design step usually results in a written document that can be shared with others interested in the issue.

Experiments are carried out in a wide range of settings. Many are inside, in labs. Others are outside in test plots or groups. For example, the basic research example of determining how digestion occurs may be in a carefully designed lab in a building. The applied research to see which feeds produce the best growth in turkeys may be on a range or in a poultry house.

5-5. A scientist is carefully examining young turkeys being used in an experiment. (Courtesy, Agricultural Research Service, USDA)

INTEGRITY

The findings of research are no better than the methods that are used. Research must be carefully planned. The appropriate methods must be followed. The researcher must be honest and truthful.

THE SCIENTIFIC METHOD

Scientists have developed a series of steps used in solving problems. The steps are known as the ***scientific method***, or scientific process. Each step is designed to insure that conclusions about scientific issues are drawn from complete and accurate information. The steps took years to develop and test.

5-6. **The effects of an insect control method in cotton production may be studied using an insect trap.**

The exact descriptions of the steps vary slightly from one scientist to another. Some scientists think of the scientific method as being linear. This is beginning at one end and working consistently to the other. Others see the method as being cyclic, with the solution of one problem leading to the rise of other problems. Some scientists describe the method as having only five steps, while others identify as many as eight. It all depends on the lines created between the steps of the method.

In this text, the scientific method has seven steps:

Step 1: **Identify a problem**

Step 2: **Investigate (research) the problem**

Step 3: **Form a hypothesis**

Step 4: **Design experimentation**

Step 5: **Conduct experimentation and collect data**

Step 6: **Analyze the data and form conclusions**

Step 7: **Report the results**

IDENTIFYING A PROBLEM

We face many problems every day. These are individual, community, and global-level problems. Issues such as obesity, environmental destruction, and population growth are challenges to individuals and our society. Although the scientific method is designed to solve scientific problems, similar strategies can be used in almost any situation.

The first step in the scientific method is to recognize a problem. A problem may be an issue that has a direct impact on a scientist. A problem may affect a local community or be one about which the scientist has special knowledge and skills. Often, scientists work on problems that are of interest to the company or agency that has engaged them.

Career Profile

PLANT GENETICIST

A plant geneticist works to develop improved crop plants. The geneticist may use traditional methods of plant breeding or may create transgenic plant material. The goal typically is to develop plants with resistance to diseases or insects, resistance to herbicides, improved nutrition, and/or improved taste and storage. A geneticist may experiment with 40,000 different plants to bring a single plant to full-scale production. It may take as long as 20 years for a new variety to become available commercially.

Education to be a plant geneticist begins in high school with science, mathematics, and agriculture classes. Most have college degrees in genetics, botany, agronomy, or a related area. Master's and doctor's degrees are in genetics or a closely related area.

Jobs for plant geneticists are with USDA agencies, colleges and universities, and private agribusinesses. This photo shows a plant geneticist examining poinsettias in an experiment.

In identifying the problem, be specific and clearly define the issue. For example, it is not efficient to describe the problem as "environmental contamination." This subject is too broad. No scientist could possibly solve the problem. If, on the other hand, the scientist defines the problem as petrochemical (such as oil) contamination of aquatic environments, then the work is much more manageable.

A problem must be accurately identified. It is more than a symptom. A symptom is evidence of a problem. For example, a dead fish floating on the water in your aquarium is a symptom or sign. The problem is why the fish is dead. Removing the fish does not solve the problem.

INVESTIGATE THE PROBLEM

The second step in the scientific method is to learn as much as possible about the problem. This gathering of information is known as a ***literature***

review. The purpose is to get a good understanding of the topic and to learn what has already been done. This saves investing time doing work that has already been done. It also helps in designing experimentation.

Literature review may lead to new problems or different aspects of a problem. This illustrates the cyclical nature of the scientific method.

HOW TO DO IT

Many approaches are used in literature review. Always keep a complete and accurate record of the sources used. This makes it easy to relocate the sources at a later date if necessary. It also makes it easy for others to access the source when they read your research report.

Many resources are available. You may choose to begin in a local public library or at a school library. Begin with textbooks or reference books that provide background information. Next, use research reports and journal articles on the specific problem that is being studied. Some people use the Internet and the World Wide Web. Interviewing authorities on a subject is also useful. Some beginning researchers give up too easily. They fail to persist in finding materials on a topic.

5-7. Using books on research methods can help sharpen your skills.

SOURCES OF INFORMATION

In general, the best sources for a broad overview or basic explanation of a topic are textbooks, reference books, news magazines, and popular literature. These are useful in beginning research and may provide sources for further investigation. They typically do not contain the technical content needed for effective research.

Other sources are the general scientific magazines. Examples include *Science News, Technology Review, Science Teacher,* and *American Biology Teacher.* These magazines have articles on a wide range of topics of interest to scientists.

The most detailed written sources are technical journals. A ***technical journal*** is a publication written by specialists in a given field for their peers. They are often published monthly, quarterly, or other times. Technical journals use specialized terms and are often difficult for laypeople to understand. They are often the best sources. In particular, each journal article typically cites other literature related to the topic, thereby allowing expanded research with each source consulted. Examples of specialized journals are the *Journal of Pesticide Reform* and the *Journal of the American Veterinary Medicine Association*.

Book catalogs, on-line book sellers, book stores, and local libraries are good sources for finding information. The information offices of agricultural experiment stations at land-grant colleges often have bulletins and research reports available free of charge.

In doing a literature review, consider other scientists actively involved in similar research. Make telephone calls, write letters, visit research facilities to interview scientists, and use the Internet. With the Internet, you may establish personal contacts with scientists around the world. Never overlook local people. They may prove one of your best sources of information as well as help in designing your study.

In the example of a scientist interested in oil and gasoline spills into aquatic habitats, a thorough review would reveal that scientists have developed a wide number of mechanical and biological strategies to remove these substances. In fact, scientists have actually designed bacteria that "eat" these substances. The Exxon Oil Company holds patents on bacteria and fungi that its scientists have developed for just this problem.

INFORMATION ASSESSMENT

All sources of information should be assessed for accuracy and being up-to-date. With the Internet, carefully evaluate each site for its accuracy and scientific value. Like the written sources previously described, sites vary significantly in their detail and depth of coverage. Sites may be maintained by organizations and individuals who seek to promote a specific point of view. These sites give opinions, which are described as being ***subjective.*** Scientists, on the other hand, try to deal only in facts. They want complete information. The sources are described as being ***objective***.

FORM A HYPOTHESIS

After doing the literature review, you should have information about the problem. Perhaps you have a different idea for solving the problem. Or, perhaps you want to try activities similar to those already done to see if you get the same results. In either case, the third step in the scientific method is formation of a hypothesis.

Biotech Connection

CLONING

Cloning has become a hot-button issue. It results in more than one individual with identical genetic make-up. Plant growers have used cloning for many years. They do not call it cloning. Several names are used: budding, grafting, layering, and rooting.

With cloning, the daughter plant is completely identical in genetics to the parent plant. The plants share the same gene composition. If they are used to produce fruit, the fruit is the same. Appearance of the plants as they grow may be quite different due to their environment.

Cloning assures similarity of the desired product. Apples grown from seed may not have the same fruit flavor, while cloned apples will. The same is true with pecans, peaches, pears, and many other fruits, nuts, and ornamental plants.

You can clone a typical houseplant. Simply cut off a small piece of stem and remove the lower leaves. Place the stem in either water or damp soil. It should grow roots in a few weeks. Similar techniques are used in horticulture to get new plants. Rooting hormones and a greenhouse environment promote cloning.

This photo shows the proper way to prepare a rose cutting for rooting. The plant that grows is a clone.

5-8. Montana students and their teacher are discussing ideas to help refine a hypothesis and protocol.

A ***hypothesis*** is an "educated guess" or a statement that gives a solution to the problem. The hypothesis usually indicates a relationship between the problem and a specific intervention by the scientist. For example, "The efficacy of naturally-occurring, oil-eating bacteria can be increased by the addition of nitrogen and phosphorus to the biosystem."

KINDS OF HYPOTHESES

A hypothesis that indicates a definite relationship between the problem and the action is a ***directional hypothesis***. Sometimes, a directional hypothesis cannot be proven (supported by data) because any number of other events could also cause the observed result. In this case, the scientist develops a null hypothesis. A ***null hypothesis*** simply states that no relationship exists between the problem and the action. The research is to disprove the hypothesis, something that is sometimes easier. For example, a null hypothesis for our oil-eating bacteria could be, "The efficacy of naturally-occurring, oil-eating bacteria will not change with the addition of nitrogen and phosphorus to the biosystem." If this statement is proven true, then the directional hypothesis stated above is false. If, however, the efficacy changes, the directional hypothesis gains validity–but it is not proven. If the null hypothesis is disproven in extensive experimentation, the directional hypothesis is acceptable.

MULTIPLE HYPOTHESES

A hypothesis should be stated clearly and in concrete terms. The hypothesis guides the remainder of the scientific method. If the problem is complex, a scientist may develop ***multiple hypotheses*** (the plural form or more than one hypothesis for a study) that are used to direct the work. If so, the multiple hypotheses would be listed in the research proposal.

DESIGN EXPERIMENTATION

Experimentation is to test your hypothesis. This requires careful planning. Generally, a single experiment is an inadequate test of a hypothesis.

THE DATA YOU NEED

The terms "prove" and "disprove" are sometimes used in testing a hypothesis. Some scientists disagree with these terms and prefer "supported by the data" or "unsupported by the data." ***Data*** are the items of information collected in an experiment. These are the facts or statistics that are tabulated. A hypothesis cannot be proven with a single experiment or activity.

Experimentation is very important in providing data. If it is done poorly, the work may fail or be non-conclusive. An effective design might lead to outside support of the project.

The design should include the following items in detail:

- A description of the problem
- A review of the pertinent literature
- Statement of the hypotheses
- A list of the necessary equipment and materials
- A description of the experiments you intend to conduct
- An explanation of the data you intend to collect
- A description of the statistical strategies you will use to analyze the data
- A timetable for the proposed activities

5-9. **Protocol may require the use of sophisticated laboratory equipment. (Courtesy, Emory University, Atlanta)**

PLANS AND PROTOCOLS

In planning experiments, scientists include variables, constants, and controls. It is important to be certain that the intended aspect is being studied, and not some other quality. For that reason, scientists create many environments that are exactly the same, and change just one thing. The "thing" that is changed is the ***variable***. A quality that is being controlled or kept the same is a ***constant***. In addition, at least one environment is not changed in any way. This is known as the ***control***.

Once the plan is developed, it must be shared with as many interested parties as possible. Others can often see flaws and issues that the scientist

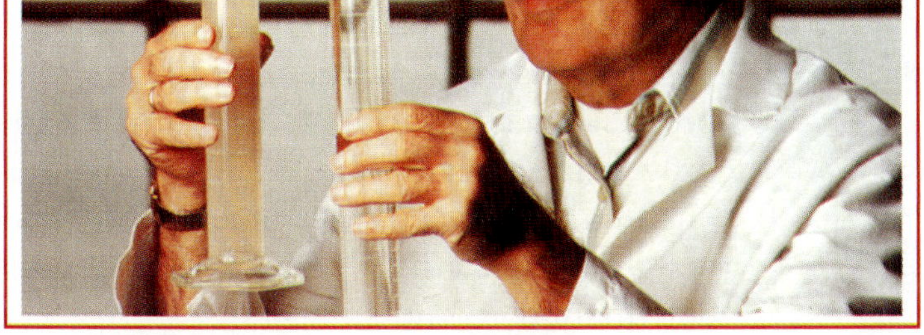

5-10. **A scientist observes a control and an experimental variable. (Courtesy, Agricultural Research Service, USDA)**

has overlooked. Outside input helps develop the strongest possible research protocol. ***Research protocol*** is how the research or experimentation is designed. It specifies the procedures that will be used.

In the example of the oil-eating bacteria, a scientist would design experimentation intended to disprove the null hypothesis. The written document would:

- describe the problem as one of environmental contamination,
- describe the journal articles and other sources of information discovered,
- list the hypothesis, and
- provide a detailed plan for testing the hypothesis.

The experimentation might be in aquariums filled with water containing the selected bacteria. The variables, which would be changed in each tank, might be additions of varying amounts of fertilizer containing nitrogen and phosphorus. All other characteristics – water source, air and water temperature, lighting, etc. – would be exactly the same, or constants. At least one tank would receive no fertilizer, thus becoming the control.

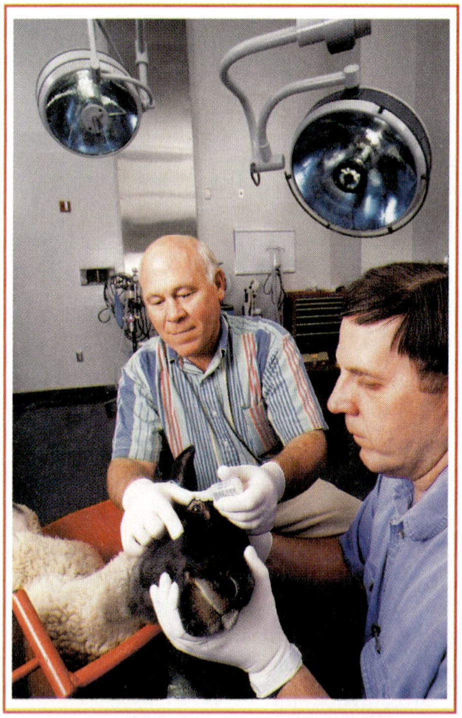

5-11. An anesthetic is applied to the eye of a sheep before taking a tissue sample. Anesthetic eliminates pain to the sheep. (Courtesy, Agricultural Research Service, USDA)

Replication is needed. ***Replication*** is running the experiment more than once. Replication avoids chance outcomes that would make the results invalid.

USE CAUTION WITH ANIMALS AND HUMANS

Experiments with animals and humans are special. The well-being of the organisms must be protected. Scientists working for companies, or at educational institutions, generally must have their proposed use of animals or humans approved by a review board. These boards ensure that the activities meet federal and state laws, are ethical, and are not unduly detrimental to the subjects.

CONDUCT EXPERIMENTATION AND GATHER DATA

Once a plan is complete, the experimentation begins. The length and complexity of the activity depends on the design of the experiments. If the design phase has been careful and extensive, the process of experimentation is simply following the plan.

Three keys to successful experimentation are safety, careful observation, and complete documentation.

Safety must be an integral part of any scientific activity. Potential risks should have been identified during the design of the experimentation, along with strategies to mitigate and/or eliminate the risk. During the work, a scientist must follow those strategies.

The introduction to this chapter mentions the concept of serendipity. Serendipity largely depends on careful observation. During experimentation, the scientist must see everything happening in the experimental environment. Unexpected results should be observed. Some scientists are blinded by their expectations. They do not see what is happening in their experiments.

Information collected during the experimentation process is data. Data must be recorded accurately, consistently, and carefully. During the design phase, the scientist will decide what information to collect and how it will be kept. Sometimes, it is recorded in writing on paper. Data are often kept on computers. Good, accurate data are essential.

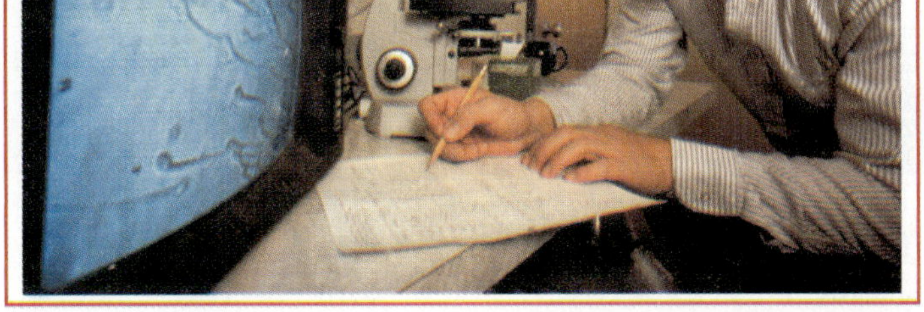

5-12. Data are being carefully recorded by this scientist. (Courtesy, Agricultural Research Service, USDA)

Careful documentation is important because it is the foundation for the decisions the scientist makes. If data are recorded incorrectly, then the whole work is without value. If data are lost, the deductions cannot be supported. Without the foundation of data, a scientist cannot proceed with the next step in the scientific method.

In our example, the scientist would set up multiple aquariums with water containing the desired bacteria. Each of the aquariums would be placed in an identical environment. A measured quantity of oil would be added to all the aquariums. At least one aquarium would not have any fertilizer added (as the control); other tanks would have varying amounts of fertilizer added. Each day, the scientist would measure the size of the oil slick in each tank as well as any other observed differences. The data would be recorded. After the determined number of days had passed, the scientist would properly dispose of the oil and water and would repeat the experiment. An experiment may be repeated many times to create the desired amount of data.

ANALYZE THE DATA AND FORM CONCLUSIONS

5-13. Computers can be used for data analysis.

Once experimentation is complete, the scientist must determine what the data mean. How do the data relate to the hypothesis?

The first step is converting the raw data into meaningful information. This is done using a science and mathematics specialty called statistics. ***Statistics*** uses numerical data to provide useful information. This enables scientists to make a ***conclusion***, or inference, about the work.

STATISTICS: A BRIEF EXPLANATION

Statistics are used in everyday life. For example, polls that measure the popularity of a political figure or a concept use statistics. One of the basic statistics, the average, is commonly used to represent a student's performance in class. Two major kinds of statistics are descriptive statistics and inferential statistics.

Descriptive Statistics

Descriptive statistics, or summary statistics, are numbers that define the data from an experiment. The number of subjects in a study is descriptive. It tells how many were involved. The number is often referred to as frequency. The weight of something is another descriptive measure, such as the weight of the tomatoes harvested comparing yields of two varieties. Another example is the oil-eating bacteria. Suppose the bacteria disposed of the oil in 5 days during one test, 4 days in another, and 4 days, 5 days, 6 days, 5 days, and 5 days in the tests that followed. We could use descriptive statistics to summarize these results.

The most common descriptive statistics are measures of central tendency: the mean, the mode, and the median. Each is a type of average. The ***mean*** (or average) is the sum of the scores divided by the number of scores. The ***mode*** is the most frequently occurring number. The ***median*** is the number in the middle if the numbers are arranged by value. Half the scores are above the median and half are below the median.

In our example, the sum of the seven tests is 34. The mean number of days is found by dividing 34 by 7, or 4.86. The mode is 5, as is the median. Each of these describes the findings of the experimentation in slightly different terms.

Descriptive statistics also consider the variability of data, the degree of difference between the measures of central tendency and the most distant results. The most common descriptor of variability is the standard deviation, which is a measure of the extent to which scores (on average) are different from the mean. Another measure of variability is the range, which is reported as the lowest score to the highest score.

The third component of descriptive statistics is correlational statistics. These are numbers that describe the strength of a relationship between the variable and some aspects of the results. In other words, correlational statistics determine if the action taken by the scientist had an impact on the results.

Inferential Statistics

Inferential statistics allow the scientist to expand the data in the experiment to predict what would happen in a larger environment. Perhaps the most important inferential statistic is the test of statistical significance. Sometimes differences occur in experimentation, but they are not great enough to justify a conclusion that the variable was the cause. At other times, the differences may appear small, but they may still indicate impact by the tested variable. Statistics use numbers to determine whether the difference is "real."

As you may have gathered, statistical methods are a very complex part of science. As with other sciences, statistics have a number of computer programs that can be used as a tool in the process. If a scientist has carefully considered statistical demands in developing the research protocol, then the work may be relatively easy. If a scientist is unfamiliar with statistics, he or she would consult other scientists or a statistician for help.

CONCLUSIONS AND RECOMMENDATIONS

Once the data have been analyzed, the scientist must begin interpretation. Do the results support the hypothesis? Why or why not? The scientist must carefully consider the numerical results and the possible explanations. Even at this stage, it is important that the scientist consider all possibilities – not just the desired one.

5-14. Carefully examine data in drawing conclusions and making recommendations. (Courtesy, Agricultural Research Service, USDA)

Sometimes drawing conclusions presents the most difficulty. The scientist must focus on the facts and be certain to maintain objectivity. Sometimes consulting colleagues helps clarify the scientist's thoughts.

In completing the work, the scientist may see additional avenues of exploration, or steps that could have been done differently. Those also are recorded, both as possible future experiments and to enhance the understanding of colleagues.

In the example of the oil-eating bacteria, perhaps the statistical analysis would find no significant difference between the experimental aquariums and the control. The scientist would be forced to accept the null hypothesis and conclude that the addition of nitrogen and phosphorus had no impact on the bacteria. The scientist might further suggest that the pH of the water or some other factor be manipulated instead.

REPORT THE RESULTS

One role of science is to add to the world's body of knowledge. If the work a scientist does is not reported, it has little value for the world. Other people do not know about it. The final step in the scientific method is preparing a report. A written report is first prepared. Journal articles and oral reports at meetings may follow.

The audience for a report varies. A company scientist might prepare it for her or his supervisors. A university researcher might provide the report to university officials and the funding agency, if any. Students develop reports for science fairs, career event displays, and their instructor.

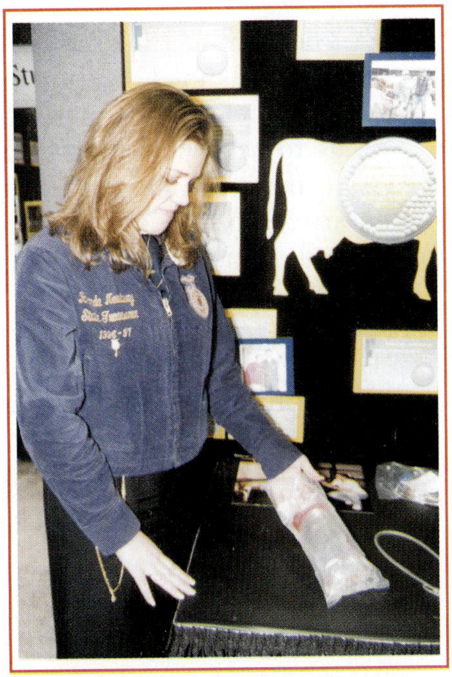

5-15. This student has an agriscience display that reports the major events in her experimentation.

REPORT FORMAT

The report describes the work in each of the steps. Most reports follow a traditional format. The following parts are often included in a report:

- cover and/or title page
- abstract (scientific summary)
- introduction (includes a statement of the problem)
- literature review
- the hypotheses
- description of the experiment design and methods used
- reported data and analysis
- conclusions and recommendations
- list of references (also known as bibliography or literature cited)

REPORT STYLE

In preparing the report, follow the standards of proper English. References on writing style should be used. These tell how to prepare various parts of a report. A common guide in agricultural education is the *Publication Manual of the American Psychological Association* (often known as the APA guide). The American Society for Horticultural Science and other associations may have guidelines for a particular area of biotechnology.

FINISHING THE REPORT

The introduction and experiment design were written when the research was being planned. It is completed during the design step. Much of the work was done before the experimentation was carried out. This part of the report needs to be reviewed and updated.

If the results are significant, the scientist may share them by submitting an article to a technical journal. Articles follow a style set by the specific journal. Journal articles are generally shorter than the written report.

Scientists also share their work with others at conferences and workshops. In general, those presentations are an oral summary of the report.

Giving good written and oral reports is important. They are a part of sharing the findings of the report. Chapter Six is devoted to these topics.

REVIEWING

MAIN IDEAS

Biotechnology research is an integral part of agriculture, medicine, and environmental study. It is one of the tools used to meet the demands of future generations. Humans, plants, animals, and the environment are all being studied using the scientific method.

It is easy to get caught up in the enthusiasm about new opportunities and to forget about the scientific method. The purpose of the method is to provide organization to scientific work. It is an important tool.

In reality, a similar strategy is useful in solving any problem you confront. Identify the problem, research it, develop a list of strategies, select one, plan its implementation, put the plan in action, and evaluate the results. Remember, the most important step is to correctly identify the problem. Be sure to go beyond a symptom or sign of something that needs study. The scientific method can be a successful model for all decision-making!

The scientific method is a series of steps that guides the work of scientists. It provides an organized and efficient way to solve problems. The scientific method is the foundation for all scientific research.

The major steps in the scientific method are identification of a problem, literature review, formulation of hypotheses, design of experimentation, experimentation and data collection, data analysis and the formulation of conclusions, and reporting the results. A scientist uses these steps to work through whatever issue he or she has selected.

The organized, methodical approach of the scientific method could successfully be modeled to solve problems in any area!

QUESTIONS

Answer the following questions using correct spelling and complete sentences.

1. List and explain the steps in the scientific process.
2. Why do scientists disagree about the number and division of steps in the process?
3. List some agricultural problems in your community that have the potential to be solved through biotechnology.
4. Compare and contrast basic research and applied research.
5. Describe the process you would use to research a specific agricultural topic.
6. Explain the difference between a directional hypothesis and a null hypothesis. Where would you use each?
7. What are some important considerations during the design stage of the scientific method?
8. What is replication? Why is it important?
9. What are some keys to successful experimentation? Explain each.
10. Why is observation so important to successful experimentation?
11. What are statistics? How are they used in the scientific method?
12. What are measures of central tendency? Provide an example that illustrates the most common three.
13. What are the major parts of a written report?
14. List some of the challenges you experience in preparing written reports and presentations.

EVALUATING

Match the term with the correct definition. Write the letter of the term in the blank that is provided.

a. conclusion
b. control
c. descriptive statistics
d. directional hypothesis
e. research
f. literature review
g. data
h. research protocol
i. technical journals
j. applied research

_____ 1. Learning about a topic of interest from books, journals, etc.

_____ 2. Written material prepared by specialists in a given field for their peers.

_____ 3. Used to solve an immediate problem.

_____ 4. Using systematic methods to answer questions.

_____ 5. A hypothesis that indicates a relationship between the problem and the action.

_____ 6. Items of collected information from an experiment.

_____ 7. Environment that is not changed.

_____ 8. Experimentation design.

_____ 9. An inference made from data based on facts gathered by experimentation.

_____ 10. Numbers that summarize research data.

EXPLORING

1. Arrange a personal or class tour of a local biotechnological business. As you tour, ask the scientists how the scientific method guides their work. Ask them what challenges they face in one or more of the steps in the method.

2. Consider an agricultural issue facing your local community that might be solved through biotechnology. Demonstrate your understanding of the scientific method by using it to pursue solutions to the problem.

3. Review a report of research prepared by a scientist. Get research bulletin reports from an agricultural experiment station, medical school, research organizations, or others in your state. Critique the report in terms of the format that is presented in this chapter. Write your critique and provide an oral report in class.

6

PREPARING REPORTS

What if George Washington hadn't written down any of his farming practices? What if Louis Pasteur had kept no records of his work on making milk safer to drink? What if large companies did not require their scientists to write down the steps they take in developing improved plant varieties?

Written reports are important. We could not have reached the technological level we are at today without written records of past work. Sharing knowledge is an important part of biotechnology.

Writing can be fun! Surprised? Once you are excited about research you will be excited about reporting it. Unfortunately, some people dislike writing. Others dislike speaking in public. Both skills are important to success in science. The best way to overcome weaknesses is with practice. You will be amazed at what you can do!

6-1. Computers are used to prepare research reports.

OBJECTIVES

The goal of this chapter is to help in understanding the importance of research reports and to provide a format to follow in preparing them. It has the following objectives:

1. Identify the kinds and importance of reports.
2. Describe the meaning and procedures in keeping a research log.
3. Explain the procedures in preparing a written report.
4. Explain the procedures in preparing and giving an oral report.
5. Describe how to prepare a visual display depicting research.
6. Identify potential problems in preparing research reports.

TERMS

abstract
appendix
bibliography
defense
display board
dissertation
figure
oral report
research log
research proposal
research report
table
technical writing
thesis
visualization
visual report

REPORTS: SHARING WHAT YOU'VE LEARNED

The report is the final step in the scientific process. It is a way of sharing the findings of the research. This includes the data gathered and the conclusions from the research. Reports may be written, oral, or visual.

WRITTEN REPORTS

The most common written report is the research report. A ***research report*** is a complete description of the activities during each step of the scientific process. Written reports are usually prepared on all research. Oral and visual reports may be used to further provide information about the research.

A research report is written in accurate, straightforward language. Bulletins from experiment stations are printed reports for other scientists or the general public. Modern methods make it possible for you to make multiple copies of a research report.

6-2. This statue of Johannes Gutenberg in Strasbourg, France, recognizes his work in developing movable printing type. This 14th century invention was important in making printing possible.

A preliminary report is the research proposal. A ***research proposal*** is a document that describes the first three steps before any experimentation. Other common reports are journal articles and presentations at conferences and research meetings. These typically summarize the work in a shorter form. Each report has a slightly different audience, but each explains the work done.

Writing a research report is not like the writing students usually do in English classes. Writing a research report is technical writing. It is different from creative writing. ***Technical writing*** is direct and factual. It reports the information gathered by the scientific method.

Career Profile

TECHNICAL WRITER

A technical writer prepares research reports. They prepare documents for a company, agency, or university. They develop proposals, protocols, procedural manuals, and other documents. Technical writers must follow the correct style. A technical writer may also do photography and prepare news releases or Internet information about research.

Education requirements include a college degree in journalism, communication, agricommunication, or related area. Some have master's degrees. Practical experience in writing is important, especially under the direction of a highly qualified individual. Some education in science and related areas is useful.

Jobs for technical writers are with large companies, universities, government agencies, associations, and others involved in biotechnology information. This photo shows a digital camera being

Two special research reports are the thesis and the dissertation. These are prepared by students seeking certain degrees from a college or university. In general, the ***thesis*** is the final project for a candidate for a master's degree. Some colleges require a thesis for a baccalaureate degree. The thesis reports research done to meet degree requirements. The ***dissertation*** is the final project of a candidate for a doctoral degree. The dissertation is generally longer than a thesis. It reports more advanced studies that use extensive experimentation.

In preparing reports, imagine you are trying to explain the process you followed to an interested friend. In reality, those who read the report are exactly that, interested friends.

ORAL REPORT

An ***oral report*** is a spoken presentation made to inform others about an experiment. Often, it does not have the same detail, precision, and thoroughness as a written report. Various media, such as computer presentation software or overhead transparencies, may be used in an oral report. Oral reports

6-3. **A variety of materials are being used in giving an oral report.**

may be given to civic clubs, industry groups, research meetings, or other events. Oral reports are sometimes given via radio and television. The report should be tailored to the backgrounds and interests of the people hearing it. Know your audience!

Planning and practice are needed to give a good oral report. Good research deserves a good oral report. Lack of preparation and practice in speaking can detract from the research. Be sure to plan and practice. (More details are given later in the chapter on oral reports.)

VISUAL REPORT

A ***visual report*** is a presentation that relies primarily on photographs, real things, and other devices to communicate information. Displays at fairs, conventions, and other events share the findings of research. Students are quite familiar with the displays prepared for science fair projects.

Since a visual report represents you, it should be carefully prepared. It should have a professional appearance. All safety and other rules should be followed. You do not want a visual presentation to detract from your research. Make it look good! (More details on visual reports are given later in this chapter.)

6-4. **A visual report prepared by a student for a science fair in Mississippi.**

RESEARCH LOG: AN IMPORTANT TOOL

Do you remember what you were doing at 7:00 p.m. last Tuesday? What about that same date two months ago? What about that date last year?

The human brain sometimes fails to remember even the most important details. For that reason, scientists maintain research logs.

A ***research log*** is a written record of the processes, actions, and observations made in a research project. All details must be recorded truthfully and accurately. A log is an important tool in preparing a report. It also is a very important part of the documentation shared with the world. A well-written log supports the claims of a scientist. A poorly written log casts doubt on the work.

6-5. **An example of a log book.**

The design of a research log is set by the scientist and those supervising the work. In many cases, the research log is still maintained in long hand in a notebook. Some scientists use a personal computer or a university or corporate network. If computer records are used, take care to back up the work to prevent loss of vital data in case of computer failure. Security from hackers is sometimes an issue.

In general, the log is dated and activities of the day are entered. In the literature review, bibliographic information and important facts from each document may be recorded. During experimentation, data are entered in a chart or graph. During the reporting phase, submissions may be documented in the log. The style and format vary depending upon the scientist and the work.

6-6. **A scientist refers to information in a log book in preparing the report.**

A carefully maintained log makes it easier to write the report. The log helps follow the scientific process in writing the report.

ASSEMBLING THE REPORT

Report preparation is the final step in research. As with other steps in the scientific process, a format for reports has been developed. Details of the format may change, but the general structure rarely changes. This format gives a complete "picture" of the work done in the study.

A research report can be handwritten, typewritten, or prepared with word processing on a computer. The use of word processors or personal computers makes it easier to prepare a report. Most reports should be neatly typed or printed on standard 8½ × 11 inch white paper.

Text format varies, but generally includes page margins of at least one inch, double spacing, and use of an easy-to-read font (type face), such as Times Roman. Institutions and corporations may have specific format guidelines, known as style. A widely used style reference is the *Publication Manual of the American Psychological Association*. It is commonly referred to as the APA style manual.

Like this book, a report is organized into sections with headings and subheadings. These divisions help the reader understand the study. Divisions are used in making the table of contents.

The traditional parts of a research report are front matter, such as the title page; introduction; review of literature; research methods; findings or results; summary, conclusions, and recommendations; and back matter, such as the bibliography (or list of references cited). Each of these sections is described in more detail on the following pages.

FRONT MATTER

The first section of the report is the front matter. The front matter gives an overview and a "map" of the work.

Items typically included are a title page, acknowledgments, the table of contents, an abstract, a list of tables, and a list of figures. One or more of these may be omitted in shorter reports. Other than the title page, items in the front matter are numbered with lower-case Roman numerals (ii, iii, iv, etc.). The title page counts as page "i," but is not numbered.

Title Page

The title page provides information about the work. It has the title, the names of the author(s), and the date of the work. It may also have information about the school, sources of funding, or corporation. In the case of theses and dissertations, the title page includes the signatures of the professors who served on the student's committee. Design the title page so it is appealing and professional in appearance.

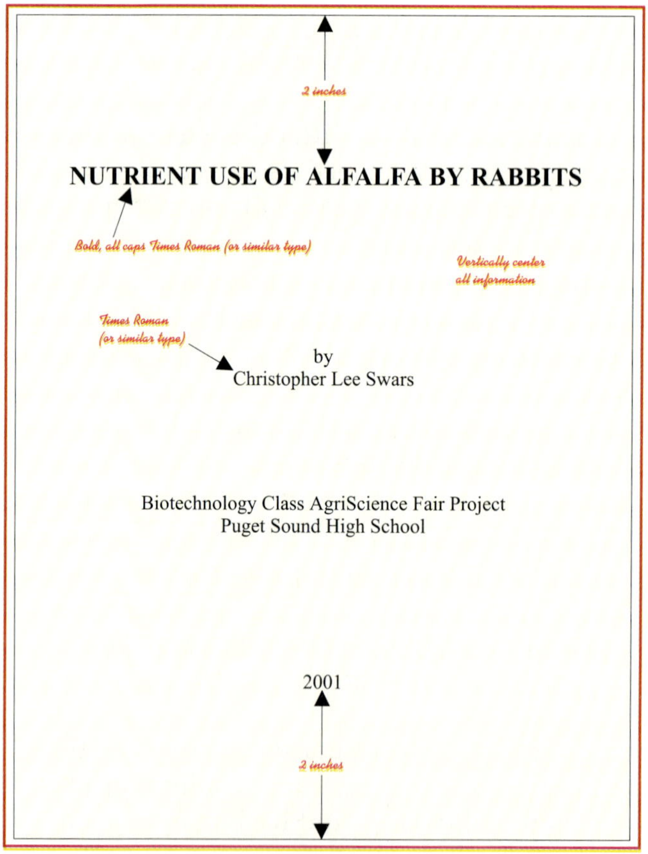

6-7. **A sample title page.**

Acknowledgments

Research is rarely done without help from others. This may be information, guidance, or other involvement. The acknowledgment page allows a scientist to recognize and thank those who have provided help.

Table of Contents

The table of contents lists the materials found in the report. Both major divisions (sections) and subdivisions may be listed. The page number where each division begins is listed.

Use 1" margins

Use of dot leader is optional

Use Times Roman or similar type

TABLE OF CONTENTS

	Page
TABLE OF TABLES	iii
TABLE OF FIGURES	iv
ABSTRACT	v
ACKNOWLEDGMENTS	vi
INTRODUCTION	1
Purpose	1
Objectives	2
REVIEW OF LITERATURE	4
Survey Methods	5
Related Studies	9
PROCEDURES	13
Background	13
Population and Sample	13
Instrumentation	14
Data Collection	15
Data Analysis	16
FINDINGS	18
Scope of Your Program	18
Planning Your Program	22
Using Environmental Resources	25
Future Needs and Challenges in Environmental Materials	27
SUMMARY, CONCLUSIONS, AND RECOMMENDATIONS	33
Summary	33
Conclusions	34
Recommendations	35
BIBLIOGRAPHY	39
APPENDIXES	
A: Questionnaire	40
B: Cover Letter	43

ii

Page number in lower case Roman Numerals

6-8. A sample table of contents.

Abstract

An ***abstract*** is a brief (no longer than a page) description of the information contained in the research report. Each of the sections of the text is touched on. Particular attention is given to the results and the conclusions.

Lists of Graphics

Tables and figures are graphics. These are within the text to provide additional information about the work. A ***table*** is a list organized into rows and

Biotech Connection

WRITING TO BRAG ABOUT

Be proud of what you write! You can learn to write good reports with practice. Get suggestions from others on how to improve. Make it your mission to be a good writer.

Here are a few pointers to help you in writing:

- Allow time for writing. Select a good time of day when your mind is fresh and clear.
- Get started. Don't put off writing. Even if the words are not flowing, start putting something on paper and most likely the words will begin to come to you.
- Chose a good place to write. It should be comfortable and quiet. Use a table or desk top and have everything you need at hand.
- Let the writing be yours. Never copy the work of another person. Copying another person's writing is plagiarism—stealing!
- Cite properly. Information from another source should be cited to that source.
- Have pride. Strive to spell correctly, punctuate properly, and have good sentences and paragraphs. Be neat in your work.

Yes, you can be a good writer. Look at other people who are writers. They began just as you are beginning. Your success is up to you!

columns. Tables often present numerical information. A ***figure*** is a graph, line drawing, photograph, or other illustration. Both tables and figures enrich a written report.

Tables and figures have separate tables of contents. These are designed the same as the main table of contents.

INTRODUCTION

The first section of the report is the introduction. This presents the case for the research. The introduction describes the problem, states the importance of the research, lists the hypotheses, defines any unusual terms, and describes limitations on the work. Each of these is listed in the table of contents.

The introduction begins with an explanation of the problem. It describes the background and setting for the work. In essence, this section sets the stage for the rest of the writing. This section clearly explains how the work could impact society or a segment of it.

The second section is a direct statement of the hypotheses. For some work, the researcher may choose to use research questions or objectives rather than hypotheses.

A section for definitions is optional. Unusual or technical terms are included. This section helps the reader understand the terms as used in the writing. For example, research about agriculture might define the word "agriculture." This is so the readers will know whether the term is being used very broadly or whether it is describing just production agriculture.

Some studies have limitations. This occurs when the conditions of the study do not apply in every situation. In that case, the researcher defines the scope of the work and explains the limitations.

REVIEW OF LITERATURE

The review of literature briefly summarizes what others have written about the research topic. It is intended to provide a foundation for the report. It helps educate both the researcher and the reader in how this study fits into a broader body of knowledge. It shows that the study is important and that it will add to knowledge of the subject.

The literature review is commonly arranged in a topical fashion. It begins with general information and moves to specific and technical information. Information from references is followed by a citation. The citation may list the author's name and date.

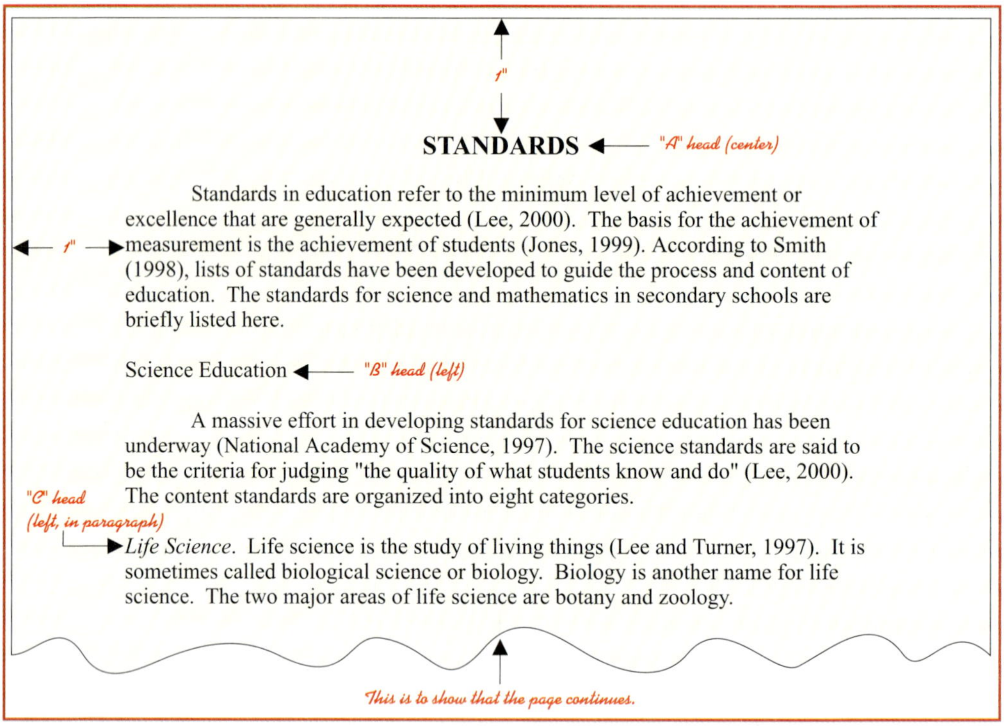

STANDARDS

Standards in education refer to the minimum level of achievement or excellence that are generally expected (Lee, 2000). The basis for the achievement of measurement is the achievement of students (Jones, 1999). According to Smith (1998), lists of standards have been developed to guide the process and content of education. The standards for science and mathematics in secondary schools are briefly listed here.

Science Education

A massive effort in developing standards for science education has been underway (National Academy of Science, 1997). The science standards are said to be the criteria for judging "the quality of what students know and do" (Lee, 2000). The content standards are organized into eight categories.

Life Science. Life science is the study of living things (Lee and Turner, 1997). It is sometimes called biological science or biology. Biology is another name for life science. The two major areas of life science are botany and zoology.

6-9. **An example of one way to cite literature. All literature is included in the bibliography (or list of references cited). Refer to a style manual for complete details.**

The literature review is easiest to write if careful records are kept of the literature research. Accurate records of the literature and the content of the references make it easy to write a report. Writing the literature review involves organizing the topics. Transitions are used to link the ideas and topics together. The topics are interpreted in terms of the stated problem.

RESEARCH METHODS

The research methods section gives the design of the research. This section includes the type or structure of the research, a list of materials and equipment, a step-by-step description of the experiment, a description of the type of data gathered, and an explanation of the statistical analysis used. The section includes the chronology of the activities. This helps readers understand the time required for the various activities.

The purpose of the methods section is to describe the planning and implementation stages of the study. It should provide an understanding of the processes followed and the reasons for each action.

In general, the description of the design should be detailed so a reader can duplicate the work. With specialized procedures, graphics may be included. Each item in the section should include an explanation of the thought process involved.

The list of materials and equipment should include sizes, quantities, and, if possible, sources for each item.

Any deviation between the planned work and the actual work should be described. In addition, a timeline (dates of major events) should be included. This may be included as a graphic or written in the text.

This section should give the reader a clear understanding of what was done and why.

FINDINGS

The findings section describes the outcome of all the work. The section begins with a summary of the activities in the experiment and the measurement techniques used. If some parts of the research were omitted, an explanation is included.

Next, the findings section presents the results as analyzed. The amount of information included varies with the size of the study. In a simple study, raw data may also be included. In a large study, data that has no impact on the hypotheses may be omitted.

The findings are usually organized by the hypotheses. They are presented in a narrative form and in a table or graph. The table or figure contains all the data and the text highlights important findings. This reinforces the information. It gives a nice presentation. In addition, some information is more eas-

6-10. **Using a computer to prepare a table.**

Table 10. Funds Budgeted for Operation of Agricultural Education Programs on a Per-Teacher Basis

Fund Source	Current Study Average Amount $	Current Study Schools Reporting No.	Current Study Schools Reporting %	Previous Study Average Amount $	Previous Study Schools Reporting No.	Previous Study Schools Reporting %
Local Funds	5,837.62	203	66.34	2,988.85	259	88.40
State/federal funds	5,128.49	141	46.08	3,060.50	136	46.42
Booster club funds	2,696.75	41	13.40	1,543.06	47	16.04
All sources*	8,677.93			4,560.59		

* The average amount for all sources was determined by calculating the total amount received by all schools on a per-teacher basis that reported funding budgeted to the agricultural education program.

6-11. A sample table.

ily understood in a graph or table. This allows readers to evaluate their perceptions of importance with those of the scientist.

Interesting questions sometimes arise during experimentation. If the data answer such questions, it is acceptable to present the information in a separate part of the section.

The findings section does NOT include any interpretation of the data nor how they impact the hypotheses.

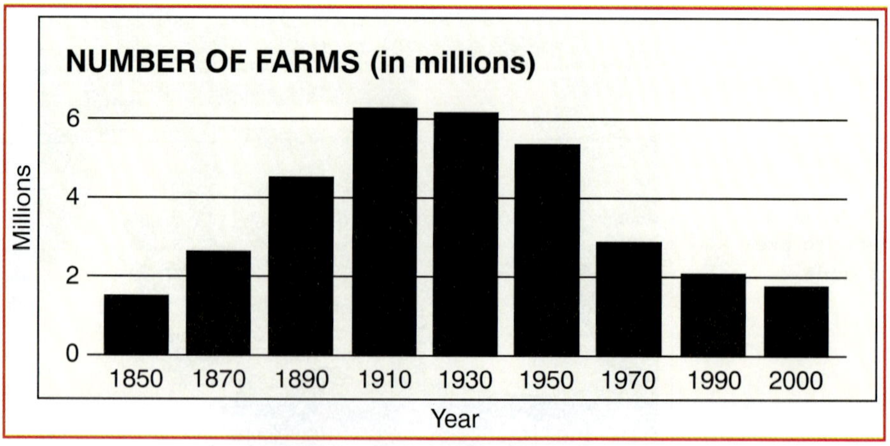

6-12. A sample bar graph.

SUMMARY, CONCLUSIONS, AND RECOMMENDATIONS

6-13. Montana students are being taught how to use a plotter in preparing graphic material for a report.

This section has subsections that summarize the study, interpret the findings, and describe the impact of the study. In addition, ideas for further study are presented.

Summary

The summary subsection generally begins with a brief overview of the problem, the experiment design, and the limitations of the work. This should enable someone to read only this section and gain a general understanding of the study.

Conclusions

The conclusions subsection provides the conclusions drawn from the study. An effective conclusion requires that the researcher objectively consider the findings and make the best possible interpretation. Sometimes the scientist will indicate that no conclusion can be drawn.

In drawing conclusions, the scientist must review the entire body of data and identify notable findings. These are findings that vary significantly from previous work. Then the scientist should consider potential explanations for each finding. In so doing, the researcher should describe how these results might be applied in real-life situations.

Recommendations

Finally, most reports present additional questions and avenues for further research. These are presented to provide direction for other scientists and acknowledge the cyclic nature of the scientific method.

BACK MATTER

Like the front matter, the back matter is not actually part of the report text. It contains items that help clarify the report, including a bibliography and, often, one or more appendices.

Bibliography

A ***bibliography*** lists all of the resources used in the literature review and in the study. This section is sometimes known as the "literature cited" section.

It should include any sources cited in the text. It may also include sources used for background or understanding, but not actually cited.

The bibliography is easy to create if the scientist has maintained accurate records during the research or literature review. The various sources are arranged alphabetically by the last name of the authors. If more than one author wrote a book, the last name of the first author is used for alphabetization.

BIBLIOGRAPHY

Books

Lee, J. S. and Turner, D. L. (1997). *Introduction to World AgriScience and Technology* (2nd ed.). Danville, IL: Interstate Publishers, Inc.

Books with Editor

Altman, A., ed. (1998). *Agricultural Biotechnology*. Hong Kong: Marcel Dedder, Inc.

Journals

Gordon, J. W. (1989). Transgenic animals. *International Review of Cytology*. 155: 171-229.

Schroth, M. H. and Hancock, J. G. Disease-suppressive soil and root-colonizing bacteria. *Science*. 216:1376

Bulletin – No Author

Technology for a Sustainable Future. (1994). Washington, DC: National Science and Technology Council.

Conference Proceedings

Mearns, A. J., Roques, P., and Henry, C. (1993). Measuring efficacy of bioremediation of oil spills. Proceedings of the 1993 International Oil Spill Conference. Washington, DC: American Petroleum Institute. pp. 335-344.

6-14. A sample bibliography with one style of entry.

A challenge in preparing a bibliography is using the correct format for citing references, such as books, journals, and web sites. Use an approved guide to clarify such issues. One of the most common is the *Publication Manual of the American Psychological Association*.

Appendices

Sometimes other materials are available that are not essential to the report, but could increase the reader's understanding. These items might include very long data tables, equipment designs and diagrams, specialty research materials, or other supporting documentation.

These materials are placed in an ***appendix***, or separate supplemental section, at the end of the report. (The plural of appendix is appendices.) The format for these sections varies.

JOURNAL ARTICLES

After finishing a research report, the findings may be prepared as a journal article. An article lets many people learn about the research. Most reports are too long to be published in journals in their entirety. In this case, the researcher prepares a shorter document using the same format, but with less detail.

The specific guidelines vary from publication to publication. Editors provide guidelines for their publication. Most guidelines are clear by looking at the journal. The guidelines will also indicate if there is a charge for publishing an article.

In general, journal articles focus on the findings and the conclusions. Other sections must be included in enough detail, however, to justify the actions described.

Journal articles may cover one part of the research and omit other areas of the study. This is done with very large studies. Similar articles may be submitted to different journals. This maximizes the possibility of publication.

6-15. Examples of journals for publishing biotechnology research. (Courtesy, Ed Osborne, University of Florida)

PRESENTATIONS

In addition to a written report, oral reports may be given on the research. Oral reports are sometimes known as presentations. They are given at conferences and other meetings. Presentations may involve a written paper. The paper follows a format similar to a journal article and is often published in records of the meeting. These records are known as proceedings.

Oral reports are much like public speaking. These allow scientists to interact with their peers and discuss their work. In addition to professional conferences, oral reports may be delivered to corporations, review boards, and other peer groups. In the case of theses and dissertations, an oral presentation is known as a ***defense***.

Some people are afraid of public speaking. Planning and practice can help overcome fear.

6-16. Practice in front of a class helps develop speaking skills.

ORGANIZING THE ORAL REPORT

Like the journal article, an oral report should follow the same format as the research report. Because of time constraints, many sections are summarized. Most of the time is given to presenting the findings, conclusions, and recommendations.

In preparing the report, find an effective way of condensing and explaining the work. During preparation of the speech, identify the most important items to discuss.

Since the work is technical, visual aids, such as overhead transparencies and multimedia presentations, may be helpful. Computer-based presentations are sometimes used. Appropriate projection and sound systems may be needed.

In the introduction, explain the problem, the hypotheses, and any limitations. Review pertinent literature, including citations (This typically follows the form of "In their 1998 article on ____________, Scientists X and Y found . . . ")

Always describe the steps in the research process, including some outline of the time line. Next, go into the findings of the study. Provide as much detail as is realistic. Finally, present the conclusions and recommendations.

6-17. **A laptop computer and presentation software are being used by this Illinois educator in an oral report.**

Oral research reports are often followed by question and answer sessions. Always respond accurately to questions. Recognize any errors or shortcomings in the research. Question and answer sessions are useful in clarifying statements in the report.

PRESENTATION STRATEGIES

Everyone involved in public speaking develops an individual style. This style comes with practice, so it's important to speak regularly and develop those skills!

Even though an oral research report is a professional presentation, use as many strategies as possible to engage and maintain the interest of the audience. Some speakers use a monotone voice and very few body movements. These lead to a very dry, boring presentation. Effective speakers are energetic and enthusiastic. They know the information and enjoy sharing it with others.

Knowing the Audience

One of the first steps in preparing a presentation is to learn about the audience and the environment. Learn the demographic characteristics of the people in the audience.

6-18. Get another person to critique your work when preparing an oral report.

Select an approach the audience can understand. For example, if your presentation is to lay people, use simple terms and avoid research terms.

Determine what is expected of the presentation. This includes its length and the availability of audiovisual devices. Some speakers like to know the physical environment in which they will be speaking, such as the size of the room.

Preparing the Oral Report

Some speakers write their speech word for word; others make an outline of the important points they want to include. Either way, speakers review their material many times before the presentation. They rehearse the mate-

To give a good oral report:

- Be proud of your research but not cocky.
- Plan and prepare ahead of time. Rehearse!
- Keep your presentation simple and concise.
- Know your subject thoroughly.
- Dress neatly and professionally.
- Relax but be enthusiastic.
- Speak clearly, at an appropriate pace, and loud enough.
- Make eye contact with the audience.
- Be polite and, if you don't know something, say so.
- Thank the audience for their attention.

6-19. Tips for a good oral report.

rial in front of a mirror, family and friends, or by themselves. Rehearsing may be videotaped and reviewed to improve on the speech.

One tool that speakers find effective is ***visualization***. They imagine themselves speaking in as much detail as possible. They see themselves at every stage of the presentation, the audience, the environment, and the reactions. This use of the imagination helps them to more effectively prepare the oral report.

Another key is to check out the setting prior to the presentation. Make certain that all equipment is working properly, including microphones and audiovisual tools.

6-20. Give your oral report with confidence.

Giving the Oral Report

Giving an oral report uses public speaking skills. Speak clearly and so all members of the audience can hear you. Use a friendly, confident voice. Use appropriate gestures and audiovisuals. Begin speaking on time and end within the time that is allowed.

Information on speaking is available from many sources. One is *Developing Leadership and Personal Skills*, a book available from Interstate Publishers, Inc. Also, get information from a speech teacher or other qualified individuals.

VISUAL REPORTS: DISPLAYS

Visual reports are displays or exhibits that provide information about research. They are sometimes known as poster sessions or they may be table exhibits in science fairs. Always know the guidelines that you are to follow.

USING DISPLAY BOARDS

Display boards are sometimes used in exhibits. A ***display board*** is a graphic report on a research project. Most display boards are on table tops about 48 inches wide and 30 inches deep. Height restrictions may also apply.

6-21. A high school gymnasium filled with science fair displays.

The display board presents information about the study. Use large lettering and organize the information so people can easily read about the research. Information on a display board includes the title of the research, purpose, hypotheses, methods or procedures, findings, and conclusions. Photographs of the research may be used. These should be enlarged for ease of viewing. Data tables and graphs may be included. Use appropriate colors in the display to appeal to the audience and to make the content realistic.

The table top should have the log book and a copy of the research report. The most effective displays have specimens and other materials used in your research. This allows the viewers to observe your findings.

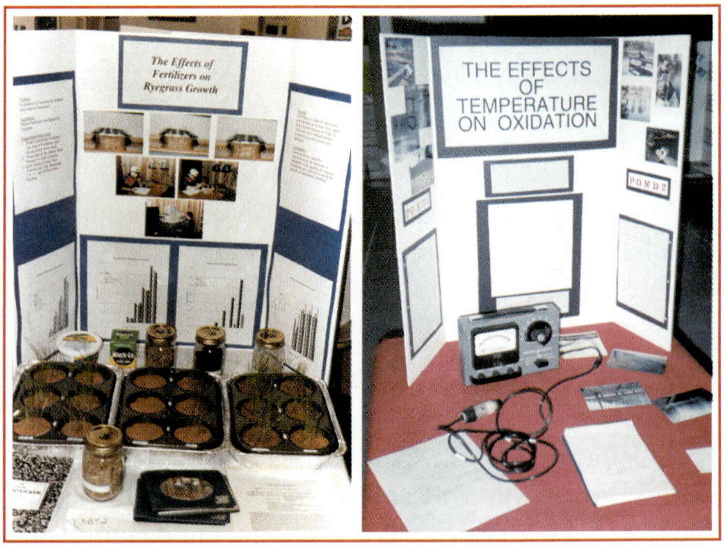

6-22. Two display board presentations.

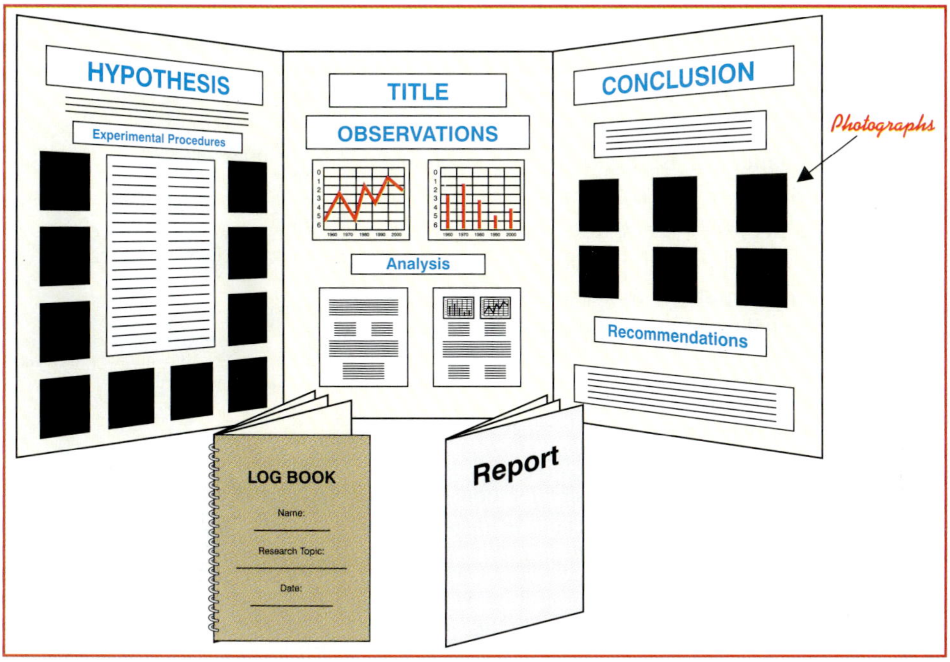

6-23. Organize a display to share the research.

KNOW THE RULES

Restrictions may apply to the use of displays. Find out the regulations before preparing a display. Here are a few questions to get answered:

- Is electricity allowed and available?
- Can engines or motors be used, if a part of the research?
- Can live organisms be displayed?
- Can animal body fluids or tissues be displayed?
- What restrictions apply to the use of chemicals in the display?
- Do I put my name on the display?
- Will my display be judged? What criteria are used? Am I to be present during judging and answer questions?
- Will the area be secure? How do I prevent the loss of materials from my display?

Students who have carried out research are sometimes disappointed to learn about restrictions. Some things cannot be put in a display, such as cer-

Biotech Connection

SCIENCE FAIRS

A common way students report research is in a science fair. Both a written report and a visual display are usually needed. The exhibits are prepared using the sections in this chapter and local information.

With computers, most students have access to quality lettering and graphics. For that reason, expectations for displays have increased. Students should prepare a high-quality display that summarizes the research: the hypotheses, the methods, the data, and the conclusions. Displays also include examples of the research materials or an experimental sample.

Some fairs also require an oral report. This presentation is closely related to the written report.

A science fair is an excellent way to develop skills in reporting research. Take advantage of the opportunities to get involved in the science fair at your school. This photo shows a student making final touches on her display at the National Agriscience Competitive Event Fair.

tain materials or animal tissues. Get the questions answered ahead of time. This will save effort and help you comply with the rules.

AVOIDING THE PITFALLS

Barriers may arise in preparing a research report. Some of these barriers are due to outside factors. Others are caused by the scientist or student. Planning ahead will help prevent some problems.

BARRIER ONE: PROCRASTINATION

A big barrier is procrastination. Proper preparation of a report requires much time. If the work is left to the last minute, the quality suffers. Writing is an integral and ongoing part of the research process. Preliminary drafts of the first sections can be done early in the study. Avoid procrastination by being on-task. Manage your time. Allot time for writing. Do not put off getting the writing done. Be excited about your research.

BARRIER TWO: INADEQUATE LITERATURE REVIEW

Inadequate literature review can be a barrier. This includes poorly kept bibliographic information. Having a good foundation for the research is important. Without the proper information, a case cannot be made for the study within the larger research picture. A well-kept research log and development of strong research skills overcome this barrier.

6-24. Use the Internet to help overcome a lack of information. This Washington student is studying aquaculture.

BARRIER THREE: POOR DATA MANAGEMENT

Poor data management is a barrier. If records are not carefully kept, data are lost, or the wrong information is recorded. The data analysis will be unsuccessful and writing an effective report is impossible. This barrier can be overcome by extensive and careful planning.

BARRIER FOUR: FAILING TO USE GOOD WRITING SKILLS

Failing to use the proper grammar, spelling, and format results in a poor-quality report. Some people have weaknesses in the areas of writing and speaking. Overcome these weaknesses with practice. Use word processing tools, such as spell-check. Another effective way to deal with this barrier is to ask for help from specialists. Teachers and others who are skilled in writing often are willing to provide help. They can review and help revise documents and presentations.

BARRIER FIVE: LACK OF COMMITMENT

People need to be committed to the task. Research reports, like the rest of the scientific method, require dedication. It is not enough to write the report or give the speech. A person must be committed to constant improvement. Initial drafts should be revised, reread, and revised again. Sloppy and careless mistakes are unnecessary. Take pride in your work. Do the best you can. Use the needed time and effort to do a good job.

REVIEWING

MAIN IDEAS

The final step in doing research is reporting the findings of the study. Reports vary, but could be broadly classified as written reports, oral reports (presentations), and visual reports.

An important tool in the report process is the research log. The log is the scientist's record of everything that happens during the research process. It may include everything from bibliographic references to raw data.

The traditional parts of a research report are the front matter; introduction; literature review; research methods; findings; summary, conclusions, and recommendations; and back matter. Each part has a role in creating a complete and accurate "picture" of the work. A short version of the report is a journal article.

The most common oral presentations of reports are to research conferences and groups involved in the research. These reports follow the same format, but typically emphasize the data gathered and the conclusions drawn from that work.

A visual report (exhibit) is using a display of graphics and real things to present the research. These are often known as poster sessions, science fair projects, displays, and exhibits.

Many potential barriers exist in preparing reports. Most of those barriers can be reduced or eliminated through careful planning and management.

QUESTIONS

Answer the following questions using correct spelling and complete sentences.

1. List and describe different types of reports.
2. Explain why preparing reports is an important part of the scientific method.
3. What is a research log? Explain its role in preparing research reports.
4. List the major parts of a report and explain the role of each within the paper.
5. What should not be included in the results and findings section?
6. Explain why a bibliography is important to a report.
7. Compare and contrast a written research report and a research meeting presentation.
8. List and explain some keys to a successful oral presentation.
9. Identify some personal pitfalls to research reporting and describe strategies you can use to overcome them.

EVALUATING

Match the term with the correct definition. Write the letter of the term in the blank that is provided.

a. appendix
b. research proposal
c. technical writing
d. research log
e. abstract
f. figure
g. bibliography
h. thesis
i. visualization

_____ 1. Final project for a master's degree.

_____ 2. Preliminary report.

_____ 3. Direct, factual writing.

_____ 4. Written record of the thoughts and actions of a scientist.

_____ 5. Brief description of the research report.

_____ 6. Chart or picture that provides research information.

_____ 7. List of the resources used in the literature review.

_____ 8. Separate section that supplements the report.

_____ 9. Imagining yourself in a setting as a way of practicing.

EXPLORING

1. Locate and read three articles from different journals about a biotechnology subject in which you are interested. Compare the style and format of the different articles. Identify the report parts in each article.

2. Attend a research conference or meeting in your area, or obtain the proceedings from such a meeting. Compare the style of the writing or speaking with other types of writings or speeches that you have seen.

3. As part of a research project, maintain and use a research log in preparing a research paper, a journal article, and/or a conference paper. Use the strategies described in this chapter to help in this process.

4. Visit a local scientist in biotechnology research. Ask about the reports he or she prepares. Ask what the biggest barriers are that the person faces. If possible, obtain samples of their reports. Give an oral report to your class about what you learn.

7

SCIENCE FUNDAMENTALS

Biotechnology is science in action. Most all areas of science are used. Some areas are quite obvious; others are less obvious. Since biotechnology often involves living things, the life sciences are important. But, what other areas are involved?

Review the definition of biotechnology from Chapter 1. A widely accepted definition is from the Office of Technology Assessment of the United States Congress. This definition says that biotechnology is "any technique that uses living organisms or substances from those organisms, to make or modify a product, to improve plants or animals, or to develop microorganisms for specific uses." That definition includes a lot of science!

Biotechnology is carried out for the benefit of humans and the world in which we live. Introducing the human element makes even more areas of science important.

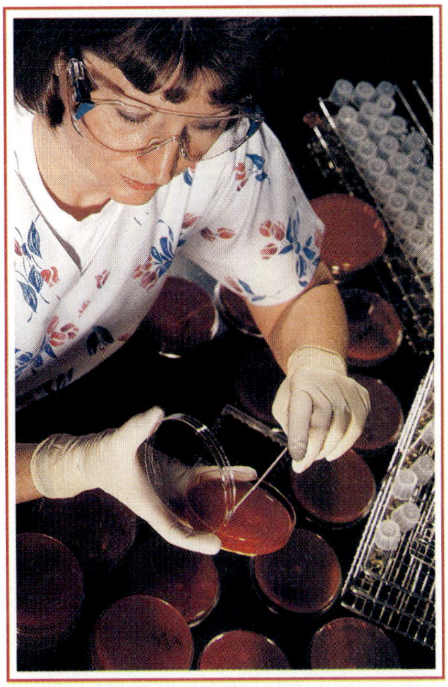

7-1. **Biotechnology work uses science in many ways. Here, cultured colonies of *Campylobacter* food poisoning are being studied using a blood agar plate. Each strain of food poisoning bacteria has a specific genetic print. (Courtesy, Agricultural Research Service, USDA)**

OBJECTIVES

This chapter introduces the role of science in biotechnology, including matter, energy, living organisms, life processes, and the structures of living organisms. It also stresses the importance of DNA, enzymes, lipids, and proteins. It has the following objectives:

1. Relate the four main branches of science to biotechnology.
2. Identify the components of matter and energy.
3. Describe the classification of living organisms.
4. Identify and explain important life processes.
5. Describe the primary structures of living organisms.
6. Explain the importance of DNA, enzymes, lipids, and proteins.

TERMS

atom
atomic number
binomial name
biology
catalyst
cell
cell specialization
chemical energy
circulation
DNA
element
elimination
energy
enzyme
gland
growth
kinetic energy
life process
life science
lipid
living condition
mathematics
matter
molecule
organ
organ system
photosynthesis
physical science
potential energy
protein
regeneration
repair
respiration
science
secretion
sensation
social science
taxonomy
tissue

THE BRANCHES OF SCIENCE

Science is knowledge about the world we live in. It involves making observations, measurements, and experiments to set up general truths or laws. A good example is gravity. Gravity is the attraction between Earth and objects on its surface. Think how different our lives would be without gravity! The law of gravity is just one of the laws that explains the nature and behavior of the living and nonliving parts of our world.

FOUR AREAS

Science is divided into four main areas or branches. These branches are life science, physical science, mathematics, and social science. Sometimes, social science is omitted. It cannot be left out in any discussion of biotechnology. You will see why later.

7-2. Students are using a model to learn the parts and functions of a flower in a biology class.

Life Science

Life science is the study of living things. This includes the structures and processes of organisms that allow them to live and reproduce. Medicine and agriculture are two areas with major interests in life science. Both use biotechnology. The definition of biotechnology most definitely includes life science: ". . .living organisms or substances from those organisms . . . " Life science is often known as ***biology***.

7-3. A horticulturist at The North Carolina Arboretum applies botany in growing ornamental plants.

Life science is further divided based on the kinds of organisms being studied. Here are a few examples:

- Botany is the study of plants. It is sometimes known as plant science. Plants are classified into the Kingdom Plantae.
- Zoology is the study of animals. It is sometimes known as animal science. Zoology is further divided into several areas, such as entomology (the study of insects) and ichthyology (the study of fish). Animals are classified into the Kingdom Animalia.
- Bacteriology is the study of bacteria. Bacteria are in the Kingdom Monera.
- Mycology is the study of fungi. Fungi include yeasts, molds, and mushrooms. These organisms form the Kingdom Fungi.
- Virology is the study of viruses. A virus is a tiny particle that lives in the cells of other living organisms. Viruses are hard for scientists to classify. They do not meet the normal conditions to be living organisms.

Physical Science

Physical science is the study of the nonliving things around us. Living things often depend on these nonliving materials for life-supporting essentials, such as water and minerals. Rocks, mechanics, and climate are three examples of nonliving things.

Areas of physical science:

- Chemistry is the study of the composition of substances. This includes how the substances relate to each other, including combining to form compounds.
- Physics is the study of energy and matter. Simply, much of physics deals with the physical nature of objects. It includes heat, light, sound, minerals, and electricity.
- Earth science deals with the study of the environment and the composition of the planet Earth. This includes meteorology, geology, geography, and oceanography.

7-4. **Properly setting up and using apparatus is important in doing science work. This shows the correct setup of hot and stir plates in a chemistry ventilation hood. Note that the Erlenmeyer flasks are securely held in position with clamps. (Courtesy, Chemistry Department, Emory University)**

Mathematics

Mathematics is the science of numbers. This includes both simple and advanced operations. The simple operations include adding, subtracting, division, and multiplication. The advanced operations include calculus and statistics. Statistics is often used in biotechnology work. Statistical applications help summarize and interpret data.

Social Science

Social science is the study of the interactions of people as individuals and groups. It includes sociology, ethics, anthropology, and other areas. In biotechnology, social science helps interpret how people will respond to new developments, such as the production of transgenic foods. Social science has a big role in the acceptance and benefits gained from biotechnology.

MANY SCIENCES TOGETHER

Biotechnology uses all four of these science branches. In research labs, chemists, biologists, physicists, sociologists, and mathematicians may work on biotechnology projects.

Here is how the sciences cooperate in studying the genes of an organism.

- Biology is used to culture and grow the organism.
- Chemistry is used to isolate the genes.

Career Profile

PLANT TECHNICIAN

A plant technician performs a wide range of activities in studying plants. They work under the direction of a scientist. The work may be in an enclosed lab or outside in a field plot. A variety of laboratory equipment may be used. This photo shows a plant technician studying a virus of the soybean plant. Ultraviolet light is being used with a stained specimen to detect strain-specific RNA. (Courtesy, Agricultural Research Service, USDA)

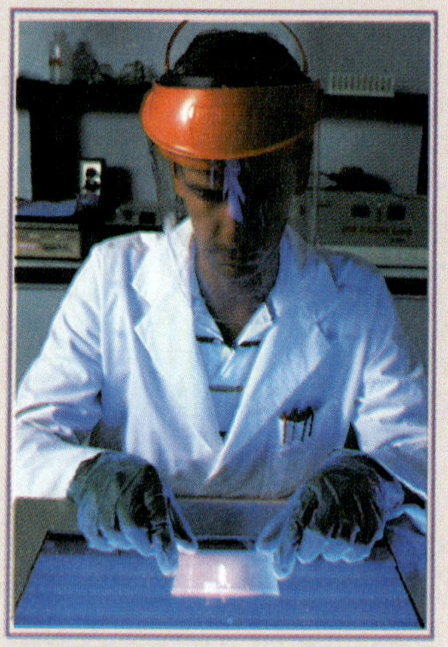

Plant technicians need a good education in science as well as skills in laboratory work. Many have bachelors degrees in areas of biological science, agronomy, botany, or a related area. Some may have masters degrees in similar areas. High school courses in biology, agriculture, horticulture, mathematics, and biotechnology are useful.

Jobs for plant technicians are found where research and development work is done with plants. Biotechnology firms, crop-research agribusinesses, colleges, and research stations are possible places for jobs.

- Physics, and its applications in engineering, is used to help create the instruments to identify and study genes.
- Mathematics is used to determine if the results obtained are trustworthy or valid.
- Computer science (applied physics) is used to collect and store data, scan and store images, and in other ways.
- Sociology assesses the results of a study in terms of human impact and acceptability.

MATTER AND ENERGY

A basic understanding of matter and energy is needed to do biotechnology work. Chemists and physicists study these areas in detail.

MATTER AND MOLECULES

Everything is composed of matter. Your desk, pencil, house, and even your teacher are made of matter. ***Matter*** is anything that has mass. The three common states of matter are solid, liquid, and gas. All three states of matter are used in biotechnology.

All matter is made of chemical elements. An ***element*** is any substance that cannot be broken down into simpler material by ordinary means. Elements are the fundamental units of a chemical substance. In their pure form, elements are made up entirely of atoms with the same number of protons. Common elements in our world include oxygen (O), nitrogen (N), carbon (C), hydrogen (H), and iron (Fe).

An ***atom*** is the smallest unit in which an element can exist. Some atoms can exist alone; others are in pairs. Copper (Cu) can exist alone. Oxygen exists as paired atoms—O_2.

Most elements are natural; others are artificial. Scientists have identified 115 elements but not all have been named. Ninety-four are natural elements.

Periodic Table of the Elements

Metals — Transition Elements — Nonmetals

	1 IA	2 IIA	3 IIIB	4 IVB	5 VB	6 VIB	7 VIIB	8	9 VIIIB	10	11 IB	12 IIB	13 IIIA	14 IVA	15 VA	16 VIA	17 VIIA	18 VIIIA
1	1 H Hydrogen 1																	2 He Helium 4
2	3 Li Lithium 7	4 Be Beryllium 9											5 B Boron 11	6 C Carbon 12	7 N Nitrogen 14	8 O Oxygen 16	9 F Fluorine 19	10 Ne Neon 20
3	11 Na Sodium 23	12 Mg Magnesium 24											13 Al Aluminum 27	14 Si Silicon 28	15 P Phosphorus 31	16 S Sulfur 32	17 Cl Chlorine 35	18 Ar Argon 40
4	19 K Potassium 39	20 Ca Calcium 40	21 Sc Scandium 45	22 Ti Titanium 48	23 V Vanadium 51	24 Cr Chromium 52	25 Mn Manganese 55	26 Fe Iron 56	27 Co Cobalt 59	28 Ni Nickel 59	29 Cu Copper 64	30 Zn Zinc 65	31 Ga Gallium 70	32 Ge Germanium 73	33 As Arsenic 75	34 Se Selenium 79	35 Br Bromine 80	36 Kr Krypton 84
5	37 Rb Rubidium 85	38 Sr Strontium 88	39 Y Yttrium 89	40 Zr Zirconium 91	41 Nb Niobium 93	42 Mo Molybdenum 96	43 Tc Technetium 98	44 Ru Ruthenium 101	45 Rh Rhodium 103	46 Pd Palladium 106	47 Ag Silver 108	48 Cd Cadmium 112	49 In Indium 115	50 Sn Tin 119	51 Sb Antimony 122	52 Te Tellurium 128	53 I Iodine 127	54 Xe Xenon 131
6	55 Cs Cesium 133	56 Ba Barium 137	71 Lu Lutetium 175	72 Hf Hafnium 178	73 Ta Tantalum 181	74 W Tungsten 184	75 Re Rhenium 186	76 Os Osmium 190	77 Ir Iridium 192	78 Pt Platinum 195	79 Au Gold 197	80 Hg Mercury 201	81 Tl Thallium 204	82 Pb Lead 207	83 Bi Bismuth 209	84 Po Polonium 210	85 At Astatine 210	86 Rn Radon 222
7	87 Fr Francium 223	88 Ra Radium 226	103 Lr Lawrencium 257	104 Unq 261	105 Unp 262	106 Unh 263	107 Uns 262	108 Uno 265	109 Une 266									

Lanthanide Series	57 La Lanthanum 139	58 Ce Cerium 140	59 Pr Praseodymium 141	60 Nd Neodymium 144	61 Pm Promethium 147	62 Sm Samarium 150	63 Eu Europium 152	64 Gd Gadolinium 157	65 Tb Terbium 159	66 Dy Dysprosium 163	67 Ho Holmium 165	68 Er Erbium 167	69 Tm Thulium 169	70 Yb Ytterbium 173
Actinide Series	89 Ac Actinium 227	90 Th Thorium 232	91 Pa Protactinium 231	92 U Uranium 238	93 Np Neptunium 237	94 Pu Plutonium 242	95 Am Americium 243	96 Cmn Curium 247	97 Bk Berkelium 247	98 Cy Californium 249	99 Es Einsteinium 254	100 Fm Fermium 253	101 Md Mendelevium 256	102 No Nobelium 254

7-5. **An example of a Periodic Table of the Elements showing 109 named elements.**

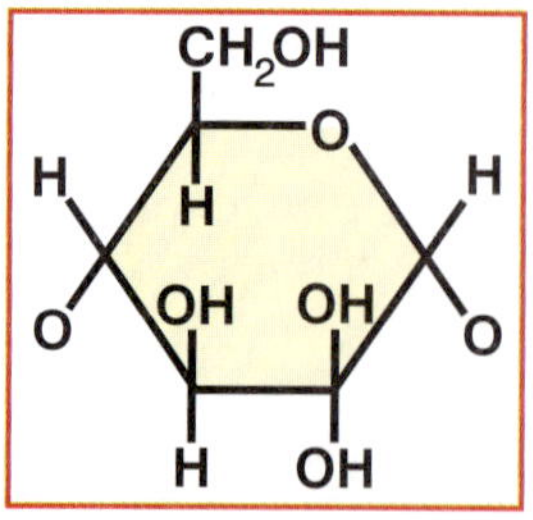

7-6. A molecule of glucose has a distinct structure. What elements are in the molecule? (Glucose is a sugar that serves as the building block of carbohydrates.)

Additional artificial elements may be identified in the future. Chemists use a Periodic Table of the Elements to list the elements and see relationships among them.

Atoms are composed of protons, neutrons, and electrons. The number of protons in the atom is its ***atomic number***. Atomic numbers establish the positions of elements on the periodic table. Each element on the table is assigned a symbol of one to three letters.

Matter is made of molecules. For example, the wood found in your desk or home is made of molecules of cellulose. Examples of molecules include cholesterol, aspirin, water, and vitamins. A ***molecule*** is the smallest group of atoms that acts together to form a stable, independent substance. The simplest molecules are composed of only two or three atoms. Water is an example. Two atoms of hydrogen combine with one atom of oxygen to form one molecule of water—H_2O.

Much of biotechnology involves the study and production of molecules that make up living organisms. A knowledge of chemistry is essential for biotechnology work.

ENERGY AND WORK

Energy is the capacity or ability to do work. Forms of energy include chemical energy, heat, sound, light, and electricity.

7-7. Engines in automobiles convert potential energy in fuel into heat energy to power the vehicle.

When a force acts on a body, work is performed and energy is expended. ***Potential energy*** is the capacity for doing work that a body has because of its position. For example, a heavy weight lifted to a certain height has potential energy because of its position. ***Kinetic energy*** is the energy a body has because it is in motion. A moving truck has the kinetic energy to move it further down the road.

In living organisms, energy is found as chemical energy. ***Chemical energy*** is present in chemical bonds that have formed molecules. This energy is produced in living organisms when molecules are broken into smaller molecules or atoms. Atoms are connected through chemical bonds to form molecules. These chemical bonds hold energy that is released when the molecules are broken down. When you eat food, your body breaks down the molecules in the food to produce energy for living.

A biotechnologist needs to understand the basics of matter and energy before beginning work with living organisms. Knowing how organisms are similar and different helps in the work.

7-8. A steer fitted with a cannula is used in feed digestibility and nutrition research. A cannula is a carefully prepared opening that allows feed samples to be entered and removed from the digestive system. The animal is normally in the chute only a few minutes to assure its well-being.

CLASSIFICATION OF LIVING ORGANISMS

How are cats and dogs alike? Different? Or, how are cats and oak trees alike and different? Living organisms can be classified based on various attributes and features. Casual observations can be used. Organisms can be classified in many ways. We could classify them by where they are found, what they use for energy, or how they move around. Most organisms have been classified using a system of scientific classification.

SCIENTIFIC CLASSIFICATION

Scientific classification is based on the careful study and observation of organisms. The classification of living things is known as ***taxonomy***. The current scientific classification system utilizes seven divisions. The divisions or stages show differences and relationships between organisms in the classifications. The seven divisions are Kingdom, Phylum, Class, Order, Family, Genus, and Species.

7-9. The mushroom and parrot are both living organisms. How would you begin to classify them? What is the kingdom of each?

The kingdom is the broadest classification. All living organisms are classified into kingdoms. A few years ago, only two kingdoms were used. The number was increased to five. They are:

1. Animalia—Animals
2. Plantae—Plants
3. Fungi—Mushrooms and yeast
4. Protista—Amoeba and paramecia
5. Monera—Bacteria

7-10. The taxonomic classification of a horse.

Scientists are now looking at a sixth kingdom consisting of what is known to them as brown plants. These organisms do not look like plants nor are they always brown. Some are quite small; others are large. Most of these organisms are forms of seaweed. They produce tremendous amounts of oxygen for the air. They also have other uses in foods, cosmetics, and medications. Scientists have known of them since their discovery in 1836. They were initially placed in the plant kingdom. In 1999, some scientists had begun talking about the need for another kingdom, according to biologist Brent Mishler of the University of California at Berkeley. A name for the new kingdom has not been identified.

After being classified into a kingdom, an organism is classified into the subsequent groups of phylum, class, order, family, genus, and species. The number of organisms in each group becomes fewer from the phylum to the species. The characteristics of the organisms increase in similarity from the phylum to the species.

NAMES OF ORGANISMS

Organisms often have common names. These names are quite familiar. Common names, such as dog or spruce tree, are the names given organisms by non-scientists. Common names vary, such as a lima bean may be known as a butter bean. This lack of agreement on name makes it difficult for people to share information.

Once the characteristics of an organism are known, the scientist gives a binomial (two-word) name that is specific to that organism. A ***binomial name*** consists of the genus and species. Binomial names are sometimes known as scientific names. The genus name is used for a few related organisms, and each organism in the genus is given a different and specific species name. Genus and species names are used together to describe a specific organism. Most people are familiar with common names, such as bullfrog, but are not familiar with the scientific name *Rana catesbeiana*.

Table 7-1. Common and Binomial Names of a Few Animals and Plants

Common	Binomial
Animals:	
Cattle (those with humps)	*Bos indicus*
Cattle (those without humps)	*Bos taurus*
Chicken	*Gallus domesticus*
Hog (swine)	*Sus scrufa*
Rainbow trout	*Salmo gairdneri*
Sheep	*Ovis aries*
Plants:	
Apple	*Malus domestica*
Carnation	*Dianthus caryophyllus*
Corn	*Zea mays*
Live Oak	*Quercus virginiana*
Rice	*Oryza sativa*
Soybean	*Glycine max*
Sugar Maple	*Acer saccharum*
Tomato	*Lycopersicon esculentum*
Wheat	*Triticum aestivum*

LIFE PROCESSES

What is life? When does life occur? In organisms, the ***living condition*** is the ability of the protoplasm in its cells to carry out complex chemical processes. When these processes stop, the organism is not in a living condition. It has died. Death results when an organism cannot carry out the activities of a living organism.

The processes for the living condition (being alive) vary with the species of the organism. A horse must be able to live as a horse—eat, drink, walk, breathe, and other activities. An oak tree must carry out the life processes of an oak—absorb water and nutrients, carry out photosynthesis, and many others.

A ***life process*** is an essential function for an organism to be in the living condition. Most organisms carry out a number of life processes. Animals, plants, and other organisms all have life processes. If those life processes stop, the organism dies.

The common life processes are growth and repair, circulation, respiration, secretion, sensation, movement, and reproduction. One of them, reproduction, is not essential for the organism to live. Reproduction, however, is essential if the species is to continue in existence. Some species have life processes unique to them. For example, all female mammals have mammary glands that secrete milk for their babies. This process is known as lactation. Milk is not essential for the mother to live, but it is for the baby mammal.

7-11. Even in the driest desert conditions the Saguaro *(Carnegiea gigantea)* gains sufficient moisture and nutrients to live and grow in Arizona.

Biotechnology uses and affects life processes. It can be used to increase processes, such as giving a growth hormone to cows to increase milk production. Or, biotechnology can be used to restore the functions of organs by stimulating the growth of new cells and tissues.

GROWTH AND REPAIR

Growth and repair occur throughout the life of organisms. These processes require food nutrients, which are changed into new living material by chemical activities within the organisms. These processes occur most rapidly in young organisms. The rate declines as organisms age. Of course, activity ceases upon death.

Growth is the process by which the cellular components and structures of an organism increase in size. The growth of individual cells and the growth of populations of cells are two separate events. The growth of a cell results in an increase in size and weight. It usually results from cell division. With division, individual cells divide into two identical cells, which may initially be smaller in size. Cell growth and division lead to population growth, which is an increase in the number of cells.

Animals and plants grow by adding new cells. The effect of growth varies with the species. The bodies of animals get larger. Plants grow longer stems and larger leaves. All growth is within the appropriate range for a species. A small plant would never grow a stem the size of an oak tree! A mouse would never grow as large as a kangaroo!

Repair is the replacement of cells. The process removes and replaces damaged cells. Cells may be damaged by activity. Suppose you go hiking in the woods. You get too close to a briar and it makes a scratch. Slight bleeding occurs. Cells have been damaged. In a few days, your body will have completely replaced the damaged cells so you no longer see the scratch.

In some cases, repair involves the replacement of parts. A good example is the lizard. If a lizard loses its tail in an accident, the tail will regrow. This regrowing of a lost body appendage is known as ***regeneration***. Unfortunately, humans do not naturally regenerate new appendages or organs, such as a per-

Biotech Connection

ENZYME FARMING

Enzymes from farm animals can help humans. An enzyme in the pituitary gland of cattle can be used to suppress human pain. Finding and having enough of the enzyme is sometimes a problem.

Scientists at Georgia Tech have been culturing the enzyme in a laboratory. The enzyme is extracted from the pituitary glands of cattle. Pituitary glands are collected in packing plants from cattle being slaughtered for human food. The glands are frozen until used in the lab.

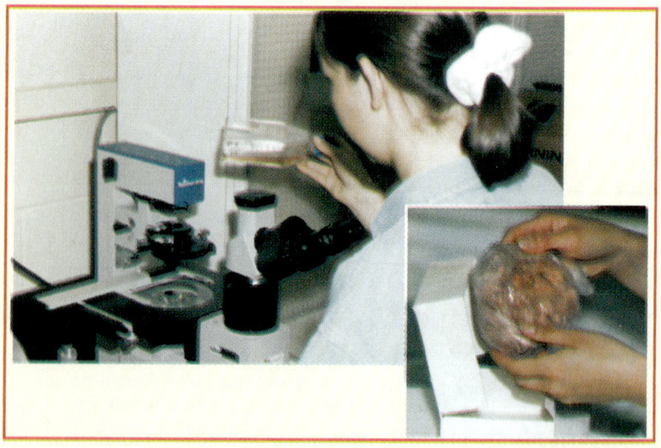

These photographs help explain. The smaller photo shows a frozen package of about 100 pituitary glands. The other photo shows a culture of the enzyme ready for study with a powerful microscope.

son who loses an arm in an accident. Under some conditions, the severed arm can be reattached by a physician. The ability to use the arm may not return to normal. With biotechnology, it may be possible to regenerate damaged or lost human organs.

Animals, plants, and microorganisms require food to grow. A lack of proper nutrients hampers growth and repair processes. In some cases, death of the organism results.

7-12. Young kittens rapidly grow and develop. These 10-day-old kittens do not have their eyes open.

OBTAINING FOOD AND NUTRIENTS

Foods are the materials or substances that provide the nutrients for an organism to carry out other life processes. They may be in liquid or solid form. Food enables organisms to grow and repair damaged cells. The substances in foods that are used for life processes are nutrients. Organisms vary in how they ob-

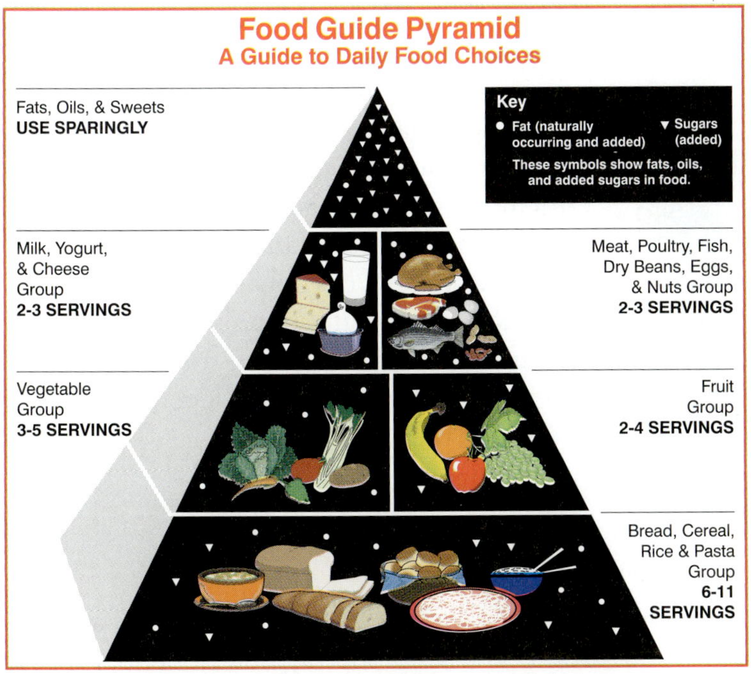

7-13. The Food Guide Pyramid helps people choose what they will have for their meals.

7-14. Fish producers use manufactured feeds to assure a proper diet for their fish.

7-15. Burley tobacco plants on this Kentucky farm are started using water float beds in a greenhouse. The medium must provide nutrients for the plants to grow for later setting in a field.

tain food. Some require food materials that are ready to use; others make their foods.

With most animals, four processes are involved in using food materials—ingestion, digestion, absorption, and elimination. Ingestion is the process of taking in food—eating. Digestion is the process of changing food materials into forms that can be used. Absorption is the process of taking in the usable forms. Elimination is the process of ridding the body of wastes. Nutrient needs vary by species, by age, and by the functions of an organism. Animals need carbohydrates, proteins, minerals, vitamins, and other nutrients. Their diets must provide these in the needed proportions. Many animals use plants for food.

Plants make their food materials. The process of food manufacture is ***photosynthesis***. The roots of plants absorb water and nutrients from the soil or medium in which they are growing. These materials are moved (translocated) to the stems and leaves. In the presence of light, chlorophyll transforms the nutrients into food. The foods are simple sugars—primarily glucose. Light provides the energy for the transformation of nutrients into food. Plant growth can be promoted by assuring that adequate nutrients are available. Soil may be chemically analyzed to determine its nutrient content (known as soil testing). Fertilizer, lime, and other soil amendments may be added depending on the results of the testing.

CIRCULATION

Circulation is the movement of substances within an organism. The process

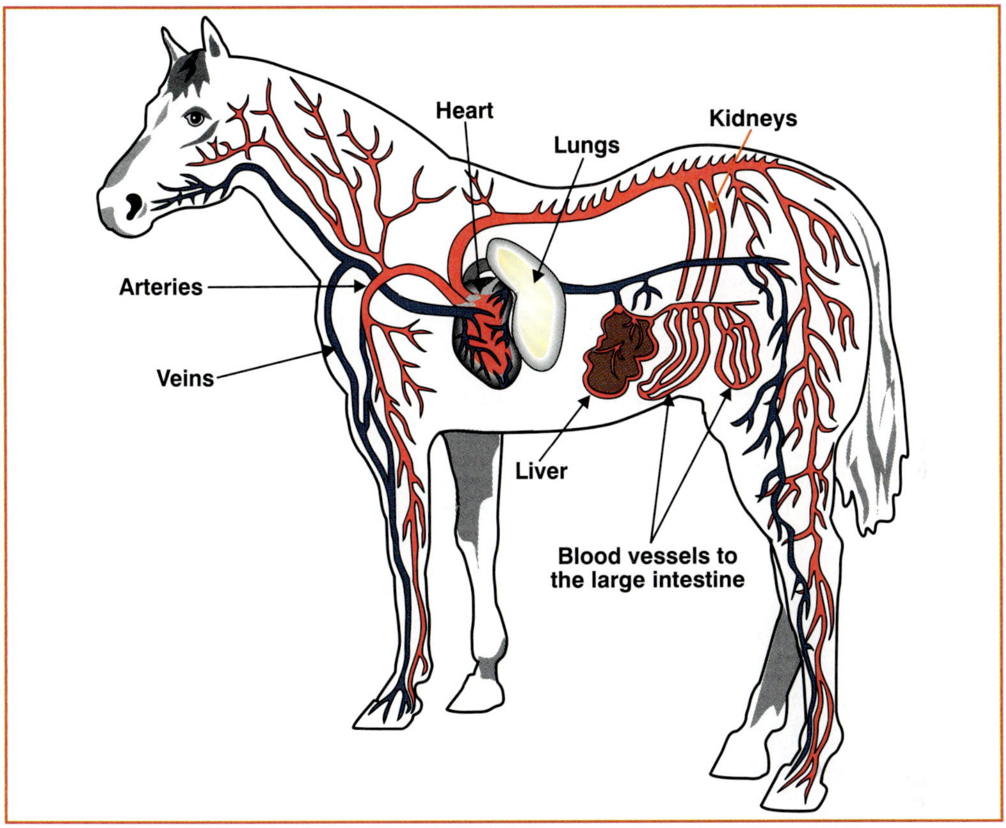

7-16. The circulatory system of a horse is similar to other animals.

varies with the species. Some organisms are more complex than others.

In animals, blood circulates nutrients throughout the organism. The heart and blood vessels form a circulatory system. It must function properly. The blood carries oxygen, digested food, and other materials. It also picks up wastes and carries them to the proper organs for elimination from the body. Injuries that cause the loss of blood reduce the circulation processes.

Plants have special tissues that carry water and food throughout the organism. These tissues form a vascular system. The two main kinds of vascular tissue are xylem and phloem. The xylem conducts water and nutrients from the roots through the stems to the leaves for photosynthesis. The phloem transports food from the leaves to other parts of the plant, including the roots. Injuries to plants may stop the flow of sap (the watery solution in plants). This results in the death of leaves, stems, or the entire plant. Careless handling causes plant injuries. These injuries can reduce the yields of crop plants, such as grapes or tomatoes.

RESPIRATION

Respiration is an oxidation process in the living cells of an organism that releases the chemical energy in foods. This makes the energy available for use by the organism in carrying out life processes. Oxidation also produces wastes. These waste products must be removed from the organism. Carbon dioxide is the primary waste. Oxygen must be provided to cells so oxidation can occur.

With animals, respiration is the process by which oxygen is taken from the environment and carbon dioxide is released. The respiratory system obtains oxygen and makes it available to the blood. The major organs in this system are the lungs (or the gills in fish). Blood vessels carry oxygenated blood from the lungs to the various organs. The reverse occurs as the blood collects carbon dioxide from the organs and delivers it to the lungs for removal from the body. Damage to the lungs (or gills) impairs their ability to function.

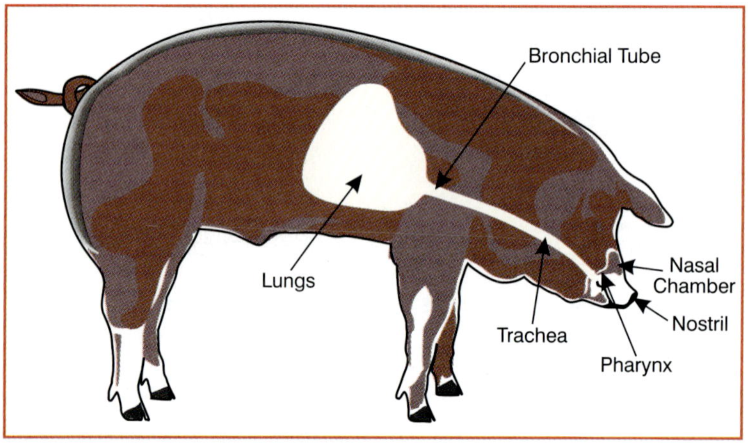

7-17. The respiratory system of a pig.

In plants, carbon dioxide is taken from the air and oxygen is released. Respiration is used to provide energy for breaking down the food. Tiny pores known as stomata in leaves allow the exchange of oxygen and carbon dioxide gases. The stomata may also give off water.

SECRETION

Secretion is the process by which the glands of organisms produce and release substances. These watery substances promote the life processes. All internal and external body fluids (except urine in animals) are secretions. In animals, tears, saliva, gastric juices, and bile are examples of secretions.

A ***gland*** is a specialized organ or tissue that produces secretions. Animals have many glands. Examples of glands are the pituitary gland, salivary gland, liver, and kidneys. Male animals may have prostate and Cowper's glands that secrete seminal fluid. The reproductive tract of female animals has glands that also secrete fluids. Female mammals have mammary glands that secrete milk.

SENSATION

Sensation is the awareness by an organism of its environment. This awareness makes it possible for an organism to respond to conditions that may be harmful. Sensation is more evident in animals than in plants.

Most animals have sense organs. These include organs that make vision (sight), taste, smell, hearing, and touch possible. A complex system of nerves is involved. Feeding your pet dog provides some examples of sensations. The dog can see you bringing the feed. It can also smell the feed and taste it when ingested. In some cases, the dog may hear you getting the food container or opening the bag or can. The sensation information is processed by the brain. The dog comes to you or its food bowl.

Plants also respond to stimuli. Light is among the most important stimuli to plants—some turn to face the sun. Plants respond to objects in their environment by growing around them, such as tree roots that grow around a rock.

7-18. Trees grow in response to large rocks in their environment. (Thigmotropism is the growth of an organism in response to contact with the surface of another object or living organism.)

MOVEMENT

Movement is needed for life processes to occur. Some movement is internal as organisms carry out life processes. The activity of an animal's digestive system is an example. Other movement is external as organisms respond to stimuli in their environment—the positions of the leaves of a plant to sunlight.

7-19. Some birds have the ability to move about by flying, such as this Western Gull.

Plants and animals move in different ways. Plants cannot move about. Roots hold them in place. Their movement occurs as they respond to their environment. Animals are much more likely to move about. Their movement is often a part of getting food, escaping danger, and seeking a mate. Some animals have strong legs or wings for rapid movement over long distances. Other animals, such as the oyster, move very little. The oyster tends to remain in the same place and wait for food to come along.

Animals have skeletal and muscular systems that are used in movement. Moving from one place to another is known as locomotion. The skeleton and muscles also give shape to the body and help protect the internal organs from injury.

ELIMINATION

Elimination is the process by which an organism removes wastes and harmful substances from its body. It is also known as excretion. Wastes include solid materials as well as fluids and gases. The nature of these materials varies with the species of the organism. Elimination varies between animals and plants.

With animals, the solid substances are undigested food materials, dead cells, and other materials. The body rids itself of waste food through the colon or large intestine. The waste is feces. The fluids include water, urine, and others. These are eliminated through the skin (as perspiration), breathe (gases, including water), and urinary tract.

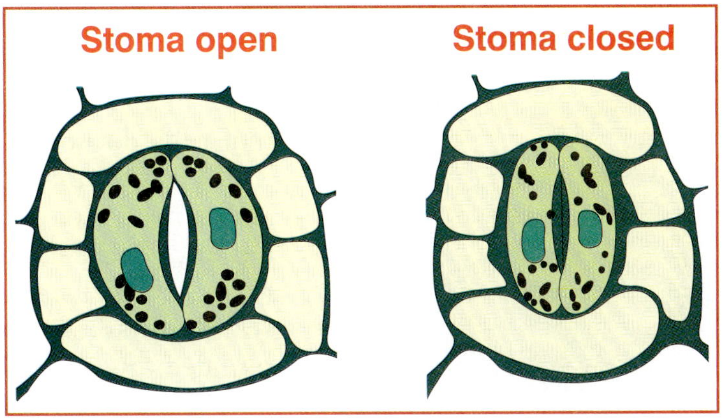

7-20. Enlarged drawings of stomata. ("Stomata" is the plural form of "stoma.")

With plants, most elimination is through the stomata in leaves. They release gases, such as oxygen and carbon dioxide, and water into the air.

STRUCTURES OF LIVING ORGANISMS

Living organisms are made of many structures. Each performs important life-sustaining processes. The overall structure of an organism depends on how its parts fit and work together.

CELLS

Cells are said to be the building blocks of living organisms. A ***cell*** is a unit of life that has a definite structure. All cells have a membrane that contains protoplasm, a nucleus, and cytoplasm. Cells vary among the species but tend

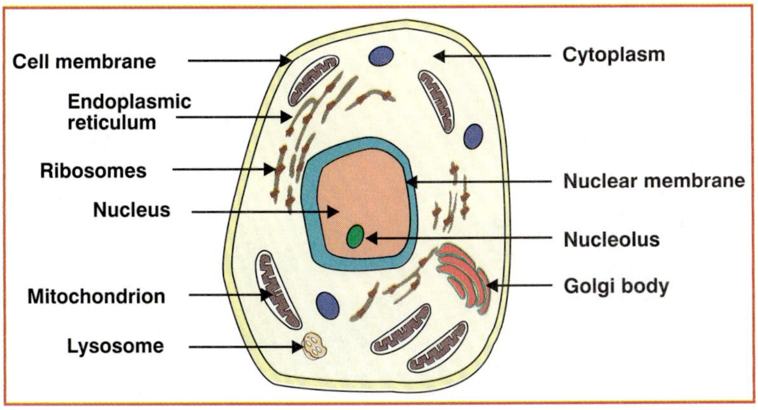

7-21. A typical animal cell. (More details in Chapter 9.)

7-22. All organisms have specialized cells. How many do you see in this desert bighorn ram?

to have the same parts. Plant cells have walls that give a rigid protective shell and allow plants to have shape and form. Animal cells do not have walls.

Cells vary in size. Some are quite small and can be seen only with a microscope. Other cells are large, such as the eggs of a bird. Some cells are long, such as those in some muscles. Other cells are flat, rounded, or square-like.

Cells are the building blocks of organisms. They may be specialized and form tissues, organs, and organ systems. ***Cell specialization*** is the difference in cells. Cells must be different to perform different functions. If all cells were alike, how would we have different organisms? Specialized cells form tissues that are eyes, skin, muscles, glands, and other body structures. (More details on cell structure and processes are presented in Chapter 9.)

TISSUES

A ***tissue*** is a group of cells that is alike in structure and activity. All of the cells are specialized. The specialization allows tissues to perform specific

7-23. The flower of a tulip poplar tree has several tissues.

functions for an organism. Even when specialized, all cells in tissues must carry out important chemical activity to keep them alive. This includes producing energy through respiration.

Tissues specialized for a particular function do not perform other functions. Skin, bones, muscles, nerves, and other tissues are examples of specialized animal tissues. Examples of specialized plant tissues include rinds, skins, flower petals, roots, and leaves.

ORGANS

A collection of tissues that work together to perform a function is an ***organ***. Organs may contain several different tissues. For example, the heart of an animal has muscles, valves, vessels, nerves, and other specialized tissues.

Both animals and plants have organs. Examples in animals include lungs, liver, eyes, tongue, kidneys, liver, heart, and testicles and ovaries. Examples in plants are leaves, flowers, stems, and roots.

ORGAN SYSTEMS

An ***organ system*** is a collection of organs that work together to perform a specific function for an organism. Organ systems are found in higher organisms. They are the major systems of animals and plants. For example, the

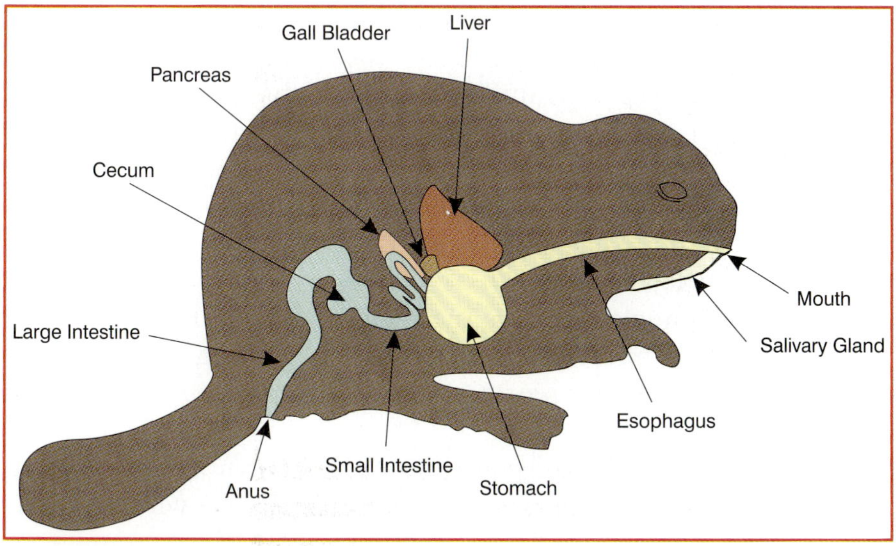

7-24. The digestive system of a beaver is made up of several organs. All are needed for the system to work properly.

circulatory system of an animal is made up of a heart, arteries, veins, and other small blood vessels.

The reproductive systems of large animals vary with the gender of the animal. A male has a system designed for delivering sperm for fertilization of an egg. A female has a system designed to produce eggs, provide a place for fertilization, and provide a place for the fertilized egg to develop.

DNA, PROTEINS, ENZYMES, AND LIPIDS

DNA, proteins, enzymes, and lipids are important in biotechnology. Each performs or promotes functions within organisms. A basic understanding is needed to begin a study of biotechnology. These are the important areas in manipulating cells and organisms. They are the major focus of genetic engineering. Each is introduced here and covered in more detail later in the book.

DNA

Cells have two types of nucleic acid: DNA and RNA. DNA is in the chromosomes in the cell nucleus. RNA is found throughout a cell but mostly in the nucleus.Knowledge of the structure of DNA makes it possible to change the genetic material in an organism. Information can be taken out or added to the DNA of an organism.

RNA (ribonucleic acid) is similar to DNA. It has long chains of nucleotide units. RNA has an important role in the formation of proteins in cells. Some RNA leaves the nucleus of a cell and carries a message about making protein (amino acids). This is known as messenger or mRNA. mRNA has genetic code information that determines the position of amino acids. This code has three nucleotides known as codons.

DNA (deoxyribonucleic acid) is the molecule in a chromosome that furnishes the genes with information for their development. The genes in chromosomes carry the genetic information (hereditary traits) about an organism.

DNA is a long polymer of repeating units known as deoxyribonucleotides. Each DNA molecule is made of two strands. This is why the term, double helix, is used in referring to the structure of DNA. Each strand has many alternating units of sugar and phosphorus. The two strands are held together by hydrogen bonds between pairs of bases.

A deoxyribonucleotide has three parts: a phosphate group, a sugar, and nitrogen-containing bases. The sugar is known as deoxyribose. The four bases are adenine (A), guanine (G), thymine (T), and cytosine (C). The genetic information is stored in the four nitrogen-containing bases. Alternating sugar and phosphate parts make up the backbone of the DNA.

When cells divide, the genetic material is replicated (copied) so that each daughter cell contains identical DNA copies. The two strands of the double helix separate in cell division. Once separated, the base-pairing process allows each single-stranded molecule to serve as a template to form a new complementary strand. Each of the two new identical double-stranded molecules consists of one original strand and one new strand.

The bases respond in interesting ways on a DNA molecule. Thymine always pairs with adenine. Guanine always pairs with cytosine. This results in the ladder-like appearance of a DNA double helix.

DNA is the molecule of intense interest in creating transgenic organisms. The structure of DNA is interesting.

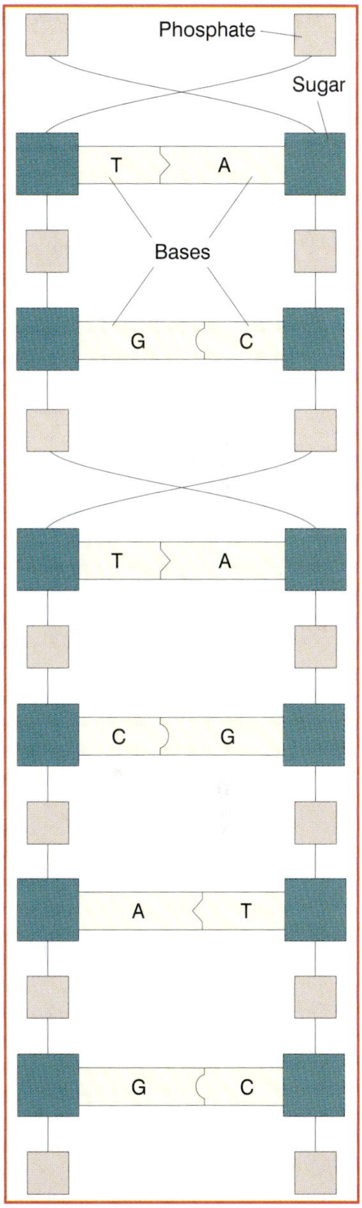

7-25. **DNA structure is simplified in this drawing.**

PROTEINS

A ***protein*** is a substance consisting of amino acids. It is important in the nutrition and growth of an organism. The sequence of the amino acids in a molecule gives proteins different properties. The amino acids, or chains, known as polypeptides may be straight or three dimensional.

More than 20 proteins have been identified; some are essential for animal growth. The essential amino acids for an animal vary with its digestive system. Ruminants (animals with more than one compartment in their stomachs, such as cattle) have fewer essential amino acids than monogastrics (animals with a single-compartment stomach, such as a dog). Ten amino acids are essential for most animals. Examples include arginine, histidine, and lysine. Essential amino acids must be provided in the food that an animal ingests. Some animals produce amino acids in their digestive systems or bodies. These are known as nonessential amino acids.

Proteins are usually made of four chemical elements: hydrogen, carbon, nitrogen, and oxygen. The arrangement of these elements in the molecule of protein determines the kind of amino acid. Some proteins also contain sulfur, iron, and phosphorus. For example, keratin is a kind of protein that contains sulfur. It forms tough tissues found in horns, hooves, hair, fingernails, and the epidermis of the skin. Keratin is known as a structural protein.

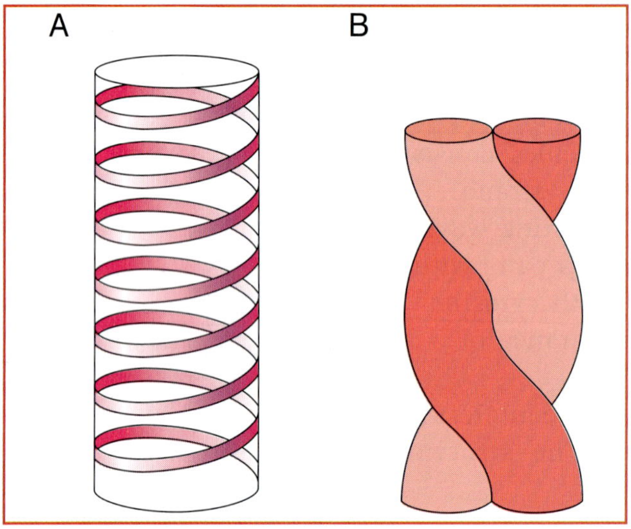

7-26. The structure of the protein keratin is illustrated with a single spiral or alpha helix (A). Spirals will wrap with another spiral to form chains or fibers (B).

Animals must have protein to grow, repair cells, and carry out life processes. A young animal needs a diet with a higher percentage of protein. For example, baby fish (known as fry) may need a diet that is 50 to 60 percent protein. Older fish weighing a half-pound or so can grow on a 28 to 32 percent protein feed.

Proteins are important to scientists interested in many areas of animal life. Nutrition and the growth of organisms, especially animals, involve protein. The antibodies in the blood of animals are made of protein. Hormones are proteins, and regulate growth and reproduction. Hemoglobin and albumin in blood are proteins. Hemoglobin is the protein that carries oxygen from the lungs to body tissues. Albumin regulates the amount of water in the blood.

Protein engineering is the design and production of non-natural proteins. This usually begins with a natural protein that is modified in some way. An example is altering the properties of an antibody. Another major use of protein engineering is altering proteins to serve as drugs.

ENZYMES

An ***enzyme*** is a kind of protein. Enzymes act as catalysts in chemical reactions needed for the living condition to occur. Enzymes are produced by living cells. Most enzymes are highly specific in their actions. Many kinds are found in nature. Some require precise conditions to function. Enzymes are not consumed in the reactions that they catalyze.

A ***catalyst*** is a substance that alters the rate of a chemical reaction without being changed itself. Many catalysts are used in the chemical industry. Several modifications of the word are used, such as catalysis (altering the rate of a chemical reaction) and catalyze (acting upon as a catalyst).

The use of enzymes is an important area of biotechnology. Scientists can make an enzyme more specific in terms of the reactions it catalyzes, how well it does them, and how long it lasts. The study of enzymes is enzymology.

Enzymes are given names and numbers. Some have common names, such as rennin and trypsin. The scientific names and numbers are provided by the Enzyme Commission (EC). All enzymes are classified into six groups. Each enzyme has a four-digit number, with the first digit being the group into which it fits. The classification is: 1, oxidoreductases; 2, transferases; 3, hydrolases; 4, lyases; 5, isomerases; and 6, ligases.

LIPIDS

A ***lipid*** is an organic compound of a fatty acid that is soluble in organic solvents but not in water. Lipids form a fairly large group of oily or fatty substances essential for good health. This means that they are important in animal nutrition. Both animal fats and plant oils are lipids. Scientists study and manipulate lipids in nutrition research and other areas related to plant and animal technology.

Lipids are classified into three groups: simple, complex, and derived. Simple lipids include fats, oils, and waxes. Complex lipids include phospholipids and glycolipids. Another kind of lipid is the derived lipid, which includes steroids.

7-27. **The seed from this large field of flowering canola in North Dakota will later be harvested for making oil.**

Lipids are concentrated sources of food energy. They are higher in calories than protein or carbohydrate. Some types of lipids help form the membranes that enclose and protect cells. They help the membrane repel water but allow certain kinds of vitamins and other substances to enter a cell. Cholesterol is a steroid lipid that is found in every animal cell membrane. Humans are well aware of the effects of cholesterol on their health, especially low density lipoprotein or LDL (the bad kind).

Humans consume lipids in margarine, shortening, salad oil, and frying oil. Plants produced for oil are soybean, palm, canola, sunflower, corn, and safflower. Some cotton oil is also used. Transgenic forms of canola have been developed with larger amounts of desired lipids. These products are beginning to come on the market in margarines.

REVIEWING

MAIN IDEAS

Science knowledge and skills are essential in biotechnology work. Simply, science is knowledge of the world around us. But, we live in a world that is sometimes quite complex. Life science, physical science, mathematics, and social science are used in biotechnology.

Biotechnologists must understand matter and energy. They need to know the life processes of living organisms. Biotechnologists need to know the structures that make it possible for organisms to carry out the processes. The structures and processes are used in areas that involve, among others, DNA, proteins, enzymes, and lipids.

Living organisms are classified into kingdoms, which are further divided into six sub-levels of increasing similarity. The species is the sub-level with the greatest similarity. Five kingdoms are now used: Animalia, Plantae, Fungi, Protista, and

Monera. Another unnamed kingdom containing species formerly thought to be plants is being discussed by scientists.

Life processes are the processes that keep an organism in a living condition. These include growth and repair, obtaining food and nutrients, circulation, respiration, secretion, sensation, movement, and elimination. Reproduction and lactation are nonessential processes for an organism to live but are important in the perpetuation of a species.

Structures of living organisms are built of blocks known as cells. Cells form tissues. Tissues form organs. Organs form organ systems. Each performs important life-sustaining functions for an organism.

DNA, proteins, enzymes, and lipids are important in biotechnology. These are important in performing or promoting functions within organisms.

QUESTIONS

Answer the following questions using correct spelling and complete sentences.

1. What are the four branches of science? Briefly describe how each is important in biotechnology.
2. What is matter?
3. What is a chemical element? How do chemical elements relate to matter?
4. What is energy? Distinguish between potential energy and kinetic energy.
5. What is chemical energy? Why is it important?
6. What are the five kingdoms in the taxonomy of organisms?
7. How are organisms named?
8. What is a life process? List and briefly explain important life processes.
9. What are the basic structures of living organisms? List and briefly explain these structures.
10. What is DNA? Why is it important in biotechnology?
11. What are proteins? Why are proteins important?
12. What is the role of enzymes?
13. What is a lipid? Why are lipids important?

EVALUATING

Match the term with the correct definition. Write the letter of the term in the blank that is provided.

a. biology	f. life process
b. matter	g. regeneration
c. atom	h. gland
d. molecule	i. catalyst
e. taxonomy	j. enzyme

_____ 1. A specialized organ or tissue that produces secretions.

_____ 2. Another name for life science.

_____ 3. The smallest unit in which a chemical element can exist.

_____ 4. Anything that has mass; may be a solid, liquid, or gas.

_____ 5. A substance that alters the rate of chemical process.

_____ 6. A function essential for an organism to be in the living condition.

_____ 7. The classification of living things.

_____ 8. The smallest group of atoms that act together to form a stable substance.

_____ 9. The ability of an organism to regrow an organ or limb.

_____ 10. A kind of protein that acts as a catalyst in important life processes.

EXPLORING

1. Job shadow a life or physical scientist. Carefully observe his or her work. Follow all safety rules. Record your observations on a regular basis for use in preparing a report on your experience.

2. Investigate the nutritional protein needs of animals found in your local community. Do this for young animals as well as mature animals. Use reference books, the Internet, or resource people as sources of information. Prepare a computer-based presentation on your findings.

3. Arrange to investigate the systems of an organism that carry out the life processes. This will vary depending on what is available in your school or community. In some cases, you may be allowed to dissect a laboratory-prepared specimen of a cat or fetal pig. In other schools, you may use a computer program for a virtual dissection experience.

8

ASEPTIC TECHNIQUE

At sometime in your life you have had a common cold. You might have sneezed and coughed. You felt miserable for a few days. You got over it. Your body helped you overcome the cold.

Another time you have probably seen mold growing on food. You didn't eat the food. You threw it away. The food was spoiled and you didn't want to run the risk of being sick.

Both of these conditions started from microbes that infected a host. In the first case, the host was you. In the second, it was the food that had grown the mold. At first, only one or a few microbes were present. The number grew into a sizeable population. Microbes may become visible to the naked eye when present in large numbers. Why did the population grow? What can you do about it?

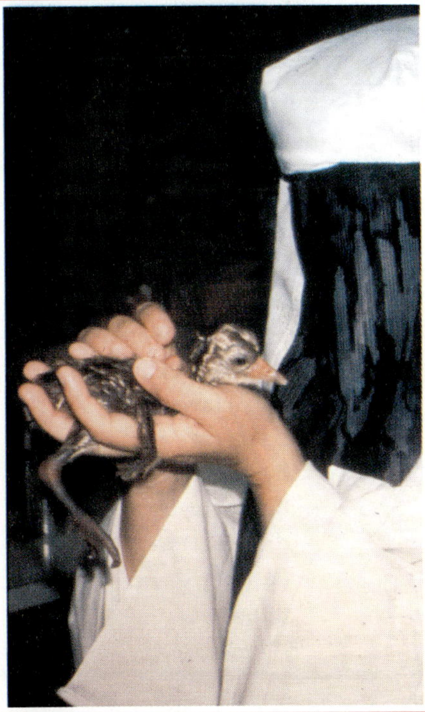

8-1. Some organisms need a germ-free environment. This shows an artificially hatched whooping crane held by a scientist. The scientist is wearing protective clothing to keep from spreading germs that might endanger the frail bird. (Courtesy, David Ellis, U.S. Fish and Wildlife Service)

OBJECTIVES

This chapter describes the meaning, importance, and procedures associated with the aseptic technique. It has the following objectives:

1. Define asepsis and aseptic technique and tell why they are important.
2. Describe the main types of microorganisms of concern in asepsis.
3. Explain how to create an aseptic environment.
4. Provide examples of aseptic conditions.

TERMS

asepsis
aseptic environment
aseptic technique
autoclave
bacteria
bactericide
bacteriostat
disinfection
flaming
fungi
germ
invasive procedure
laminar flow hood
microorganism
sanitization
sterilant
sterilization
virus
yeast

ASEPTIC ENVIRONMENTS: "PEOPLE VS. GERMS"

Asepsis means that something is without germs. It is sterile. In some areas of biotechnology, work must be done in an aseptic environment. Other areas also require asepsis. (Note: Asepsis is a noun; aseptic is an adjective.)

A ***germ*** is a microorganism. Many people think of them as organisms that produce disease. But, a germ is any microbe that can cause problems with biotechnology, food keeping, and other areas. Many do not cause disease. Microbes are not all bad. They perform many useful roles in our environment.

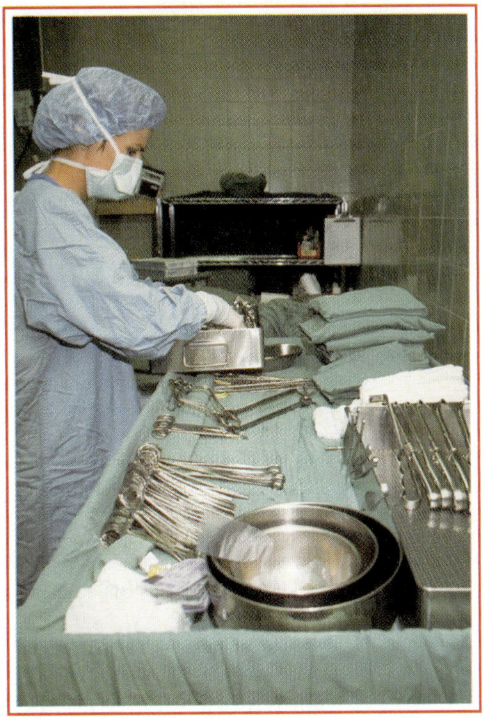

8-2. Asepsis is extremely important in surgery.

Medicine is an example of where asepsis is used. Physicians and veterinarians use sterile techniques. These are to prevent infections among their patients. Surgical devices that cut or penetrate the skin must be sterile. If not, microbes may be introduced. Germs on instruments can transfer diseases from one patient to another.

ASEPTIC TECHNIQUE

Aseptic technique is taking action to have an environment that is free of germs. In many cases, it is the practice of removing microbes from an environment. It is to assure asepsis.

Microbes are the primary sources of contamination in lab work. Aseptic techniques are used to reduce microbe populations so they are not factors in the work. The goal is to have a sterile environment, which means that no microbes are present. That is often difficult to achieve.

A tolerance level for microbes may exist in some work. How many microbes can be present before they pose a problem? This depends on the risk we are willing to take. It is obvious that medical facilities where surgery is

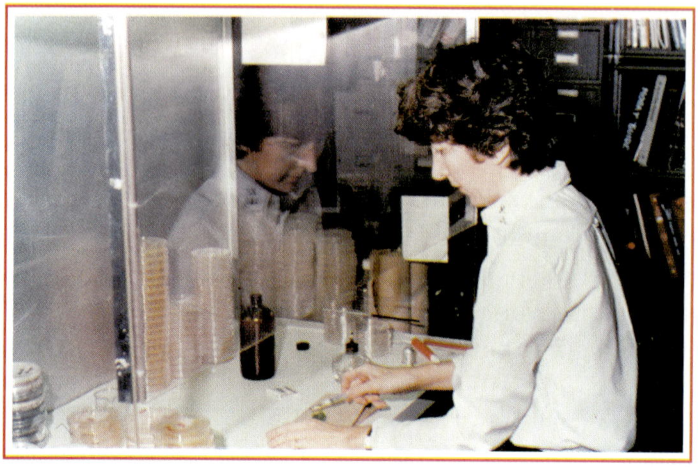

8-3. Tissue culture isolations are being prepared in an aseptic environment in an Auburn University lab. The work is using sterile media dishes for the cultures. The hands of the scientist were disinfected with alcohol. (Courtesy, Jackie Mullen, Auburn University)

done must have a very low tolerance for microbes. Facilities where plant tissue culture is done may not be as strict as the medical field. In general, the work area must be free of microbes.

With aseptic techniques, sterilization, disinfection, and sanitation are used.

Sterilization

Sterilization is the destruction of all microbes. It is used to have an environment or instrument that is free of microbes. Sterilization is most widely used in human and veterinary medicine.

Two principal methods of sterilization are physical and chemical. Physical sterilization is using heat. The heat may be dry or wet, as with steam. The most widely used method is heat.

Heat, at a high enough level, destroys any and all microbes that may be present. It is used when no microbes can be tolerated. An autoclave may be used to heat instruments to assure that their surfaces are sterile. Dry flames are used with edges and small areas of metal instruments. Metal and glass instruments can be sterilized with heat. Plastics tend to melt or warp when heated.

Chemical methods use sterilants. A ***sterilant*** is an agent that kills microbes. Sterilants may be gases or liquids. Ethylene oxide gas is effective against all types of microbes. It can be used with plastic materials. Common liquid sterilants are alcohols and bleaches. Liquid materials often do not thoroughly sterilize instruments. Instruments are typically sterilized in covered trays. Exposure time should be a minimum of 24 hours. Plastics and in-

8-4. Sponges used in labs and kitchens can harbor many germs. Most are killed by placing a wet sponge in a microwave oven set on full power for four or so minutes.

struments with cemented parts cannot be sterilized with liquids. Only people with proper training should use chemical methods.

Disinfection

Disinfection is removing potential disease-causing microbes. In some cases, disinfection merely lowers the number of microbes to a point where they do not pose a hazard. Various approaches are used in disinfection, such as commercially available disinfectants.

Sanitization

Sanitization is reducing the level of microbe contamination to an acceptable level. It is done with cleaning solutions, hot water, or chemical disinfectants.

Sanitization is used where a sterile environment is not essential. Some microbes are tolerated.

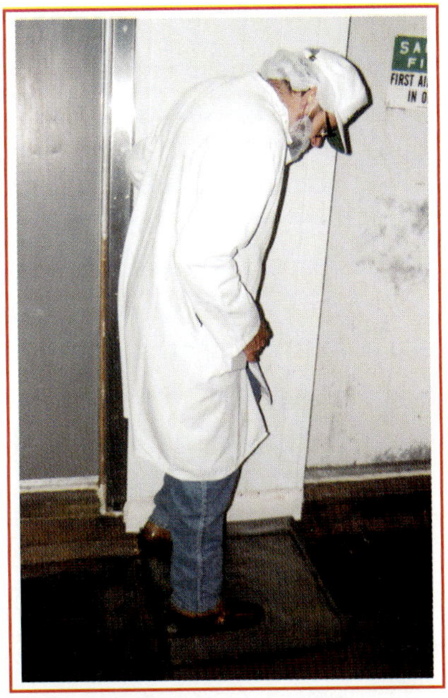

8-5. Sanitation may involve dipping the soles of shoes in a disinfectant solution. Note that a hat, beard net, and clean outer garment are worn.

IMPORTANCE

The importance of asepsis depends on the nature of the work that is being done. Three general areas of concern are given.

Career Profile

LABORATORY TECHNICIAN

Laboratory technicians perform a variety of lab work. They help maintain a safe lab, process data, and communicate with researchers. Technicians must carry-out tasks in a detailed, precise manner, and follow instructions exactly as written.

A high school diploma with emphasis in science is needed for entry-level jobs. An associate degree from a tech school or community college is beneficial. A college degree in biology or a related field is needed for the better jobs. Some employers require state certification of skills.

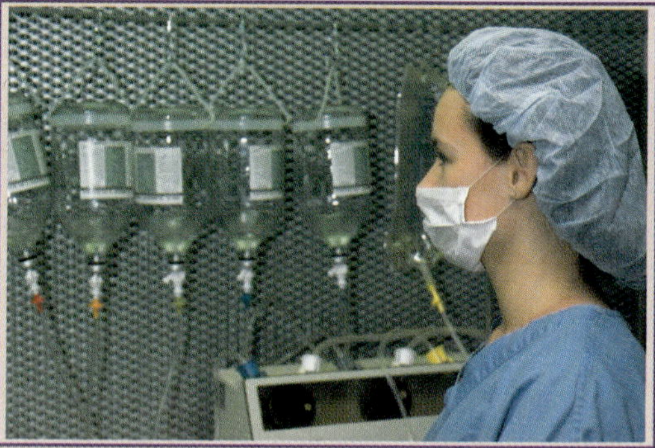

Jobs for laboratory technicians are in a number of settings: biotechnology research labs, hospitals, veterinary diagnostic clinics, and food processing facilities. Colleges, government agencies, and testing labs may hire laboratory technicians.

Worker Protection

In some labs, asepsis is needed to protect the workers. Workers can be exposed to hazardous microbes. Some would likely produce no ill effects. Others can create disease and impair the health of the worker. For example, scientists who do research in human pathology use sterile techniques. This is to contain the pathogens. They do not want to get the disease nor release it.

Environment Protection

Some microbes can damage the environment if they were accidentally released. Aseptic techniques help assure security of microbes and protect the environment. Scientists call this containment. The microbes are kept in the lab where they belong!

Experiment or Product Protection

Microbes can have a negative effect on experiments or products. Microbes can cause an experiment to fail or have an inaccurate result. This means that microbes could alter the outcome of the experiment. Aseptic techniques also protect products from injury due to microbes. An example is tissue-cultured plants. If procedures are done in an environment that is not aseptic, the tissues may be attacked and fail to develop.

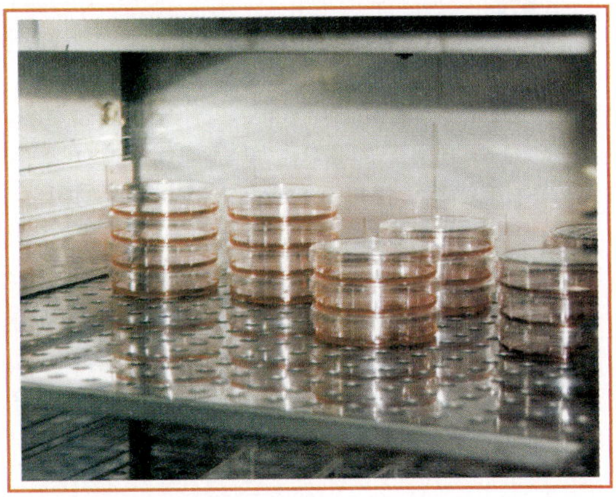

8-6. Cells being cultured in a protective environment.

Researchers in microbiology, people doing plant tissue culture, and those doing other work must keep contaminants down. This protects the fragile cultures that they work with. Delicate tissues may be attacked by microbes and fail to grow. The medium used with tissue cultures is often quite fertile for microbe growth. Food processors strive to have sanitary conditions to protect food quality and increase shelf-life.

MICROBES

A ***microorganism*** (microbe) is a living organism that is not normally seen with the naked eye. A microscope is needed to see them. Masses or colonies are sometimes visible without a microscope.

Some microbes are beneficial; others are harmful. Some cause disease infections. Others are used to create food products like cheese, wine, and yogurt. Most microbes that are sources of contamination can be placed in three groups: bacteria, fungi/yeast, and viruses. Protozoa and algae are also microbes but are not described here. More information about all microbes is presented in Chapter 12.

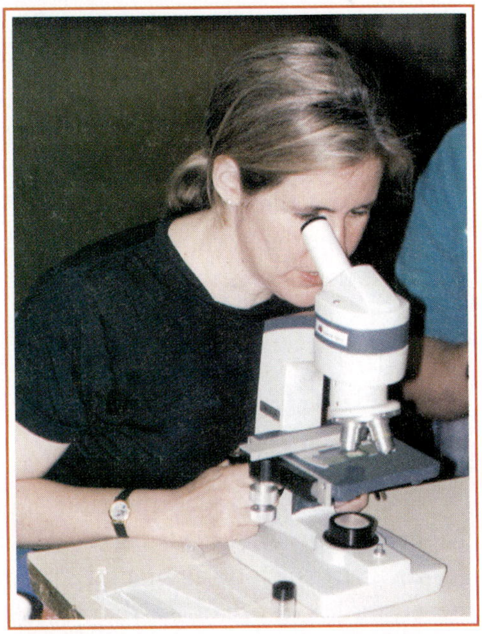

8-7. A microscope must be used to see microbes.

BACTERIA

Bacteria (plural of bacterium) are single-celled prokaryotes with a single chromosome and no cellular organelles. Prokaryotes do not have a cell nucleus. Bacteria live by digesting healthy cells and producing a toxin as a by-product. Bacteria are in the kingdom Monera and phylum Schizomycota.

Most bacteria reproduce by cell division. However, some bacteria (primarily the *Bacillus* and *Clostridium* genera) form endospores. An endospore is a special reproductive structure formed within the cell of a bacterium. Endospore forms are more difficult to control than the bacteria.

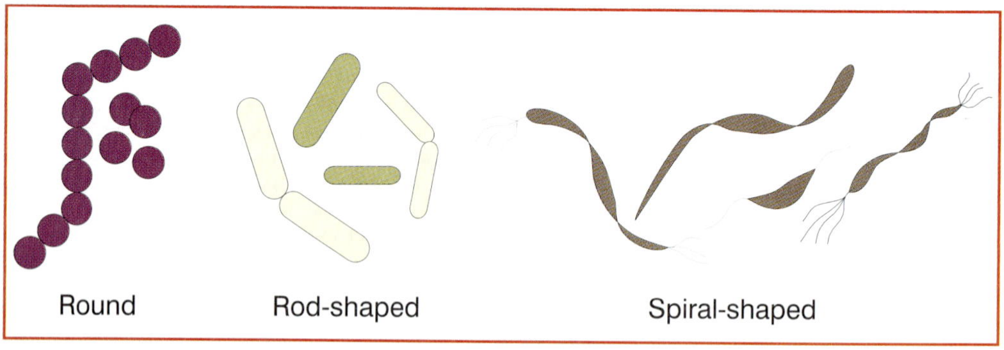

8-8. **Common shapes of bacteria are round, rod, and spiral.**

Bacteria can be destroyed and controlled in several ways. These depend on the nature of the organisms that are involved. Bactericides and bacteriostats can be used. A ***bactericide*** is a chemical or physical agent that kills some kinds of bacteria. They work best on those that do not form endospores. A ***bacteriostat*** is an agent that prevents the formation of bacteria.

Some bacteria are useful; others are harmful. For example, bacteria are used in cheese and yogurt production. Other bacteria are responsible for diseases, such as anthrax.

FUNGI

Fungi (plural of fungus) are complex, eukaryotic organisms (meaning their cells have a nucleus). Fungi spread by spores and may form long, branching chains of cells known as hyphae. The large hairy mass you see forming on food is black bread mold, *Rhizopus stolonifer.* Bread mold is not confined to bread. It also forms on fruits, such as strawberries, grapes, and plums. The drug penicillin is produced by the fungus *Penicillium notatum*.

Individual organisms may be quite small and require a microscope for study. Some fungi are larger, such as mushrooms. Examples of fungi include molds, mildews, rusts, smuts, and mushrooms.

Yeast is a group of single-celled fungi of the phylum Ascomycota. They reproduce by fission or budding. Some yeasts have the ability to convert sugars into alcohol and carbon dioxide. This is a useful feature in baking bread that "rises." Yeasts are used in industrial fermentation processes.

8-9. Cheese that is not stored properly will mold. This cheese is not fit to eat.

VIRUSES

A ***virus*** is a very small particle composed of a core of nucleic acid surrounded by a protein coat. Viruses reproduce only inside of another cell. They do not have the properties of living things.

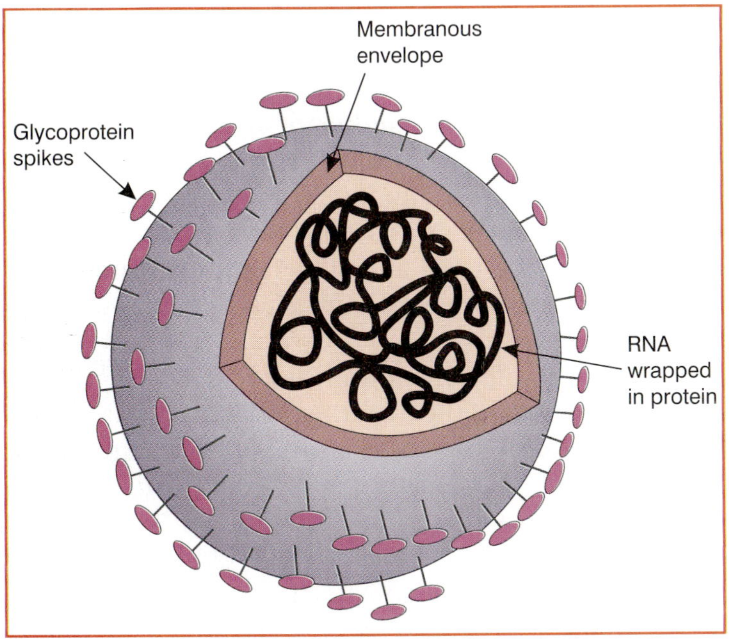

8-10. One kind of virus is said to be enveloped.

Viruses are not visible with an ordinary microscope. An electron microscope is needed. Viruses alter the functions of the cells they invade. They are responsible for many serious human diseases, including AIDS (acquired immune deficiency syndrome), hepatitis, and chicken pox.

ASEPTIC ENVIRONMENTS

An ***aseptic environment*** is an area free of microbes. The size and location of the area may vary considerably. In some cases, an entire room may be aseptic. In other cases, special chambers or work areas may be aseptic. These range from laminar hoods to plastic bags in which work is performed. In some cases, only surfaces are made aseptic, such as the surface of a microscope slide or the tips of forceps or tweezers.

CREATING ASEPTIC ENVIRONMENTS

All instruments and media should be free from the presence of any living microbe. The air should provide a good environment to prevent the contamination of work with microbes.

Approaches in creating aseptic environments are included here.

8-11. Bleach and Lysol are common disinfectants. The label tells how much to mix in water. Make no more than will be used in a few hours.

Washing Surfaces

Wash surfaces with 70 percent ethanol alcohol, bleach (sodium hypochlorite) in water solution, Lysol, or another disinfectant. Use care with alcohol near any open flames.

The Lysol solution varies with the product used. Common household Lysol may be mixed 50 percent Lysol and 50 percent tap water. Lysol concentrate may be from 1 to 2 percent with tap water.

The bleach solution is typically 10 percent bleach and 90 percent water. This solution can be made stronger by increasing the amount of bleach. Remember that bleach will damage clothing and similar materials.

Always wear rubber gloves and protect your skin from exposure to bleach and Lysol solutions.

Autoclave Instruments

8-12. A pressure cooker can be used to sterilize instruments.

Reusable tools and instruments should be autoclaved. Those that cannot be reused should be properly disposed of. Do not leave them in the work area. They could be a source to reintroduce contamination to an area that has been aseptic.

An ***autoclave*** is a steam sterilizer that uses pressure to exceed normal water and steam temperatures. Some people say it is a type of pressure cooker.

An autoclave uses both heat and pressure to destroy germs. Heating water in an airtight container allows a temperature greater than boiling water. Some bacteria can survive boiling water, but few survive temperatures above boiling. Holding instruments in an autoclave at 121°C and 15 psi (pounds per square inch) for 15 minutes is recommended. Most all microbes are killed in these conditions though there are exceptions, such as *spongiform encephalopathic* agents.

Always use caution with autoclaves. They are hot and can quickly cause severe burns. Follow the instructions in using them. Open an autoclave safely after it has cooled. Special autoclave gloves may be used if instruments are handled while hot. It is best to let the autoclave cool before trying to use the instruments.

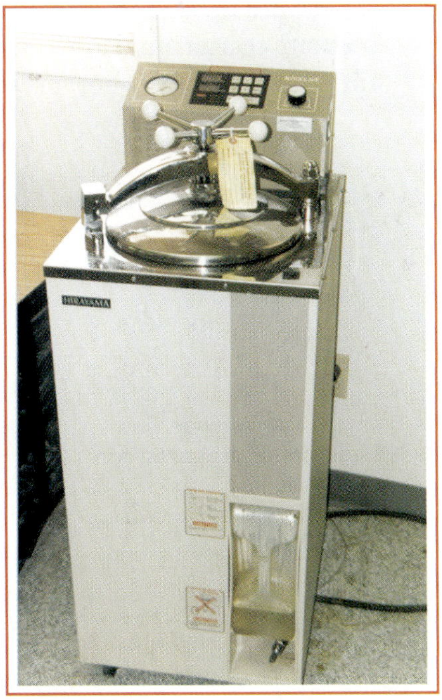

8-13. A stand-alone autoclave.

Laminar Flow Hoods

A ***laminar flow hood*** is a small work environment that helps provide aseptic conditions. The work area inside the laminar flow hood allows experiments and procedures in a protected area.

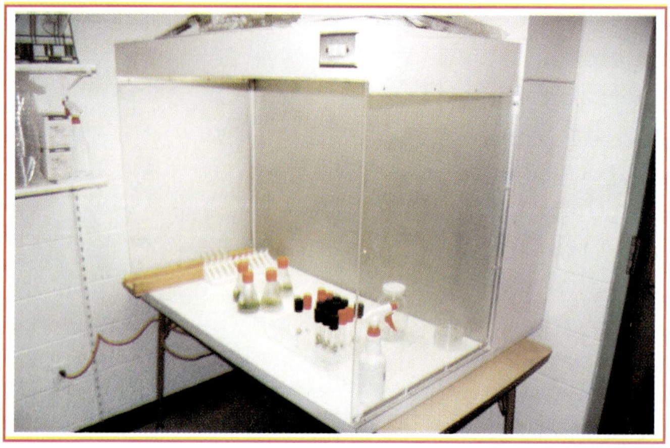

8-14. A tabletop laminar flow hood.

Laminar flow hoods have fans that blow filtered air into the work area. This creates a slight increase in air pressure inside, and this keeps outside air out while it is open. Filters remove dust, some microbes, and other materials from the air. The filters must be regularly serviced to assure that they are working properly. The inside surfaces of laminar flow hoods are wiped or sprayed with a disinfectant. The hoods may be small units designed for sitting on a tabletop or larger permanently installed units.

All instruments and materials used in a flow hood should be sterile to prevent contamination. Laminar flow hoods are used in making tissue cultures and doing other preparations and experiments.

8-15. A laminar flow hood being prepared for use.

Flaming

Flaming is exposing the tips or surfaces of instruments to a flame to destroy any microbes. A small Bunsen burner flame or an alcohol lamp is adequate for the tips of tweezers or the surfaces of microscope slides. Surfaces only need to be flamed for a few seconds. Only metal or glass can be flamed. Plastic materials would be damaged by a flame.

Use caution with flaming. Since fire is involved, no flammable materials should be nearby. Be sure to protect skin and clothing from the flame. The burner should be on a se-

8-16. Flaming is being used to sterilize the tips of these forceps (tweezers). A small alcohol lamp provides the flame.

Biotech Connection

OPERATING ROOM PREPARATION

Aseptic technique is used in hospital operating rooms. The room, equipment, instruments, and people are aseptic to the fullest extent possible. It also includes the procedures used in the surgery. If good procedures are not followed, a hospital can be a source of disease.

Hospitals should use procedures to keep from being a source of germs. The presence of disease and human tissues creates conditions where germs may thrive. Some germs are resistant to drugs and hard to treat. Sterile operating rooms are essential to prevent infections.

Aseptic technique is used to sterilize operating tools, work surfaces, linens, and floors. It is also used in the operating procedures by surgeons, assistants, and nurses. The goal is to prevent contamination of open wounds and invasive practices.

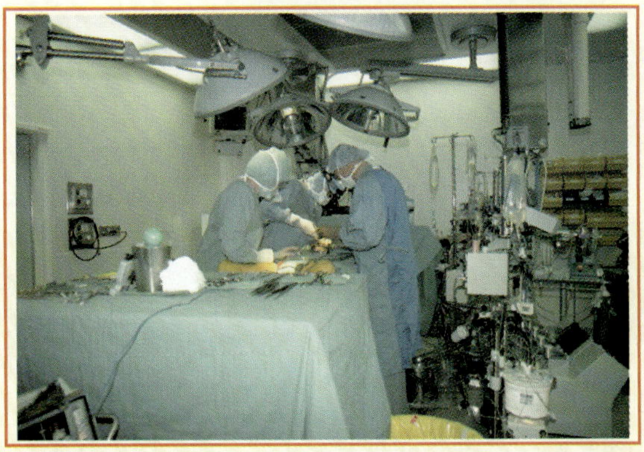

cure surface so it does not turn over. Always have a fire extinguisher nearby. Know the procedures in case of a fire emergency.

Air Filtration

Air filtration may be used in rooms or lab areas. These systems should vent outside. They should not be connected with other rooms or areas. Keep the filters clean. The system should work properly. When properly installed, these systems work well in rooms that are airtight. Doors and windows should be opened no more than is essential. The air flowing into the room creates a slight increase in pressure. This tends to keep outside unfiltered air out.

Dress Appropriately

People who are present should wear clean laboratory clothing and wash their hands. In some cases, special lab clothing may be required. Lab clothing should not be worn from the lab area. Sterile latex gloves should be worn. Some places require hats, hair nets, and facial hair coverings.

8-17. **Not reusing pipette tips after a procedure is part of proper protocol.**

Follow Protocol

Follow proper protocol in all work. Contamination can result when the work does not follow an approved and accepted sequence. Instruments should be used properly. In some cases, disposable pipette tips should be used and properly disposed of afterward.

No Eating or Drinking

No food or beverages should be eaten in any laboratory. Food materials may introduce microbes to a lab. They also serve as good places for microbes to grow. Just a few crumbs or scraps provide the food for thousands of microbes!

Eating or drinking in a lab also runs the risk of infecting a person with a microbial disease. Handling food or otherwise exposing it to microbes can contaminate it.

Safety Rules

Know and follow all safety precautions with organisms and in personal protection. Post the rules in or near the lab area. Train all people in the lab in safety.

CONSIDERING THE ENVIRONMENT

Consider the environment that is to be aseptic. It includes the work space, equipment, people and their clothing, and work surfaces.

- Transfer Area/Room—The room should be free of foreign particles. If possible, ventilation systems should not be shared with other rooms. Maintaining positive air pressure in a room is good. It decreases the number of microbes that can enter through doorways and other openings.

- Work Surfaces—Floors should be constructed of concrete or an approved tile. They should be mopped daily with a disinfectant. Likewise, all work surfaces should be continually wiped down with 70 percent ethanol, bleach (sodium hypochlorite) solution, or other disinfectant. (Note: Always wear rubber gloves and protect yourself in other ways when using disinfectants.)

- Equipment and Media—All equipment and culture media should be sterilized in an autoclave before each use. Media should be poured in a laminar flow hood. Contaminated cultures should be autoclaved or incinerated if the containers are not reusable.

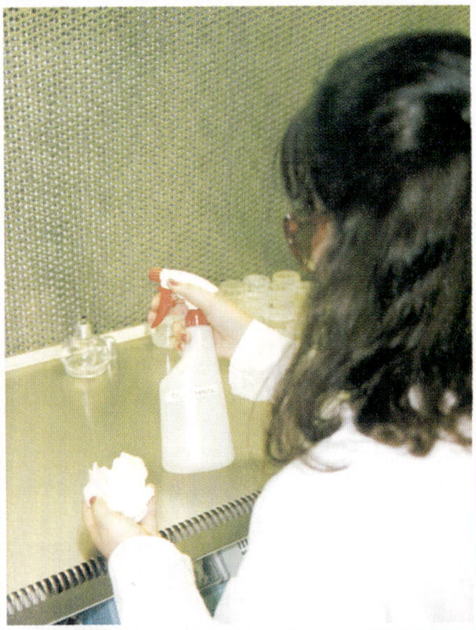

8-18. Spraying the surface of a laminar flow hood work area with 70 percent ethanol helps assure asepsis.

8-19. Mopping floors with a disinfectant solution reduces microbe populations.

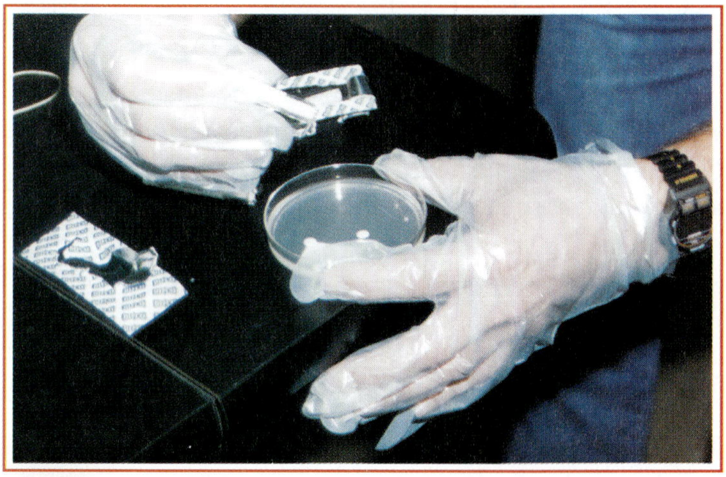

8-20. Gloves help protect from exposure to hazardous microbes, such as *E. coli.*

8-21. It may possible to do work requiring a reasonable degree of asepsis inside a sterile plastic bag.

- Personnel—Depending on the containment level needed in the lab, people will prepare differently. However, it is always recommended that hair be tied back, lab coats be worn, and shoes be surface sterilized. Individuals working with plant tissue culture will surface sterilize their hands and arms with a 70 percent ethanol solution. Individuals working in most medical situations will wear disposable medical-grade gloves.
- Cultures/Samples—Microorganism samples should be isolated using appropriate microbiological practices. Plant tissue should be surface sterilized prior to culturing. After culturing, all samples should be autoclaved and possibly incinerated. This depends on the hazard level of the organism/culture.

ANIMAL ASEPSIS

Animals are sometimes exposed to invasive actions. Treatment and research methods should be used that consider the well-being of the animal. An ***invasive procedure*** is entering the body of an animal with a needle, catheter, or other medical instrument.

Preparation of the animal is essential. Needles are used to give injections and are less invasive than surgical scalpels. All invasive procedures should

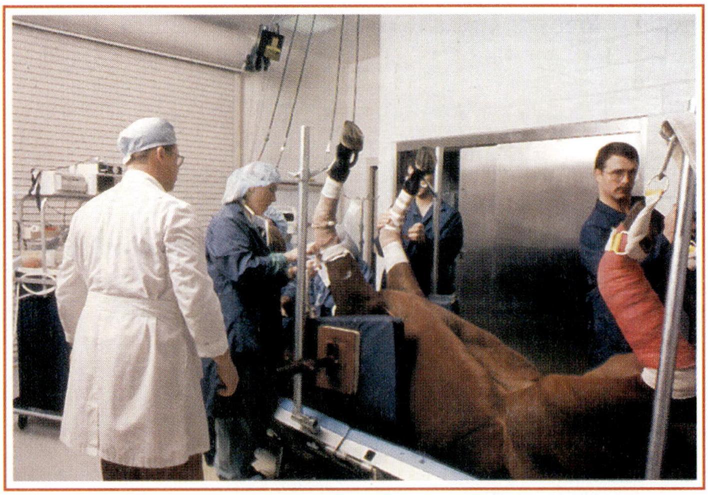

8-22. Surgery on a horse requires asepsis as well as restraint of the sedated animal. (Courtesy, Veterinary Medical Center, Texas A&M University)

control microbes and prevent entry into the system of the animal. With needles, the immediate area around the injection site may be sterilized with alcohol on a cotton ball.

Surgery usually requires the sedation of the animal. Anesthesia is used. Anesthesia is induced with drugs to prevent the sensing of pain and make the animal immobile. Once anesthetized, hair in the area around the surgical site is removed with clippers. A vacuum is used to remove the clippings of hair. Skin preparation solutions, such as chlorhexidine or iodophors, are applied with a sterile sponge. In 2 to 4 minutes, the skin solution is removed with a 70 percent alcohol solution. Drapes are used to isolate the surgical site. This involves covering the animal (except for the surgical site) and the surgical table. All people in the operating room should wear gloves, caps, masks, and surgical garb to reduce the risk of microbe infections.

People involved with veterinary medicine are trained in the proper procedures. These should always be followed. Any compromise may place the animal at risk of post-surgical infections.

EXAMPLES OF ASEPTIC CONDITIONS

Microorganisms are classified by their ability to infect individuals and cause disease. Laboratories should be designed for the work that is to be done.

Laboratories in which potentially harmful microbes are used should be designed to contain the microbe. They should offer protection from the microbe to people and the environment. Laboratories are classified on the basis of the hazard level of the organism(s) being worked with.

Table 8-1. Classification of Laboratories Based on Microbe Hazards

Level and Risk Associated with Microbe	Lab Specifications	Lab Practices	Examples of Labs
1. Unlikely to cause human disease	none open bench	general aseptic technique	plant tissue culture lab
2. Possible hazard to workers; none to community	open bench and safety cabinet/hood	general aseptic technique, protective clothing, biohazard sign	most biotech and genetic engineering labs
3. Some hazard to workers; possible hazard to community	separate room, safety cabinet for all activities	general aseptic technique, special protective clothing, controlled access, biohazard sign	pharmaceutical research labs
4. Serious hazard to workers; high risk to community	all in level 3 plus pressure gradient air control, safety cabinet	all in level 3 plus air lock entry, special waste disposal, and shower exit	disease lab, such as the Center for Disease Control

Some laboratories are used for producing new products, such as tissue cultures. These labs need to provide a good, aseptic environment for the work. The lab environment should protect the procedures from contamination. The extent of effort made to have asepsis varies with the criticality of the work.

REVIEWING

MAIN IDEAS

Asepsis is being without germs. A germ is a microbe, but commonly thought of as those that cause disease.

Aseptic technique is taking steps to remove microbes from an environment. This decreases or eliminates the chance of contamination. Creating aseptic environments is a process that must include decontamination of all components. Sterilization, disinfection, and sanitization are used. Sterilization is best. It eliminates all microbes. Many professions require the use of aseptic techniques.

Microbes targeted with aseptic techniques are bacteria, viruses, and fungi/yeasts. Bacteria and fungi are living organisms. Viruses are nonliving particles with some attributes of living organisms. Bacteria and fungi are observed with a microscope. Viruses require powerful microscopes because of their extremely small size.

Aseptic techniques vary with the nature of the work and its environment. Regular use of disinfectants helps reduce microbe populations. Always follow safety procedures with aseptic techniques.

QUESTIONS

Answer the following questions using correct spelling and complete sentences.

1. What is a germ? How do they pose threats?
2. What is aseptic technique?
3. Distinguish between sterilization, disinfection, and sanitization.
4. What are the three areas of protection provided by asepsis? Why are they important?
5. What are the three types of microbes that cause contamination? Briefly explain each.
6. What is an aseptic environment?
7. How are surfaces cleaned? What materials are used?
8. What is an autoclave? How does it work?
9. What is animal asepsis? When is it used?
10. How are aseptic procedures used to protect humans and the environment from microbes?

EVALUATING

Match the term with the correct definition. Write the letter by the term in the blank that is provided.

a. bacteria
b. fungi
c. yeast
d. virus
e. autoclave
f. germ
g. asepsis
h. sterilant
i. flaming
j. laminar flow hood

_____ 1. A machine used for sterilizing instruments.

_____ 2. Organisms that invade cells and alter cellular function.

_____ 3. Sterilizing a surface by exposing it to a flame.

_____ 4. A condition without germs.

_____ 5. A single-celled prokaryote.

_____ 6. A single-celled fungi.

_____ 7. A chambered working area that creates an aseptic environment.

_____ 8. A complex eukaryotic organism that is spread through spore production.

_____ 9. An agent that kills all types of microbes.

_____ 10. The general name given to microorganisms.

EXPLORING

1. Tour a hospital or veterinary clinic. Make a list of the areas where aseptic technique is being used. Interview lab technicians to find out what types of aseptic techniques they regularly use. Prepare a report on your observations.

2. Prepare an aseptic environment for tissue culture or other activities. Follow proper protocol as well as safety procedures. Prepare a report on your activities.

3. Use an autoclave to sterilize instruments. Get a demonstration on how the autoclave is used. Read instructions on its use. Follow all safety rules.

9

BASIC CELL STRUCTURE

When you were small, did you build towers out of building blocks? You probably used one block at a time. Soon, you had a tall tower. That is much the way a building is constructed. What materials does it take to make a foundation? To frame the building? To make the interior comfortable in any type of weather?

Just as bricks, boards, and other materials are used in building, cells provide the structures for living organisms. The cell is the basic building block of life. It is the common starting point for much biotechnology work.

Since the cell is so important in biotechnology, it is essential that you understand its parts and the roles they play. You also need to know how cells reproduce and how we have gained our understanding of them. Learning about the cell will make a big difference as you continue your study of biotechnology.

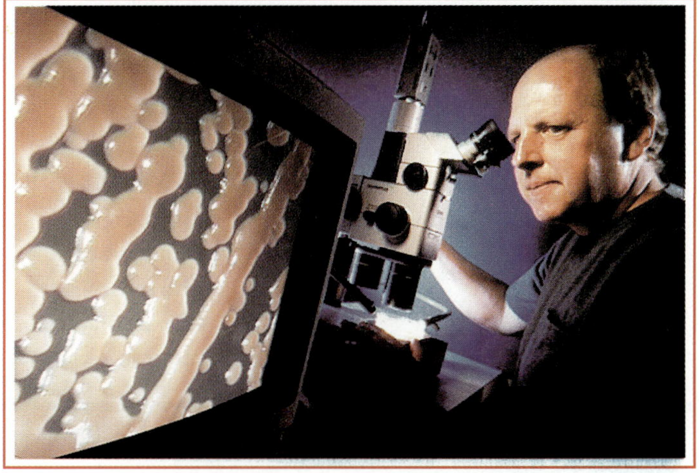

9-1. Magnified yeast cells are being studied for enzymes used in making ethanol. (Courtesy, Agricultural Research Service, USDA)

OBJECTIVES

This chapter reviews basic cell structure. It describes the parts of the cell and the functions of each. It is intended to provide the background for understanding transgenic biotechnology. It has the following objectives:

1. Name and distinguish between two types of cells.
2. Identify the major parts of the cell and their functions.
3. Describe some of the major processes of the cell.
4. Explain the use of technology to view and manipulate cells.
5. Explain the processes by which cells reproduce.

TERMS

cell membrane
cell theory
cell wall
chloroplast
cytoplasm
diffusion
diploid
electron microscope
endoplasmic reticulum
equilibrium
eukaryote
fission
Golgi bodies
light microscope
lysosome
meiosis
microscope
mitochondria
mitosis
nucleus
organelle
osmosis
prokaryote
ribosome
stereomicroscope
vacuole

THE CELL: BUILDING BLOCK OF LIFE

The cell is the smallest living unit of an organism. Individual cells grow, reproduce, use energy, adapt, and respond to their environment. All of these are characteristics of living things.

Most cells are so small they cannot be seen with the naked eye. Yet, the typical cell contains the information needed to build another organism like the one it is or represents!

It is difficult to describe cells because they have many functions within living systems. Some cells are so small that specialized technology must be used to "see" them. Other cells may be more than a meter long. A cell may be an entire organism or it may be one of the billions of cells and hundreds of types of cells making up a single organism.

Cells may be divided or classified in many ways. Since an almost uncountable number of cells perform a wide variety of functions, it is impossible to create a classification system that works in every case.

One common method of classification is by the presence or absence of a cell nucleus. The ***nucleus*** is a relatively large structure within a cell. The nucleus directs cell activities. Some very simple single-celled organisms lack this structure. They are known as prokaryotes. Cells that have a nucleus are eukaryotes.

9-2. An ostrich egg is among the largest cells. (Ostriches themselves consist of millions of cells!)

PROKARYOTES

A ***prokaryote*** (also spelled procaryote) is a single-celled organism in the Kingdom Monera. There are many kinds of prokaryotes. A prokaryote lacks a cell nucleus, which means its internal structure is less organized than other cells. The genetic information it needs is within the confines of the cell.

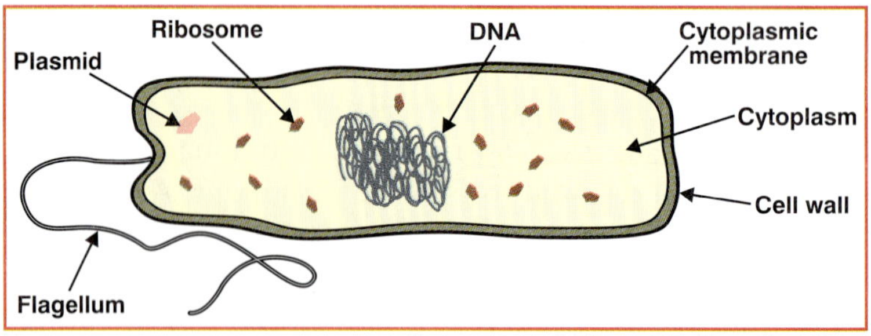

9-3. A typical prokaryote cell.

Two of the most common prokaryotes are blue-green bacteria and true bacteria. Blue-green bacteria have the ability to produce their own food like plants. Most bacteria get their food from other sources.

Although prokaryotes are less complex than eukaryotes and do not have a nucleus, they still have many of the basic internal characteristics. Common prokaryotic structures include the cytoplasm, chromosomal material, cell membrane and wall, and the ribosome. Some have a flagellum for mobility. These structures are described in more detail later in the chapter.

EUKARYOTES

A ***eukaryote*** (also spelled eucaryote) is an organism made up of cells that have a nucleus. A eukaryote may be single-celled, like the members of the Kingdom Protista, or may have billions of cells, like those in the plant and animal kingdoms.

Examples of single-celled eukaryotic organisms include paramecia, euglena, and diatoms.

Eukaryotic cells, in addition to the presence of a nucleus, typically are highly organized. They have an infinite number of shapes and purposes. The human body, for example, has at least 85 different cell types. All eukaryotic cells, however, have a number of structures in common.

COMMON CELL STRUCTURES

The cell may be compared to a factory. Like an industrial environment, it has many different parts, each of which fulfills a specific function. Smaller parts of the cell with specific functions are known as ***organelles***.

The cell and its structures are made of the molecules of life: proteins and fats. It is fueled by carbohydrates, and vitamins and minerals help with spe-

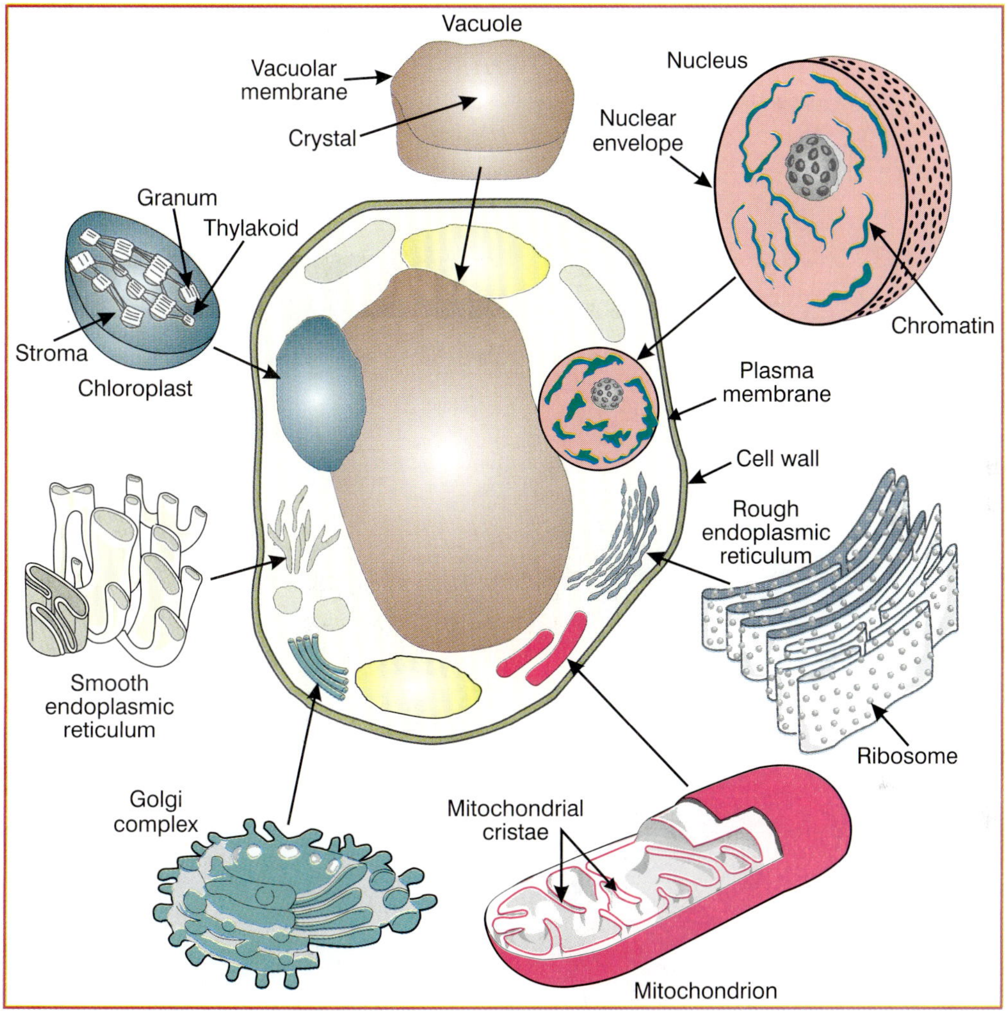

9-4. A typical plant cell.

cific functions. The nutrients needed by an organism are the same as those required for cell survival.

Each cell is divided from others by a ***cell membrane***. The membrane acts as a gatekeeper, controlling what enters and leaves the cell. Plants, algae, and bacteria typically have a tough outer structure known as the ***cell wall***. A cell wall provides additional structure and protects the cell from pressure caused by the movement of water. Animal cells do not have walls.

Within the cell, the organelles are surrounded by a jelly-like material known as ***cytoplasm***. The primary component of cytoplasm is water, which explains the importance of water to all living organisms.

In a typical cell, the largest and most easily identified part is the nucleus. Other organelles include ribosomes, mitochondria, Golgi bodies, the endoplasmic reticulum, lysosomes, vacuoles, and, in plants, chloroplasts.

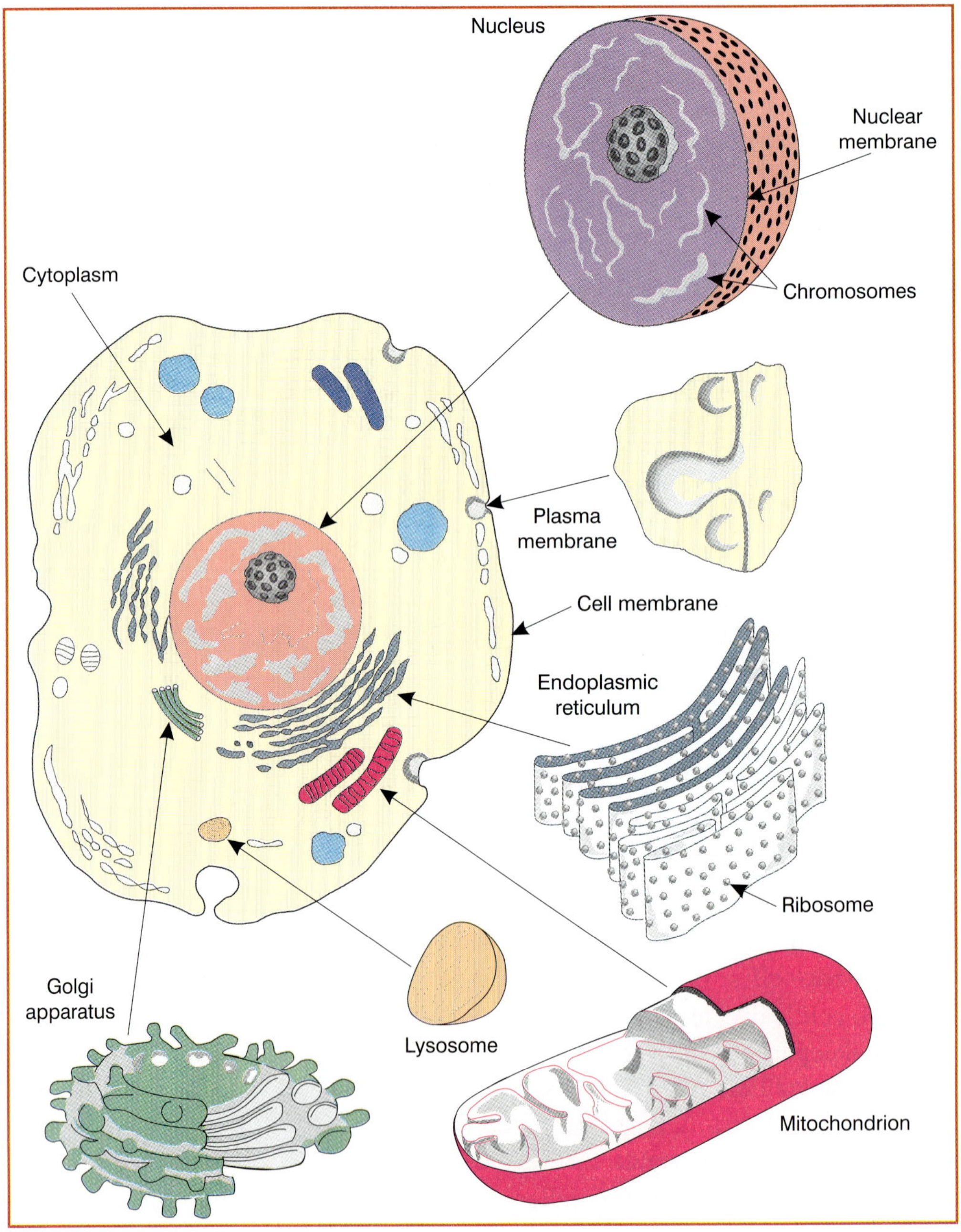

9-5. **A typical animal cell.**

THE NUCLEUS

The nucleus is frequently referred to as the "brain" of the cell. It is the site of much of the manipulation done in biotechnology. This is because it controls the activity of the cell.

The nucleus is a round or oval structure typically found in the middle of the cell. It often appears darker than the surrounding material.

The nucleus is surrounded by a membrane that, like the cell membrane, controls the passage of materials into and out of the area. One or more smaller structures known as nucleolus may be found in the nucleus.

The nucleus contains DNA, deoxyribonucleic acid, the chemical instructions for life. DNA is located on chromosomes. Most organisms are ***diploid***,

Biotech Connection

SLIDES IN REVIEW

You may need to prepare a slide to view cells with a light microscope. Several kinds of slides are used. Generally, the wet-mount slide is most common. You have probably made these before. Here is a quick review.

Steps in making and viewing a wet-mount slide to view one-celled organisms in water:

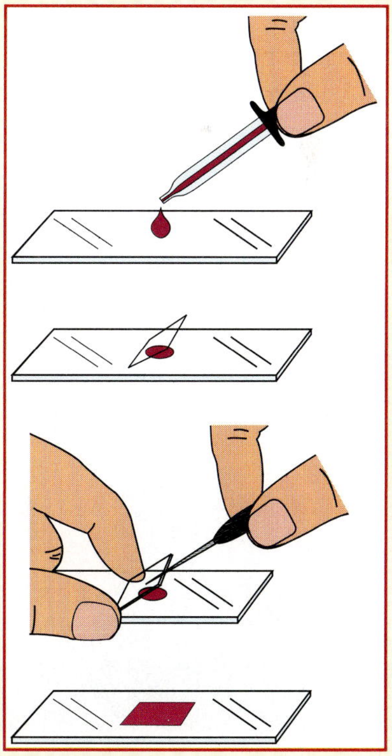

1. Place a drop of water on a clean slide.
2. Place one edge of the cover slip at the edge of the drop of water and move it toward the drop until it adheres to it.
3. Lower the top edge of the cover slip onto the water drop with a dissecting needle. (This helps keep out air bubbles.)
4. Place the slide on the stage of the microscope and examine under low power.
5. Slowly move the slide under low power to find shapes.
6. Change the microscope to medium power, refocus it, and observe the specimens.
7. Change the microscope to high power, refocus it, and examine the specimens.
8. When finished, properly clean and store the equipment.

that is, having two chromosomes that control a single trait. Humans have 46 chromosomes, 23 pairs. Goldfish have 90 chromosomes, 45 pairs. In different stages of a cell's life, the degree to which chromosomes and DNA can be seen varies greatly.

Since DNA and its counterpart RNA (ribonucleic acid) are central figures in biotechnology, they are explained in more detail in the next chapter.

CYTOPLASMIC ORGANELLES

Although the nucleus controls the activity of the cell, many other organelles are required for proper cell function. Cells may have only one of each organelle or many.

A ***ribosome*** is an extremely small, round organelle. Several ribosomes may be free floating in the cytoplasm or attached to the endoplasmic reticulum. They make proteins for cellular use and communication.

The ***endoplasmic reticulum*** is a layered, membranous organelle. It may be found surrounding the nucleus and/or as a separate organelle within the cytoplasm. The endoplasmic reticulum and the attached ribosomes make and transport proteins.

A ***vacuole*** is a large organelle without a specific shape. Its function is to store any possible particles. Vacuoles may store waste or the raw materials used in synthesis of proteins.

Golgi bodies, or the Golgi apparatus, appear similar to the endoplasmic reticulum. They look something like a thick elastic band that has been folded many times. The Golgi adds modifications to unfinished proteins and makes lysosomes.

A ***lysosome*** is a small, round organelle. Lysosomes store enzymes that break down food into chemical compounds. They may also be involved in destroying cell organelles and the cell itself.

The ***mitochondria*** are the "energy factories" of a cell.

9-6. Chlorophyll in the leaves of plants promotes photosynthesis. This is needed for plants, such as the hydrangea, to have flowers.

They are rod-shaped. The mitochondria change food into molecules that can be used for energy. Mitochondria also contain DNA material that provides instructions for their functions. Mitochondria can replicate themselves.

In plants and some other organisms, cells also have chloroplasts. A ***chloroplast*** is a green body containing chlorophyll. Chlorophyll is an important chemical in the photosynthetic process.

Although each of these organelles has a specific function, it is the cell, working as a whole, that provides the basis for living organisms.

CELLULAR PROCESSES

Since the cell is the smallest unit of life, it must be capable of completing certain processes. Some processes are specific to certain types of cells within an organism, while other processes may be carried out by almost all cells. Some of the most common processes include respiration, osmosis and diffusion, and, in plants, photosynthesis.

RESPIRATION

Respiration is the process by which cells break down carbohydrates and other molecules to produce energy. It is known as respiration in a cell because it is similar to the process of an organism taking oxygen into the body and releasing harmful gases.

One of the most common forms of respiration is the oxidation of glucose. This process is breaking a simple sugar molecule into water, carbon dioxide, and energy. The energy is used for the work of the cell, such as the synthesis of new molecules.

OSMOSIS AND DIFFUSION

A second important process is the transport of raw materials into the cell and waste and synthesized materials out of the cell. These requirements are by osmosis and diffusion. ***Diffusion*** is the movement of molecules, especially gases and liquids, from areas of high concentration to areas of low concentration. ***Osmosis*** is a specific type of diffusion, the movement of water through a membrane from areas of high concentration to areas of low concentration.

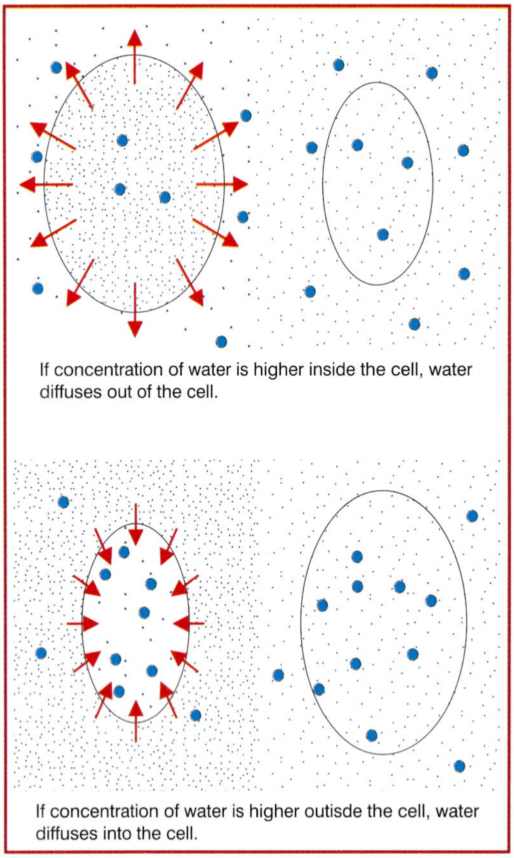

9-7. **How osmosis works.**

Osmosis and diffusion function because of nature's preference for equilibrium. ***Equilibrium*** refers to a system that is balanced, or equal. Molecules will move into and out of cells to achieve that balance.

If one cell has little water, but the surrounding cells or environment has more, then water would move by osmosis into the cell with less. In the same way, if a cell has a high concentration of sugar molecules and the surrounding cells do not, those molecules will disperse until equilibrium is reached.

PHOTOSYNTHESIS

A third and more specific process is photosynthesis. As covered in Chapter 7, photosynthesis is a food-making process. In the process, organisms (primarily plants) use sunlight, water, and oxygen to manufacture their food. It is roughly the reverse of respiration (which plants also do).

In photosynthesis, special organelles known as chloroplasts gather water and oxygen molecules. A special pigment, chlorophyll, absorbs the appropriate wavelengths of light and uses the energy provided to create chemical bonds. Those bonds lead to the formation of glucose molecules. That glucose can then be used by the plant as a source of energy or it can be converted to more complex sugars for storage.

$$6CO_2 + 12H_2O \xrightarrow[\text{chlorophyll}]{\text{light}} C_6H_{12}O_6 + 6O_2 + 6H_2O$$

9-8. **The chemical process of photosynthesis.**

HOW DO WE KNOW?

Since most cells are so small and can't be seen with the naked eye, how do scientists know they exist? How do they know what happens within the cell? The effort to explain the structure and functions of living organisms has challenged the scientific community for centuries. One of the first obstacles was finding a way to see and observe a cell.

EARLY SCIENTISTS

Have you ever used a magnifying glass? What about binoculars? Most of us today are very comfortable with the idea that a curved lens alters the appearance of the object. The lens, a curved piece of glass, is an important component of many scientific tools, including the microscope. A ***microscope*** is any instrument that uses lenses to enable the observation of very small objects.

History does not have a definitive record of the first scientist to invent a microscope. It is known, however, that Anton van Leeuwenhock, a 16th cen-

Biotech Connection

POND SCUM: ANYBODY HOME?

Some of the earliest observations using a microscope were of aquatic organisms. What do you think is living in the water around your home?

Visit a local pond or stream and collect a water sample. You might use a plankton net to help collect material from the water. Or, you can dip up some water in a clean jar or bucket. Take the water or samples back to school and look at it under a microscope. You might be amazed to see the results! (Be sure to follow all safety rules around water.)

As you look at the organisms, try to identify as many of the cellular structures as possible. Use a research log to record your work, and include sketches of the organisms. Then use taxonomic keys to identify the organisms.

This photo shows a plankton net being used to collect a sample of materials from water. The water may contain solid matter and living organisms. (Courtesy, Forestry Suppliers, Inc., Jackson, Mississippi) tury fabric merchant's apprentice in Holland, mounted a lens in a metal plate and saw some of the first miniature creatures.

Shortly after, an English man named Robert Hooke used a similar tool to look at cork. In so doing, he saw many small compartments, each completely separate from the others. In describing those compartments for his peers, he used the term "cells."

CELL THEORY

Work on identifying these new structures has continued ever since. In the 1800s, Matthias Schleiden, a German botanist, and Theodor Schwann, a German physiologist, stated that cells were the basic units of life. Schleiden and Schwann are widely accepted as the authors of the ***cell theory***, a theory used today.

Cell theory states:

- All living things are made of one or more cells
- Cells live and are the basic building blocks of life
- Cells reproduce; new cells come from old cells

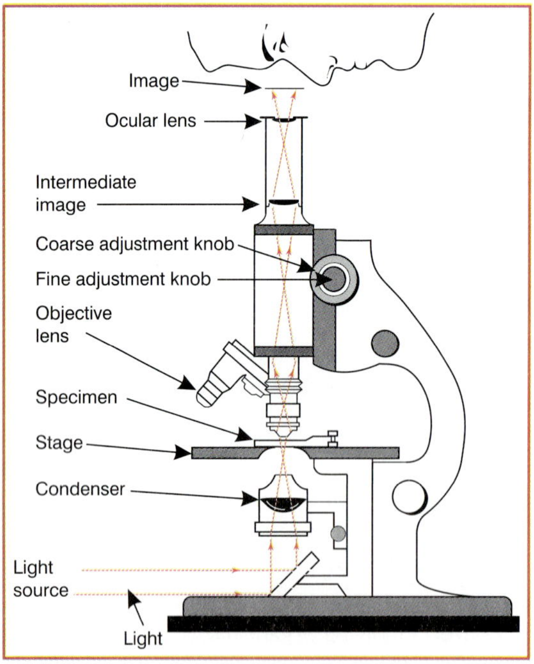

9-9. Major parts of a compound microscope.

TOOLS OF TODAY

Today some scientists continue to seek new ways to observe the microscopic world, while others use the tools developed to work within that world. Some of the most common tools used in their work are the light microscope, the stereoscopic microscope, and the electron microscope.

Light Microscope

The most commonly used microscope in the scientific community is the ***light microscope***. This microscope passes light through an

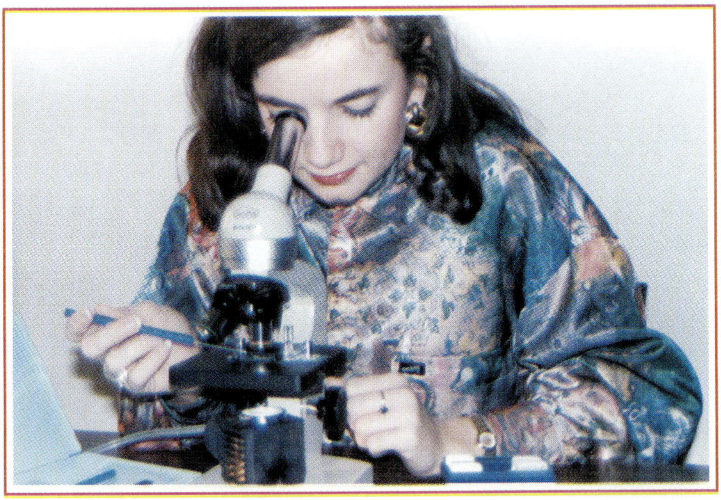

9-10. **A compound light microscope is easily used to observe cells.**

object and then through at least two lenses: the objective lens and the eyepiece. Since each lens magnifies the object, the most powerful light microscopes can magnify objects about 500 times.

The degree of magnification allows the observation of most cells and of many of the organelles within the cell. It does not, however, allow the observation of molecular structures or of viruses. At magnification greater than this, the details of individual objects blur and cannot be distinguished.

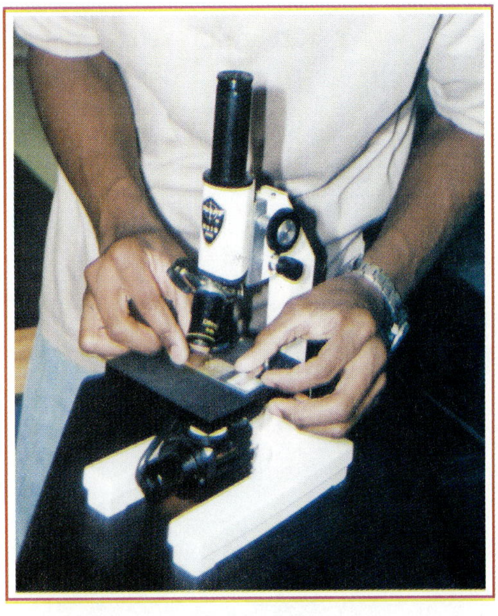

9-11. **A slide must be properly positioned on the stage of a microscope.**

Stereoscopic Microscope

Sometimes scientists need to observe parts of large organisms. In addition, some organisms are opaque and an adequate amount of light does not pass through for successful use of the light microscope. A ***stereomicrosope*** provides an image that is three-dimensional. The image is not reversed, as with some other microscopes. No preparation of a specimen is needed for some specimens. The scopes have two eyepieces. (The name, stereomicroscope, is a shortened form of stereoscopic microscope.)

9-12. Using a stereomicroscope.

The stereomicroscope magnifies from 10 to 60 times normal and allows scientists to make observations in these situations.

9-13. Using an electron microscope. (Courtesy, Agricultural Research Service, USDA)

Electron Microscope

The third type of microscope used in scientific work is the ***electron microscope***. The electron microscope operates because other types of radiation can be manipulated in the same way light radiation is bent by glass lenses.

An electron microscope operates by shooting beams of electrons through an object and using magnetic fields as lenses.

To prepare objects for observation, however, the samples must be preserved, dehydrated, and very thinly sliced. The objects are then placed in a vacuum (a space void of matter, including air). Electrons move much more easily through a vacuum than through air. The image is then projected onto a special screen.

An electron microscope enables scientists to see the structures of the cell in much more detail. It has the potential to magnify objects more than 500,000 times.

The greatest drawback is the expense and complexity of the work. Electron microscopes are not practical in a high school setting.

CELLULAR REPRODUCTION

We know that cells are the basic units of living things and that they fulfill all the functions of life. Where, then, do new cells come from? How do cells reproduce?

Cells reproduce three different ways. Prokaryotic cells reproduce through fission, while eukaryotic cells reproduce through mitosis. Eukaryotic cells also produce gametes, or sex cells, which are used in organismal reproduction.

Cells divide in response to their environment. They may divide to replace those damaged or destroyed, they may divide to increase the total mass of cells, or they may divide in response to chemical signals from neighboring cells.

FISSION

Prokaryotic cells reproduce through ***fission***. The cell simply grows larger and makes a second copy of its DNA (nucleoid). Then at some point along the cell membrane, a small division begins. This division is made by the growth of a transverse septum.

The division grows inward from either side of the cell until two separate cells exist. Each cell contains all of the structures necessary for life. Because two cells are formed, the process is sometimes known as binary fission. The two new cells are known as daughter cells.

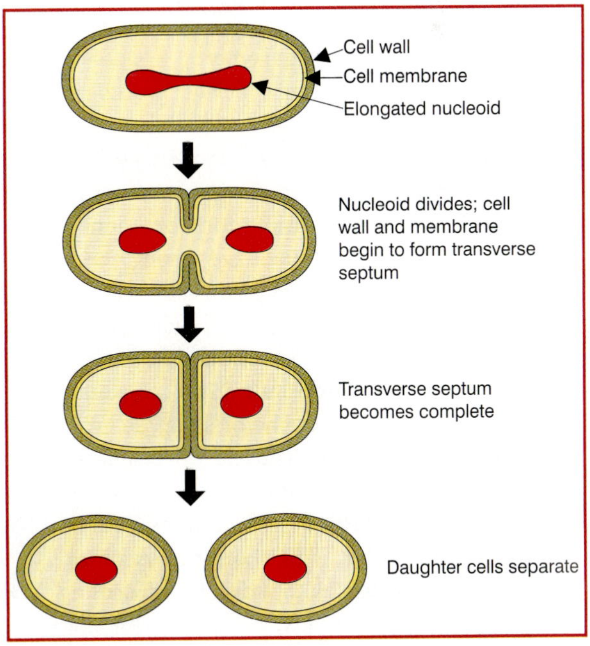

9-14. **The fission process in prokaryotic cells.**

Since fission is so simple, reproduction among prokaryotic cells can occur very quickly. Prokaryote populations can grow quite rapidly if the right conditions are present.

MITOSIS

Cell division for eukaryotes involves more steps. The division of a eukaryotic cell into two identical "daughter" cells is known as ***mitosis***. The number of stages in mitosis may vary among scientists. In this book, six stages are used: interphase, prophase, metaphase, anaphase, telophase, and cytokinesis.

Stage 1 is interphase. It is the time between cell divisions. Interphase consists of two rest periods interrupted by a phase in which the DNA of the nucleus is duplicated.

The second stage of mitosis is prophase. In animal cells, small organelles known as centrioles appear just outside the nucleus just prior to prophase.

Career Profile

MARKETING SPECIALIST

Once a bioengineered crop has been developed, producers need to be convinced to grow it. A marketing specialist works to get producers to plant the variety. The specialist may visit farms, deal with issues, set up field days, and monitor compliance with licensing agreements. (A license is needed to plant most bioengineered crops.) The marketing specialist also projects sales and gathers data used in planning. The specialist generally is responsible for documenting the work he or she does.

The specialist must have strong verbal and written communication skills. A good understanding of crops and bioengineering processes is essential. Computer skills are also important. Most marketing specialists have a bachelor's degree in agriculture, communications, or business.

Jobs are with seed companies and other businesses with bioengineered products to sell. This photo shows a marketing specialist with a display advertising agricultural products. (Courtesy, Agricultural Research Service, USDA)

During prophase, the centrioles migrate to opposite sides of the cell. In plants, no centrioles exist, but polar regions form. From the centrioles or polar regions, small tubes form. These tubes grow through the cell and attach to the chromosomes, and the nuclear membrane dissolves. The chromosomes thicken and become visible. Pairs of chromosomes are joined by structures called centromeres. The chromosomes then move to the center of the cell

The third stage, metaphase, begins when all the chromosomes are aligned. At this point, each set of chromosomes is visible as a distinct pair. Metaphase is often the shortest stage of mitosis.

The fourth stage is anaphase. The pairs of chromosomes split and move to opposite sides of the cell. When all the chromosomes have reached the centrioles or poles, anaphase is complete.

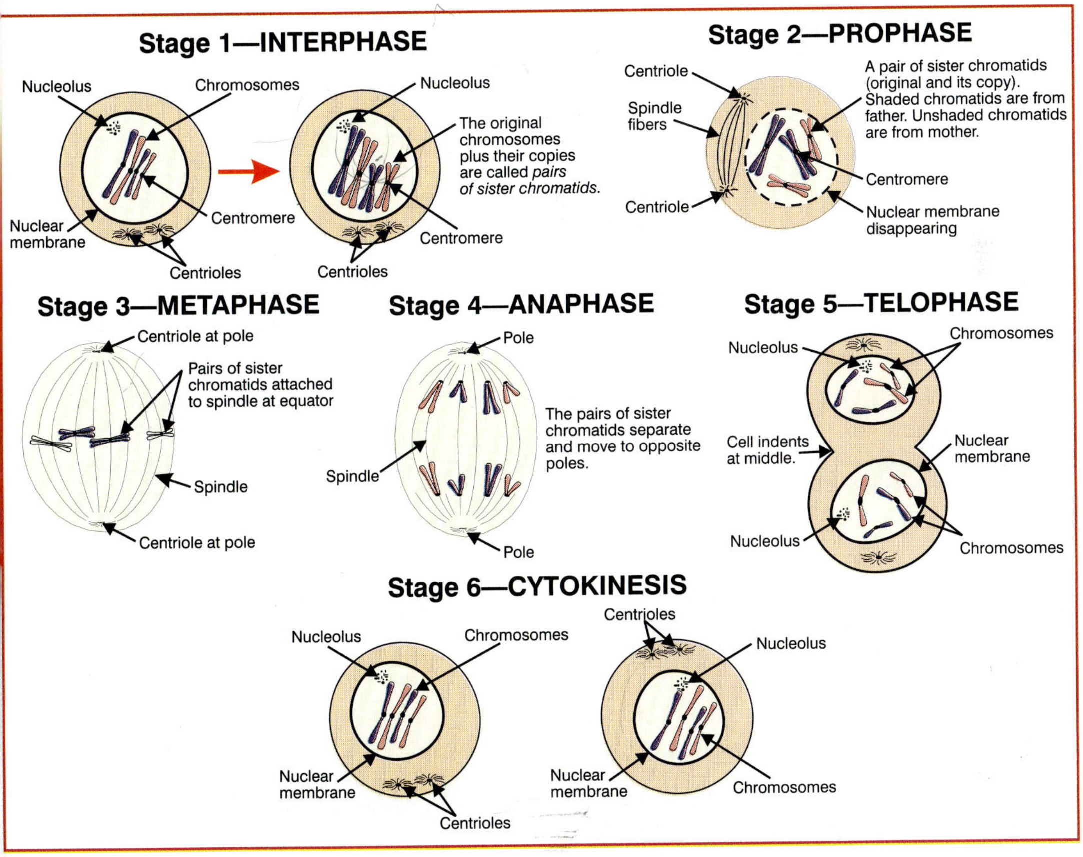

9-15. The six stages of mitosis. (Some scientists list four stages and exclude the first and last stages shown here.)

The fifth stage of mitosis is telophase. During telophase, the chromosomes reorganize into a group. The tubes disappear, and a new nuclear membrane forms around each group. When the nuclear membrane is complete, mitosis is finished.

The final stage is cytokinesis. In this stage, in most cells the cytoplasm and cytoplasmic organelles are divided to produce two cells. In plants, cytokinesis occurs through the formation of a cell plate. In animal cells, this occurs through the division of the cell membrane.

MEIOSIS

The third type of cell division is ***meiosis***, division through which sex cells or gametes are formed. Meiosis occurs in two divisions, with the same four phases of mitosis occurring in each division. The difference is that no duplication of DNA occurs between the two divisions.

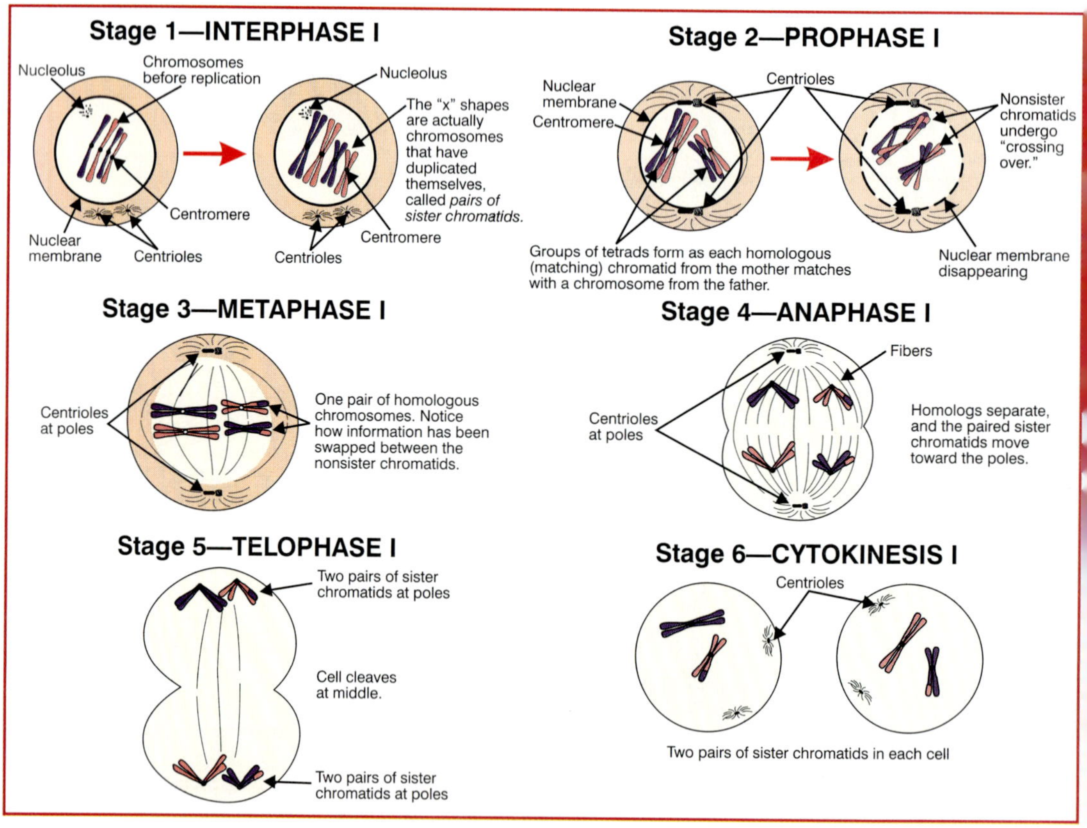

9-16. **The stages of meiosis I.**

The stages of meiosis are meiosis I: interphase I, prophase I, metaphase I, anaphase I, telophase I, and cytokinesis I; and meiosis II: prophase II, metaphase II, anaphase II, telophase II, and cytokinesis II.

The major difference between the first half of meiosis and mitosis is that chromosome pairs match up. This allows all four chromosomes that control similar traits to be together. A tetrad is formed. During the division, each cell receives one chromosome from each pair randomly. The basic process of division (formation of tubes, dissolution of the nuclear membrane, etc.) does not change.

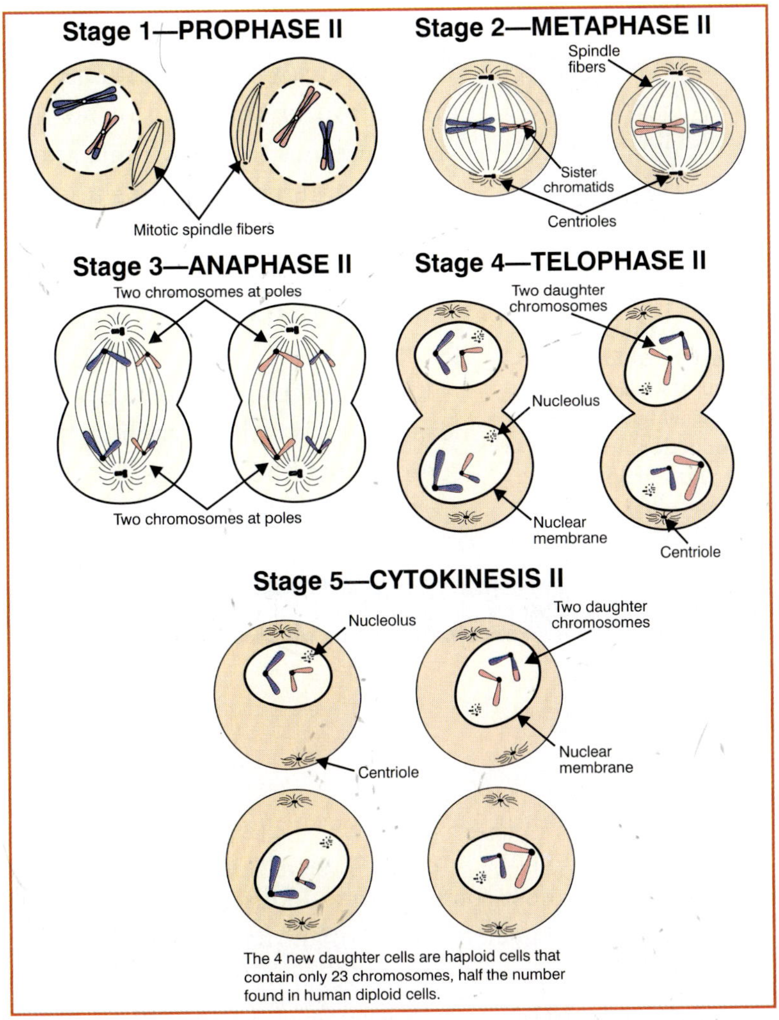

9-17. **The stages of meiosis II.**

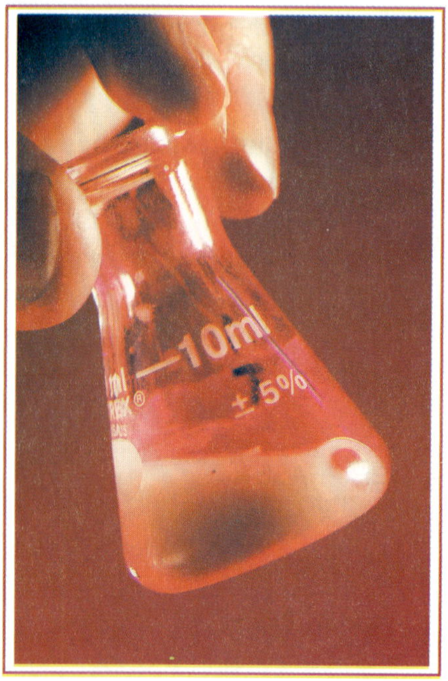

9-18. **This flask holds several mL of turkey semen. The semen has male sex cells–a product of meiosis. (Because today's turkeys have been bred for high meat production, natural mating is nearly impossible. Most all female turkeys used for egg production are artificially inseminated. A small drop of semen has many sperm cells!)**

During meiosis II, the two cells divide again. Not all cells in meiosis I go through meiosis II. Those cells that do, become either sperms or eggs, depending on the gender of the organism. In meiosis II, the cells do NOT duplicate the DNA prior to the second division. The cells follow the same process of division, except that this time each pair of chromosomes separates, so that the two cells formed have half the normal complement of cells. Each cell has half of the instructions necessary to create an organism. When two sex cells eventually merge in reproduction at the time of fertilization, the chromosomes from the male and female parent are combined to form a unique offspring.

In reality, all four resulting cells rarely become functional gametes. Many lose their cytoplasm to the sister cell and form useless polar bodies.

The outcome of meiosis is to have at least one sex cell or gamete that is used in reproducing the organism.

REVIEWING

MAIN IDEAS

The cell is the basic unit of life. Cells occur in an almost infinite number of sizes, shapes, and environments. Organisms may be made of single cells or of billions of cells. One of the most common ways to distinguish between cells is the presence or absence of a cell nucleus. A prokaryote has no nucleus; a eukaryote has a nucleus.

Cells are made up of smaller parts known as organelles. The cell is surrounded by a membrane and filled with a jelly-like material known as cytoplasm. Probably the most important organelle is the nucleus, which contains DNA, the genetic instructions for life.

Cells carry out a number of basic life functions. These life functions include respiration, osmosis and diffusion. Plants also use the process of photosynthesis to produce their own food.

One of the most important tools within cellular studies is the microscope. The most common microscope used is the light microscope. The microscope that provides the highest level of magnification is the electron microscope.

Cells reproduce through fission, mitosis, and meiosis. Prokaryotic cells reproduce through a simple division or fission. Mitosis is the process by which two identical cells are formed, while meiosis is the process through which sex cells are made.

QUESTIONS

Answer the following questions using correct spelling and complete sentences.

1. Why is a cell important?
2. Compare and contrast prokaryotic and eukaryotic cells.
3. What is the nucleus? Why is it an important structure?
4. Explain how the food you eat is important on the cellular level.
5. What is cytoplasm?
6. Using the organelles of the cell, explain why some people compare it to a factory.
7. Explain the role of respiration, osmosis and diffusion, and photosynthesis in cellular survival.
8. How does the cell theory impact biotechnological activities?
9. Compare and contrast light, stereoscopic, and electron microscopes.
10. Why must cells reproduce?
11. Describe the six stages of mitosis.
12. What is interphase?
13. What are the differences between mitosis and meiosis?
14. How many functioning cells are typically formed from mitosis? Meiosis?

EVALUATING

Match the term with the correct definition. Write the letter of the term in the blank that is provided.

a. equilibrium
b. electron microscope
c. mitochondria
d. cytoplasm
e. eukaryote
f. mitosis
g. osmosis
h. ribosome
i. nucleus
j. cell theory

_____ 1. Operates by shooting electrons through an object and using magnetic fields as lenses.

_____ 2. Division of a eukaryotic cell into two identical cells.

_____ 3. Special type of diffusion where water moves from areas of high concentration to areas of low concentration.

_____ 4. Balance; equal levels.

_____ 5. A larger structure within a cell that directs cell activity.

_____ 6. States that cells are the basic units of living organisms.

_____ 7. Jelly-like material in the cell.

_____ 8. Rod-like energy organelle.

_____ 9. Cell that has a nucleus.

_____ 10. Small, round organelles that make proteins.

EXPLORING

1. The microscope is a very important tool in biotechnology. If you haven't worked with one, gain access to a microscope. Have someone teach you to use the microscope correctly. Observe different organisms and as many types of cells as possible. (Many schools have prepared slides that can be used to observe cell structures.)

2. Using common household materials, construct a three-dimensional model of a cell. Try to use materials that represent the function of the organelle as well as the shape (such as batteries for the mitochondria.) Label the organelles, or ask your classmates to identify them using their knowledge about cellular structure.

3. Two important cellular processes are the Krebs cycle and photosynthesis. Research these two processes and prepare a written report or a poster that explains, compares, and contrasts them.

4. Obtain a commercially prepared slide of tissue that is rapidly dividing (such as the root tip of an onion.) Use the knowledge you have gained to locate cells in each of the stages of mitosis.

10

GENETICS

Inside every cell, including every cell in your body, there is a special code—your genetic code. You inherited this code from your mother and father. This "code" determines many things about you. These include your appearance, your predisposition to a healthy life, and it impacts many of your natural abilities. You can probably think of qualities that you share with your parents and other relatives.

Every species of plant and animal has a genetic code. Similarities exist among the same species of plant or animal. Within a species, minor differences are found. Biotechnologists are trying to manipulate that code to create new varieties, improve the organism, or maximize production.

Opinions differ about changing the genetic code of an organism. Once you learn more, you will be able to justify your feelings.

10-1. Careful study and selection are used with flowering plants to promote desired genetic qualities. (Courtesy, The North Carolina Arboretum)

OBJECTIVES

This chapter provides basic information on genetics. This includes a comparison of Mendelian and molecular genetics, the structure of DNA and RNA, protein synthesis, and techniques in genetic manipulation. It has the following objectives:

1. Explain the meaning and importance of Mendelian genetics.
2. Predict genetic outcomes using Punnett square and chi-square.
3. Describe the meaning and importance of molecular genetics.
4. Name and explain examples of DNA science tools.

TERMS

allele
chi-square
DNA isolation
DNA profiling
DNA sequencing
dominant
F_1
F_2
gamete
gel electrophoresis
gene
gene expression
genotype
heterozygous
homozygous
phenotype
polymerase chain reaction
Punnett square
recessive
selection
transcription
translation

MENDEL UNRAVELS MYSTERIES

Why are we what we are? Genetics helps answer such questions.

Genetics is the study of the laws and processes of biological inheritance. Parents have traits that are passed to offspring. This passing of traits is known as heredity. All organisms receive their traits from their parents and will pass their traits to their offspring. The combinations from different parents result in unique offspring. Even offspring of the same parents are different, but also resemble each other in some ways.

10-2. These two dachshund puppies are from the same litter (same father and mother and birth date). How are they similar? Different? Why?

EARLY EXPERIMENTS

Gregor Mendel carried out a variety of plant breeding experiments during the mid-1800s. He used garden peas to study the inheritance probabilities of a variety of traits. Mendel made crosses of parent lines that consistently expressed certain traits. Examples of traits he studied were round or wrinkled seed shape, white or yellow flower color, and long or short stem lengths.

Mendel discovered that if a cross (hybridization) occurred between two individuals possessing two versions of one trait, such as round and wrinkled seed shape, the resulting offspring would all express only one version of the trait. This first generation of offspring is known as the F_1. (The second generation is known as F_2.) All individuals that are from the F_1 population are expressing the dominant trait. In this case, round seeds are dominant over

10-3. Growing garden peas similar to those used by Mendel.

wrinkled. If two individuals from this F_1 generation are bred together, 75 percent of the individuals will still express the dominant trait (round), while 25 percent will express the recessive trait (wrinkled). This generation is F_2.

Today, the efforts of Mendel form an area known as Mendelian genetics. The principles set forth by Mendel help explain genetics and heredity.

GENES AND SELECTION

A ***gene*** is a segment of a chromosome that contains the heredity traits of an organism. Genes are the basic units of biological inheritance. They are in pairs, just as chromosomes are in pairs. Genes contain information for making proteins. Genes vary. This variance is in alleles. An ***allele*** is a different or alternative form of a gene. Alleles occupy the same place on homologous (similar) chromosomes. Some forms are quite similar; others are different.

Gene expression is the process of making DNA information available to a cell. The process is highly regulated by a segment of DNA known as a promoter. With higher order organisms, coding information is shared among segments known as exons. Exons are separated from each other by introns. The number of introns in a gene varies greatly.

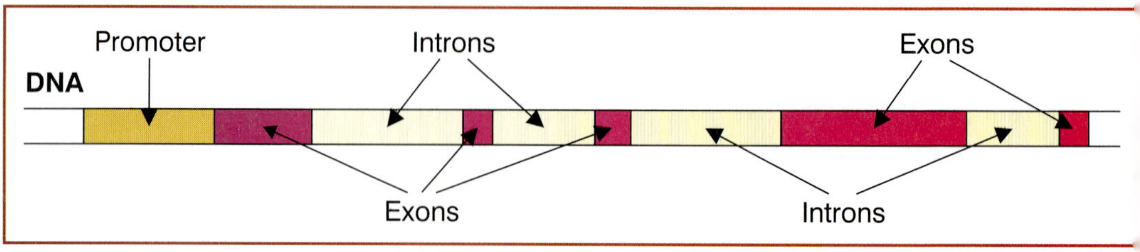

10-4. The structure of a gene.

People often select plants or animals because of certain traits. ***Selection*** is choosing parents of the next generation. The chosen parents have the desired traits, such as a hog that has large hams or peas that grow an additional pea in each pod. When reproduced, these traits tend to be present in the offspring.

Selection is not always a function of human intervention. Natural selection occurs when only the strongest animals or plants reproduce. The most aggressive bull bison may sire more calves than a less aggressive bison. A large tree may produce more seed than a small tree. Further, a large tree may shade the ground and keep new, young trees from growing.

Through selection, farmers have been able to improve the quality of their livestock and crops. Offspring do not always reflect the desired traits. However, they would have the desired traits more often than offspring of parents that did not have the traits. Some traits are dominant; others are recessive.

10-5. The parents of this lamb were selected for breeding to achieve a desired conformation in their offspring.

DOMINANT AND RECESSIVE TRAITS

Offspring inherit two alleles for each trait. One is from each parent. The alleles (traits) an offspring inherits may be alike or different.

If alleles are different, the trait that can be observed in an organism is said to be ***dominant***. Those traits that cannot be observed are ***recessive***. Dominant alleles mask the expression of the recessive alleles. Some people say that recessive traits are hidden. A recessive trait appears in an organism only when a dominant gene for that trait is not present.

If both alleles for a trait are the same, they are ***homozygous***. If they are both recessive alleles, the trait is said to be homozygous recessive. A recessive trait is masked by dominant traits except when it is homozygous.

If the alleles for a trait are different, the trait is said to be ***heterozygous***. Both male and female parents contribute to homozygosity and heterozygosity.

10-6. **Limousine cattle are typically red. A recessive gene, however, results in a black Limousine occasionally being born—the product of parents who are both red (phenotype) but have recessive black genes (genotype).**

PHENOTYPE AND GENOTYPE

The appearance and makeup of an organism are known as type. Type is one of the factors used in selection, such as a horse may be selected on the basis of its type. In biological heredity, two kinds of type are important: phenotype and genotype.

Phenotype is the outward or physical appearance of an organism. The appearance of a person, a tree, or a fish is its phenotype. Phenotype is a product of genotype and the environment in which the organism lives.

Genotype is the genetic makeup of an organism. Genotype is never observable as is phenotype. Genotype may be expressed or latent. Latent means that a characteristic is genetically present but is not expressed. Combine genotype with the environment of an organism and the product is the phenotype.

10-7. **A tiny oak may become a giant oak tree or a scrubby sapling. Its genetics are the same either way. The tree's environment for growing will have a big influence on its phenotype.**

The environment of an organism promotes the expression of its biological heredity. What we see is its phenotype. Plants are good examples of the importance of the environment. Trees that grow in good soil, have plenty of moisture and nutrients, are sheltered from prevailing winds, and have the

Career Profile

GENETIC COUNSELOR/ HUMAN GENETICIST

Human geneticists help people with hereditary health issues. People who have family histories that express a pattern of genetically inherited disease are counseled in terms of their own risk of contracting the disease. They may inform parents of the risks their children may face. Geneticists may also provide medical and emotional counseling to people who are afflicted with genetically inherited diseases.

Most human geneticists have completed advanced degrees in genetics and/or human medicine. It is a fairly new career area driven by information from the U.S. Human Genome Project. It promises to be an area of tremendous growth in the coming years.

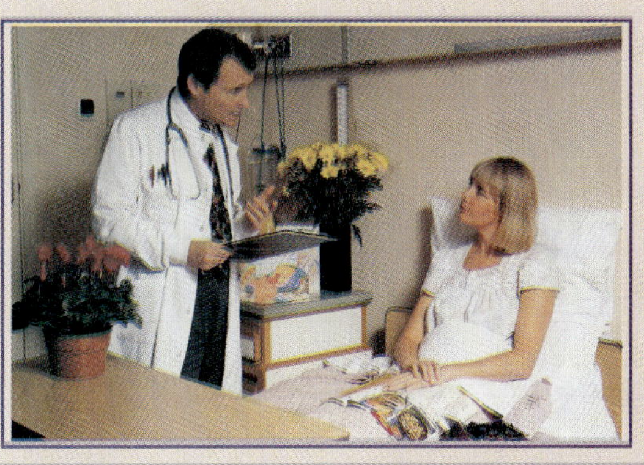

Jobs are with hospitals, clinics, government agencies, and in private practices.

space to develop may grow large and straight trunks. Those that grow in poor soil, with little space and strong prevailing winds, may be short, crooked, and gnarled.

AN ALPHABET EXPLANATION

So, why are we what we are? Each offspring gets a random sample of one-half the genes of each parent. Letters of the alphabet in upper and lower case can be used to explain. This explanation gives more meaning to meiosis. (You learned about meiosis in Chapter 9. That information is useful here.)

What we are is also related to meiosis (sex cell production) by males and females. We must consider the sex cell produced by each parent. Sperm are produced in the testes of the male parent. Eggs are produced in the ovaries of the female parent. Both sperm and eggs carry biological heredity information.

A good example is the trait of tall versus short. A **T** can be the symbol for tall. A **t** can be the symbol for short. Testes and ovaries produce **T** and **t** gametes. (A ***gamete*** is a sex cell capable of uniting with another sex cell to produce a new individual.)

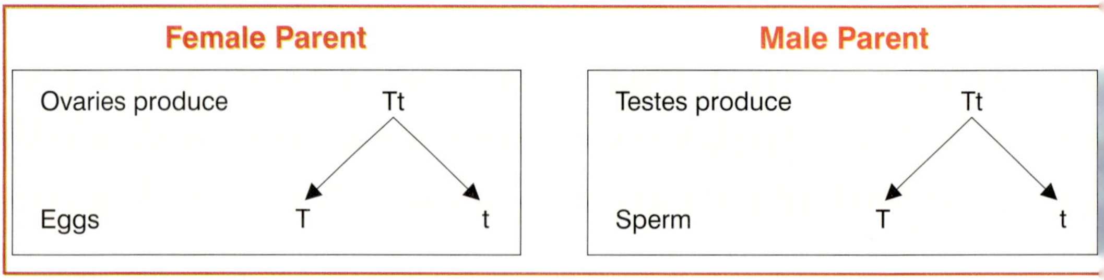

10-8. Gametes and phenotype of male and female parents for the trait of tallness. (T=tall; t=short)

When mating occurs, four phenotypes for tallness are possible: **TT**, **Tt**, **Tt**, and **tt**. The presence of a **T** produces a tall phenotype. Only **tt** results in a short phenotype. The phenotypes have a 3:1 ratio for tall versus short. The next section of this chapter will add more detail to genetics.

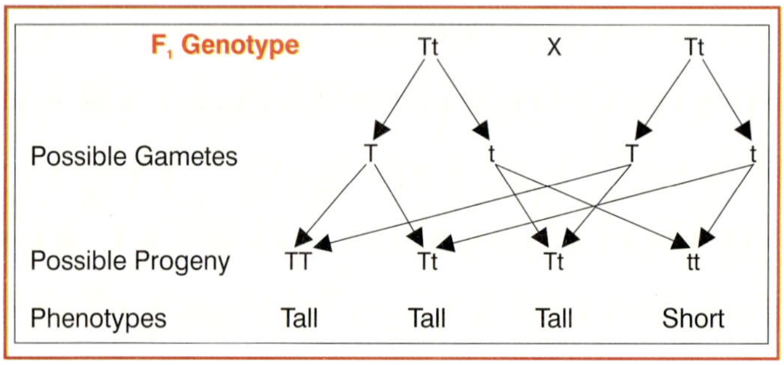

10-9. Possible pairs resulting from male and female gametes. The offspring is a product of a random sample of half the genes of each parent. (F_1 is the first filial or first generation.)

PREDICTING GENETIC OUTCOMES

Natural breeding results in offspring with random traits of their parents.

USING THE PUNNETT SQUARE

Mendel's theories can be applied to all types of complex genetic predictions. Predictions are calculated using the Punnett square. The ***Punnett square*** is a simple tool that gives a way to manipulate and visualize genotypes. The Punnett square involves assigning every trait a double set of letters. (You were introduced to this earlier in the chapter.) The letters represent that trait as alleles.

Single Traits

It is easier to predict single traits than multiple traits. The number of possible traits in offspring is fewer. The predictions begin with alleles.

Allele pairs occur in three forms: homozygous dominant, heterozygous, and homozygous recessive. The following represents what the alleles would be like for Mendel's round versus wrinkled seed hybridization:

RR	homozygous dominant	round seeds
Rr	heterozygous	round seeds
rr	homozygous recessive	wrinkled seeds

The parent cross for this classic experiment would be calculated in this manner:

round-seeded parent (**RR**) x wrinkle-seeded parent (**rr**)

The phenotypic ratio for this cross (Figure 10-10) would be 4:0. This means that all offspring would have round seeds.

The genotypic ratio for this cross would be 4:0. This means that all offspring would have one genotype (**Rr**).

	R	R
r	Rr	Rr
r	Rr	Rr

0-10. F_1 predictions for rounded and rinkled seed using the Punnett square. R=rounded seed; r=wrinkled seed)

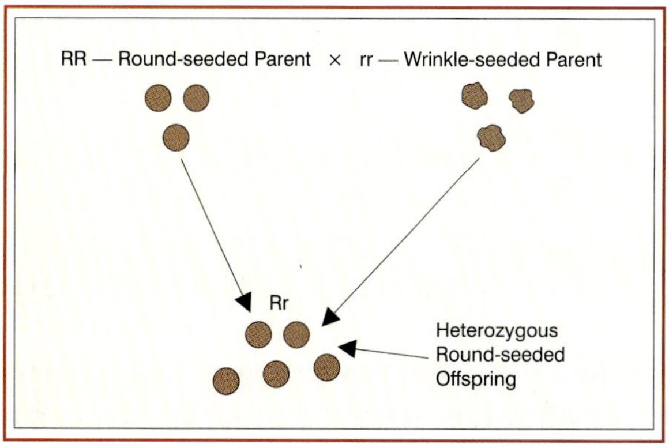

10-11. An alternate way of showing the Punnett square prediction from Figure 10-10 is shown here.

If two F_1 offspring were bred, the following would represent the most probable outcome for the F_2:

round-seeded offspring (**Rr**) x round-seeded offspring (**Rr**)

The phenotypic ratio for this cross would be 3:1, meaning 75 percent of the offspring would have round seeds and 25 percent would have wrinkled seeds.

	R	r
R	RR	Rr
r	Rr	rr

10-12. **F_2 predictions using the Punnett square and mating F_1 offspring.**

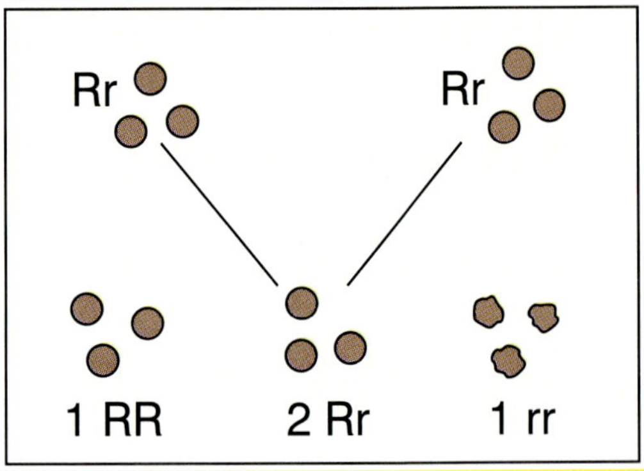

10-13. **An alternate way of showing Figure 10-12**

The genotypic ratio for this cross would be 1:2:1, meaning 25 percent of the offspring would be homozygous dominant (**RR**), 50 percent of the offspring would be heterozygous (**Rr**), and 25 percent of the offspring would be homozygous recessive (**rr**).

Multiple Traits

The Punnett square can also be used to look at multiple trait inheritance. An example of multiple traits would be flower color and seed traits. If two parent lines have the following characteristics and genotypes, P = yellow flowers & wrinkled seeds (**Yyrr**) x P = white flowers & round seeds (**yyRR**), then the phenotypic ratio for this cross would be 8:8 (1:1) or 50 percent yellow-flowered with round seeds: 50 percent white-flowered with round seeds.

The genotypic ratio for this cross would be 8:8 (1:1) or 50 percent (**YyRr**) and 50 percent (**yyRr**).

The number of cells in the Punnett square increased when the number the number of traits was increased.

Yyrr yyRR

	Yr	Yr	yr	yr
yR	YyRr	YyRr	yyRr	yyRr
yR	YyRr	YyRr	yyRr	yyRr
yR	YyRr	YyRr	yyRr	yyRr
yR	YyRr	YyRr	yyRr	yyRr

10-14. **F_1 two-trait prediction using the Punnett square for flower color and seed shape. (Y/y=flower color; R/r=seed trait)**

USING CHI-SQUARE

Genetic predictions can also be made by using a mathematical formula known as chi-square. ***Chi-square*** is an inferential statistic that compares observed frequencies with expected frequencies. The formula requires information about the number of times an event has actually occurred. The "number of times" an event has occurred is known as frequencies. In genetics, each time a trait occurs is a frequency. The symbol for chi-square is χ^2. Chi-square is sometimes known as the "goodness of fit" test.

A null hypothesis is used with the chi-square. A null hypothesis is a type of hypothesis that states that there is no relationship between the observed frequencies. The value obtained with chi-square is used to accept or reject the null hypothesis. The symbol for a null hypothesis is H_O.

$$\chi^2 = \text{sum}\left[\frac{(\text{observed} - \text{expected})^2}{\text{expected}}\right]$$

usually written as $\chi^2 = \Sigma \dfrac{(O - E)^2}{E}$

10-15. **A formula for chi-square.**

Once the calculations have been made, a table of chi-square values is needed. This table will reflect rows and columns based on probability and degrees of freedom (df). Details on the use of chi-square are available from statistics books and selected web sites.

MOLECULAR GENETICS UNRAVELS MORE MYSTERIES

Prior to the 1970s, genetics was evaluated based upon the physical expression of traits. This method of genetic evaluation is Mendelian or "qualitative" genetics. This was because the predictions were based upon the qualities that individuals possessed.

Since the 1970s, new technology has allowed scientists to evaluate genetic qualities from a much more precise, molecular level. This group of techniques has created the area known as molecular or "quantitative" genet-

Biotech Connection

FORENSIC SCIENCE

You may have seen the TV show, *The X Files*. In this series, Dana Scull serves as a forensics specialist for the FBI. Frequently on the show you will see blood and tissue samples being taken to provide DNA evidence. This is not science fiction. Law enforcement agencies use DNA science to build criminal cases on a daily basis.

Your DNA is unique to you. With advances being made every day in molecular science, DNA information becomes more and more reliable. It is more commonly used to prove innocence than to prove guilt, but it can be an important part of a prosecutor's case. Currently, DNA information is being used in rape, murder, theft, and paternity cases. (The photo shows a human DNA specimen being prepared for analysis.)

ics, because the actual genetic code of every living individual can be mapped and evaluated.

DEOXYRIBONUCLEIC ACID

DNA (deoxyribonucleic acid) is the molecule that codes the genetic information for all living things. Long pieces of DNA code for genes and each

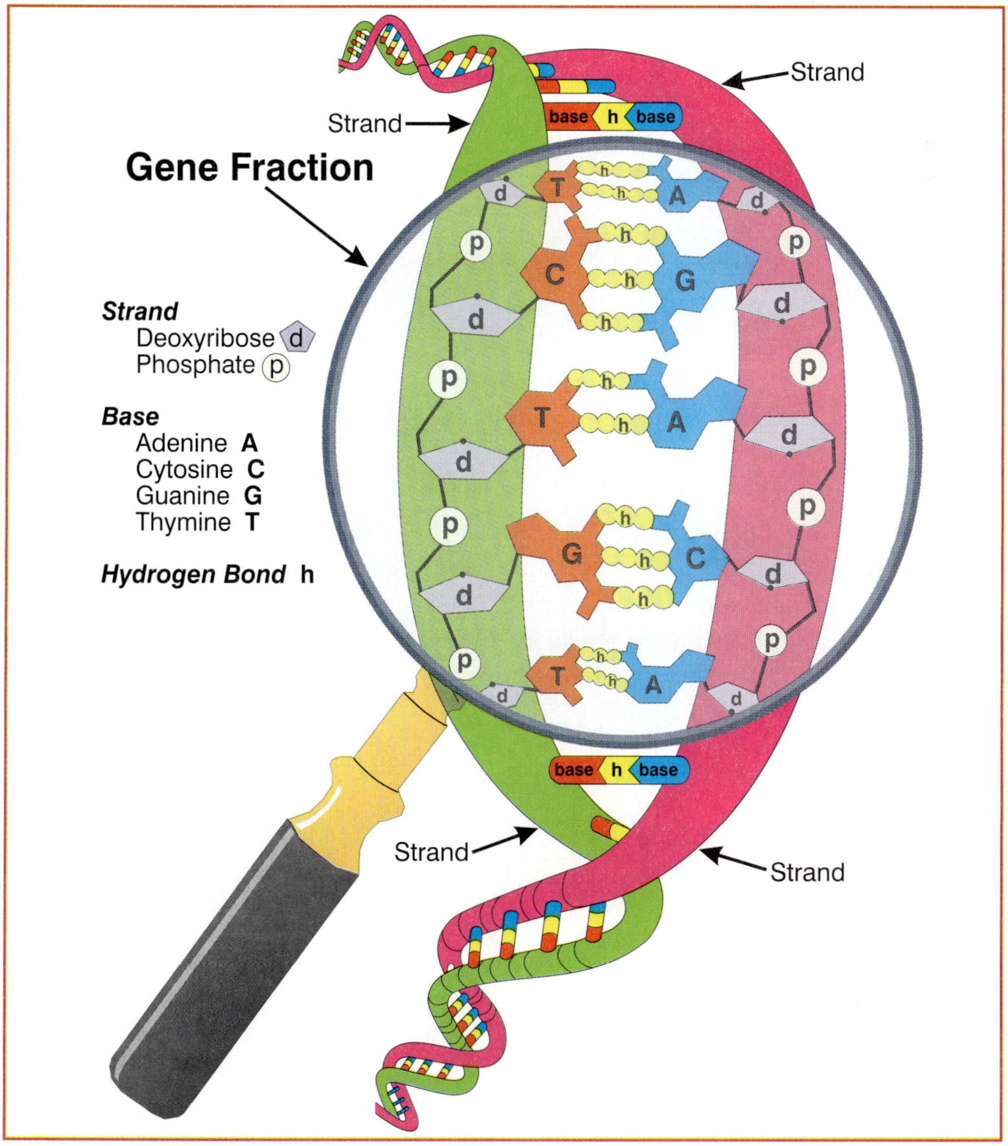

10-16. A DNA model.

chromosome may contain several genes. DNA was introduced in Chapter 7. A short review will be useful.

The DNA molecule is spiraling, double-stranded, and consists of a ribose sugar and phosphate backbone on each side. At the core of the molecule are two nucleotide bases. There are four nucleotides that DNA is composed of, and it is the arrangement of these nucleotides that determines the genetic code for all beings. The nucleotides are adenine (A), thymine (T), cytosine (C), and guanine (G). At the core of the molecule, adenine always pairs with thymine and cytosine always pairs with guanine. Figure 10-16 shows an example of DNA including hydrogen bonds, deoxyribose strands, and phosphate strands.

Protein Synthesis

Proteins are responsible for cellular function and development. They are essential for all living things. DNA codes for the manufacture or synthesis of proteins. There is a central dogma (system of principles) to describe this process:

$$\text{DNA} \xrightarrow[\textit{transcription}]{} \text{mRNA} \xrightarrow[\textit{translation}]{} \text{Proteins}$$

Transcription is the process of making an RNA that is complimentary to a strand of DNA. During transcription, in the cell nucleus, enzymes split the DNA molecule in half at the nucleotide bond, each single strand is known as RNA or ribonucleic acid. When this occurs, the nucleotide thymine (T) changes to uracil (U). One of these strands will code for protein synthesis, it is known as mRNA or messenger RNA.

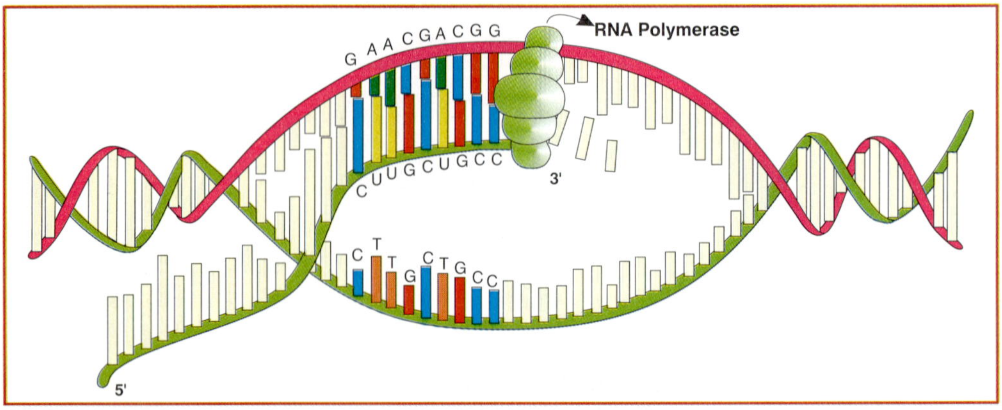

10-17. **Transcription of DNA into RNA.**

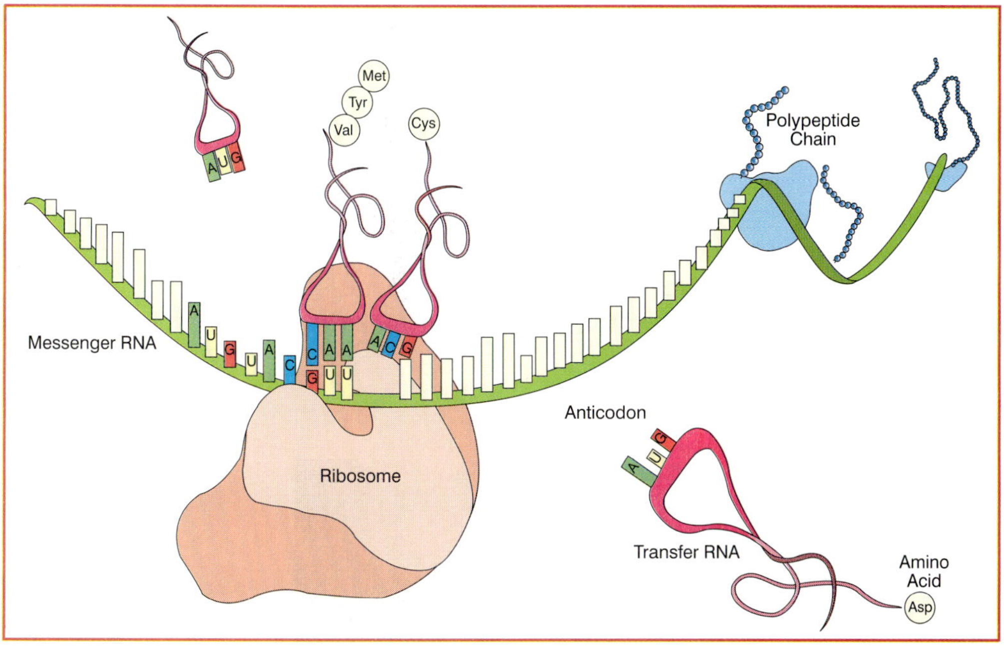

10-18. Translation of RNA into protein.

The messenger RNA carries DNA information from the nucleus to the ribosomes. When the mRNA reaches the ribosome, translation begins. ***Translation*** is the process of a cell beginning to build a protein (amino acid). During translation, a three-base pair unit binds to a complimentary unit on the mRNA. This nucleotide sequence is known as tRNA or transfer RNA, and it serves as the acceptor/bridge for amino acid production.

For every possible RNA three-unit nucleotide combination, there is a corresponding amino acid. Long chains of amino acids bind to then become proteins.

TOOLS OF DNA SCIENCE

The DNA code of any organism can be analyzed. This includes plants, animals, bacteria, and other organisms. Most of these procedures begin with DNA isolation.

DNA ISOLATION

To study DNA, it must be isolated. ***DNA isolation*** is the process of extracting and separating DNA from all other cell materials. Different procedures can be used to isolate "clean" DNA. The procedures vary with the cells/tissues being used. Wheat germ, bovine thymus glands, and other tissues are used in school labs. It is likely best to use cells that have large nuclei. Most of these procedures follow the same basic steps.

Step 1. Break open the cell wall and/or membrane. This is done with liquid nitrogen and/or grinding. This step varies with the source of the DNA. A mortar and pestle may be used to crack and grind seed cells, such as wheat germ. A food blender may be used with animal tissues, such as a thymus gland.

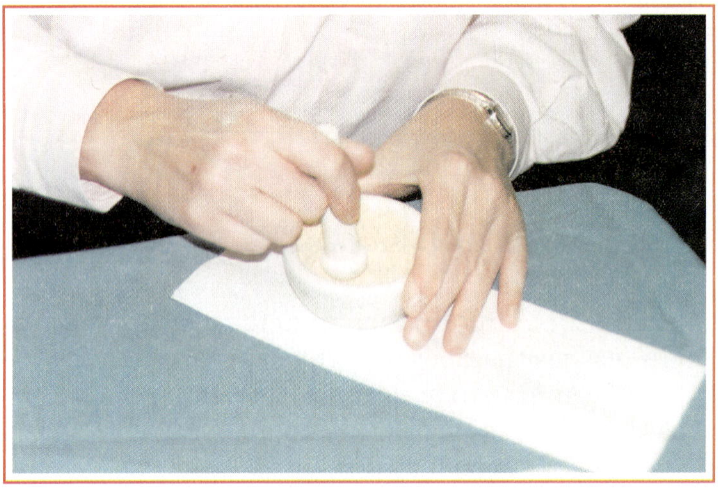

10-19. A mortar and pestle are being used to crush wheat germ for DNA extraction.

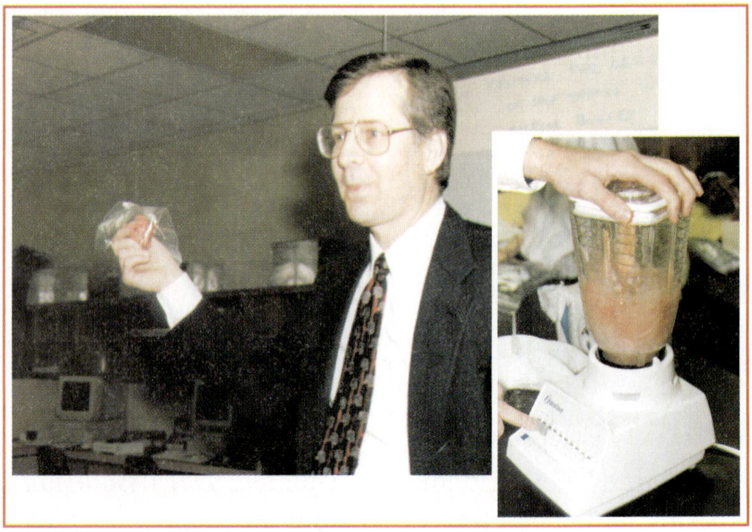

10-20. A calf thymus from a meat packing plant yields good visible DNA. The thymus is ground in a food blender that also contains a sucrose solution.

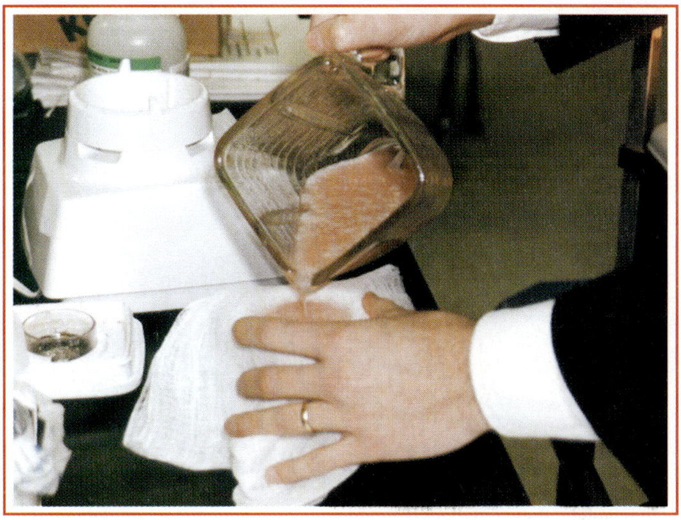

10-21. **The blended calf thymus solution is strained through a cheesecloth to remove clumps and gristle. (Low-speed centrifuging may help settle the nuclei to the bottom of a test tube and leave the cellular material in suspension.)**

Step 2. Digest the cellular components. This is often done by heating with a detergent. With wheat germ, a small amount of detergent solution is mixed with the ground powder. The wheat germ solution is heated to 60°C for 10 minutes. With the thymus solution, a small amount is put into a small test tube. EDTA solution is added to weaken membranes and inactivate DNA-digesting enzymes.

Step 3. Separate the polar compounds. This involves using a detergent solution, such as sodium dodecyl sulfate (SDS). The detergent dissolves the lipids (fat) in the nuclear membranes. Salt, such as NaCl, may be added to the solution. Gently stirring the solution with a slow circular motion promotes the process. Procedures may vary slightly depending on the source of the DNA. Some involve using chloroform or ethanol.

Step 4. Extract and precipitate the DNA. Remove the top aqueous layer with a pipette and place it into cold absolute alcohol. In demonstrations, the white mass of DNA may be "spooled" or collected onto a glass stir rod for observation.

POLYMERASE CHAIN REACTION (PCR)

Polymerase chain reaction is a procedure that uses controlled temperatures and the enzyme, *taq* polymerase, to replicate pieces of DNA. This technique allows scientists to make many copies from a few target DNA mole-

10-22. Spooled DNA extracted from the cells of a thymus gland is being removed from the test tube.

cules. Essentially, by multiplying the molecules exponentially, DNA can be made visible and, thus, analyzed.

Taq polymerase is the DNA replication enzyme found in bacteria that live in hydrophilic vents in the ocean (*Thermus aquaticus*). In its natural environment, the bacteria work at very high temperatures. Temperature is used to control PCR reactions.

PCR is a three-step process performed in a piece of equipment called a thermocycler. The machine alters the temperature at each step of the process and the process is repeated many times.

The following is what occurs at each step of the PCR process:

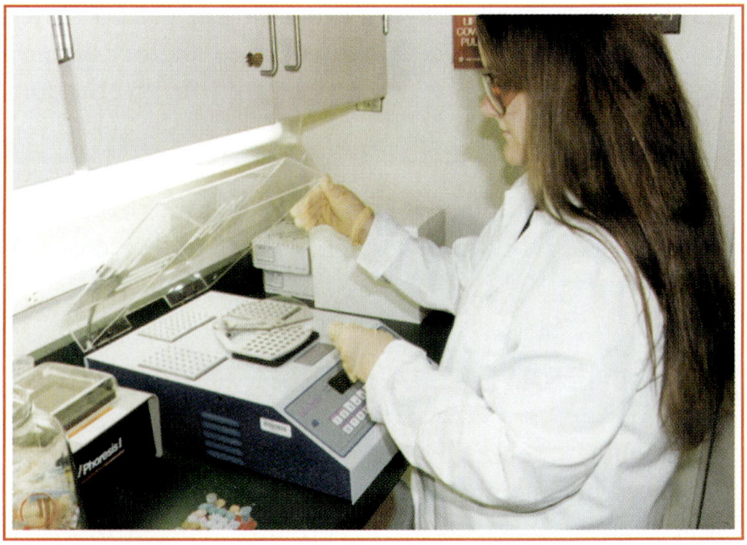

10-23. Preparing to use a thermocycler in performing PCR.

Step 1. Separation—Heat to 95°C to separate the DNA strands.

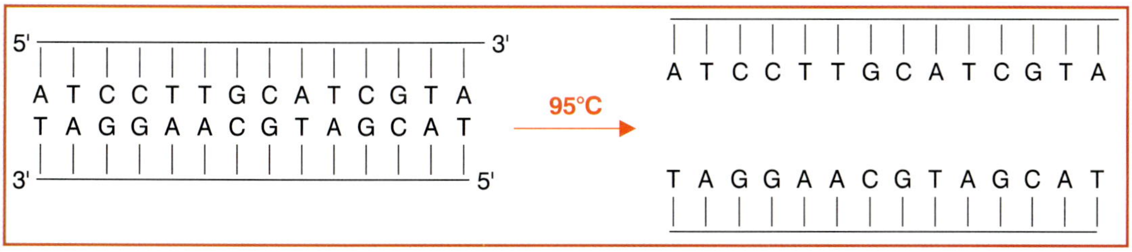

10-24. **Separation of a DNA strand.**

Step 2. Annhealing—Cool to 35-58°C for the primers to bind to complimentary DNA regions.

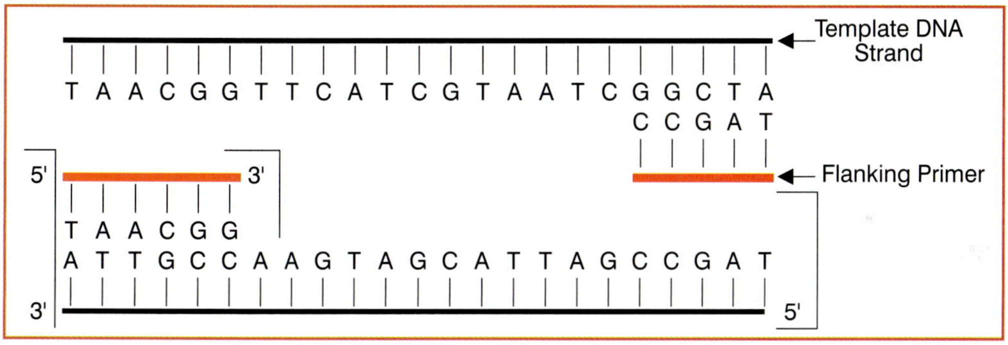

10-25. **Annhealing is used to bond complementary single strands of DNA.**

Step 3. Extension—Warm to 72°C for the *Taq* polymerase to build a new DNA strand from primed regions.

10-26. ***Taq*** **polymerase builds a new strand of DNA.**

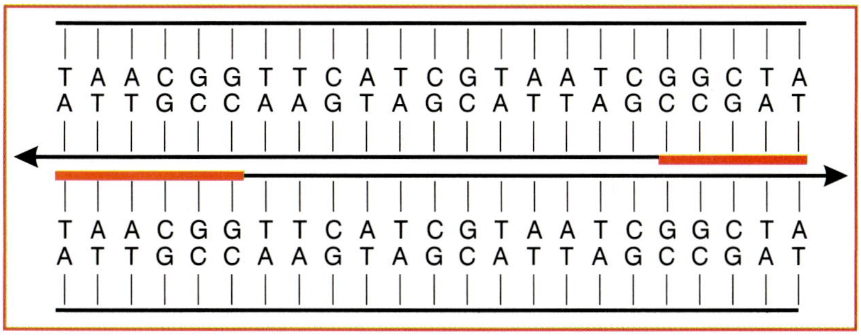

Each cycle represents an exponential step in DNA production:

(1)

i.e. 1st Cycle = 2 = 2 Copies

(5)

5th Cycle = 2 = 32 Copies

(20)

20th Cycle = 2 = 1,048,576 Copies

GEL ELECTROPHORESIS

Gel electrophoresis is a process of using an electrical field in agar to separate DNA and RNA molecules based on molecule size. First used with RNA, it is now a powerful tool for visualizing DNA products.

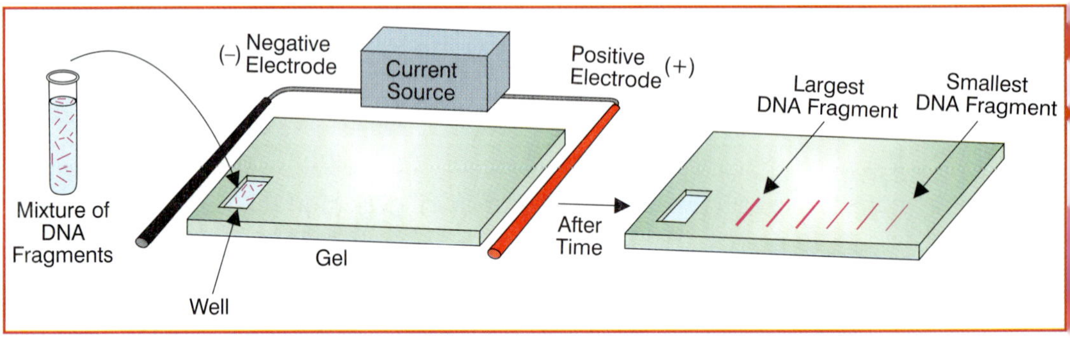

10-27. **A gel electrophoresis model.**

Gel electrophoresis provides a sequence of the DNA fragments. ***DNA sequencing*** is the process of determining the order of the nucleotides in DNA fragments. The chain termination sequencing is most commonly used. The chain has three steps: DNA synthesis, gel electrophoresis, and DNA detection.

The DNA is synthesized using an extraction process, as described earlier in this section. The DNA samples are loaded into a gel matrix, an electric current is applied, and the molecular fragments separate as they are pushed through the gel. A polyacrylamide gel was used in the first gel electrophoresis. The gel has been replaced with agarose, which is a highly purified agar. A fluorescent dye is used to stain the DNA fragments. Electrodes at each end of

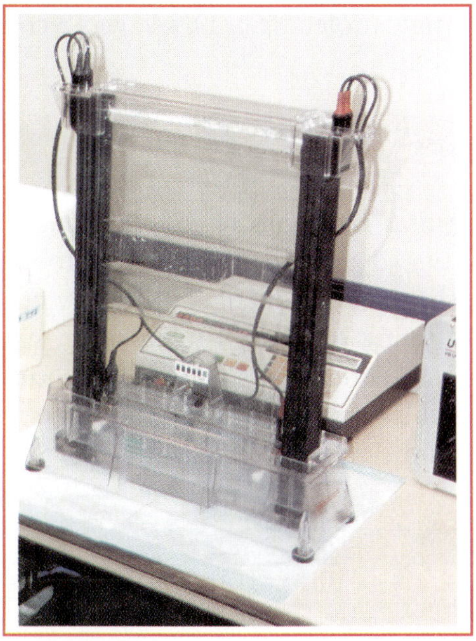

10-28. **Apparatus used in gel electrophoresis.**

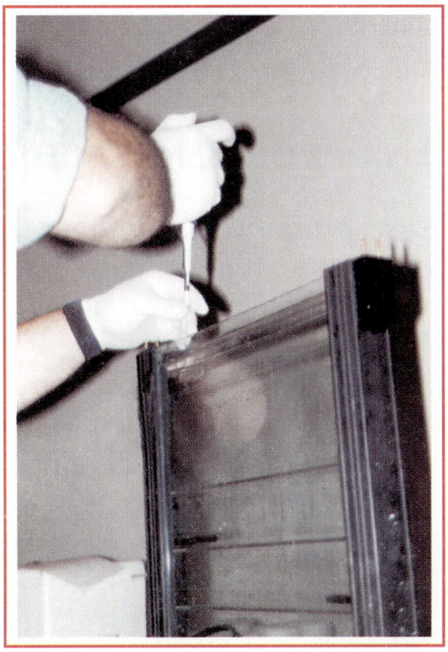

10-29. **Loading extracted DNA samples into gel electrophoresis apparatus.**

the gel create a current across the gel to move the DNA particles. The dye makes it easy to monitor the migration of the DNA particles across the gel. It also makes distinctions easier to identify based on distance of movement. Ultra violet light may be used to view the results.

Since DNA is negatively charged, it travels from the negative electrode toward the positive electrode. Heavier molecular fragments move slower than lighter weight fragments, so smaller pieces of DNA travel farther through the gel. Both distance and direction of travel are of interest in gel electrophoresis. Each band represents a molecular point of interest that can be compared to other samples, analyzed, or sequenced.

10-30. **The DNA sequence of a cow is being studied. (Courtesy, Agricultural Research Service, USDA)**

The DNA seen in the gel is not one single molecule of DNA. It is a collection of millions of identical DNA molecules.

DNA PROFILING TECHNIQUES

DNA profiling is identifying an organism based on regions of DNA that vary greatly from one organism to another. It is most widely used today in identifying people who cannot identify themselves, such as murder victims. DNA profiling is also known as DNA typing or DNA fingerprint.

No two individuals have identical DNA sequences. The only exception is with identical twins. This makes it possible to identify nearly every person on Earth using DNA. Table 10-1 lists some profiling techniques.

Table 10-1. Examples of DNA Profiling Techniques

Technique	DNA Isolation	PCR	Electrophoresis
RFLPs—Restriction Fragment Length Polymorphisms: restriction enzymes cut DNA into large pieces	X		X
RAPDs—Random Amplification of Polymorphic DNA: short 10 base pair primers bind to DNA. Segments between binding sites are aplified via PCR.	X	X	X
Microsatellites—15 base pair primers bind to reoccuring non-genic DNA regions.	X	X	X
AFLPs—Amplified Fragment Length Polymorphisms: A combination of RFLP's and RAPD's, provides enormous amounts of information.	X	X	X
Southern Blot—Used to analyze DNA in combination with RFLPs; radio or phosphorescent probe is applied; sample is blotted to a nylon membrane.	X	X	X
Other Molecular Profiling Techniques			
Northern Blot—Used to study RNA			
Western Blot—Used to study proteins			

REVIEWING

MAIN IDEAS

Genetics is the study of the laws and processes of biological inheritance. Gregor Mendel performed early experiments with peas and established several principles of genetics. Important areas are dominant and recessive traits and phenotype and genotype.

Mendelian genetics is said to be qualitative. This means that it makes genetic predictions based upon the traits of ancestors, the Punnett square, and chi-square.

Quantitative or molecular genetics is based upon actual DNA profiles of organisms. Deoxyribonucleic acid (DNA) provides the genetic code for all living organisms. Tools of DNA science include DNA isolation, polymerase chain reaction (PCR), and gel electrophoresis. A variety of DNA profiling techniques can be used in combination with PCR and electrophoresis to analyze DNA molecular differences.

QUESTIONS

Answer the following questions using correct spelling and complete sentences.

1. What is genetics? How does genetics relate to the traits of offspring?
2. What is gene expression?
3. What is selection? When is it used?
4. When does a recessive trait appear in an organism?
5. How does the environment of an organism promote the expression of its biological heredity?
6. What is the Punnett square? How is it used?
7. What is the role of proteins in a cell?
8. What is DNA isolation? What general process is involved?
9. What is PCR? Why is it used?
10. What is gel electrophoresis? How does the process sequence DNA?

EVALUATING

Match the term with the correct definition. Write the letter of the term in the blank that is provided.

a. homozygous
b. F_1
c. phenotype
d. gene
e. dominant
f. gamete
g. transcription
h. DNA sequencing
i. PCR
j. genotype

_____ 1. The physical expression of a trait.

_____ 2. A DNA profiling technique that uses restriction enzymes.

_____ 3. A trait that can be observed in an organism.

_____ 4. The genetic makeup of an organism.

_____ 5. A first generation cross.

_____ 6. Occurs when both alleles for a trait are the same.

_____ 7. A segment of a chromosome that contains the heredity material of an organism.

_____ 8. A procedure used to replicate pieces of DNA.

_____ 9. The process of making RNA that is complimentary to a strand of DNA.

_____ 10. A sex cell capable of uniting with another sex cell.

EXPLORING

1. Use colored marshmallows and toothpicks to create a 3-dimensional DNA model. Use white marshmallows for the structural backbone and a different color marshmallow for each type of nucleotide.

2. Interview a person working with forensic medicine. Find out the techniques of DNA profiling they use and what situations they use them in. Prepare a report on your findings.

3. Extract the DNA from wheat germ, onion skin, thymus gland, or other source. Research the procedures to follow. Be sure to observe all safety precautions. Prepare a report on your experiences.

11

GENETIC ENGINEERING AND TRANSGENIC ORGANISMS

People are living longer. Why? Is it because of better food and improved health care? Those are most likely important factors. A challenge is to continue making gains. Each new generation wants to do better than the previous one.

We are going to need to use new approaches. Some of the new approaches are on the horizon. A few that represent a vast number of possibilities are now being used. These use important new technologies. Genetic engineering is one of the new technologies. With some people, the mention of genetic engineering brings up notions of monsters and mutations. But, that is not what it is about. Genetic engineering deals with helping people have a better quality of life. What are the facts? Do a little investigation. You will find that in the next decade, genetic engineering will affect all of us on a daily basis. We need to learn more about it.

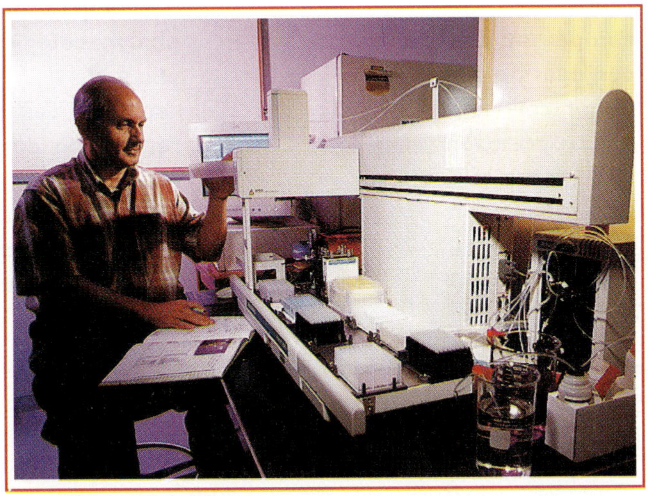

11-1. A robot is being used in searching for new wheat genes. (Courtesy, Agricultural Research Service, USDA)

OBJECTIVES

This chapter provides information on the meaning and use of genetic manipulation to produce transgenic organisms. It has the following objectives:

1. Explain why genetic engineering and transgenic organisms hold promise for the future.
2. Identify practical applications of genetic engineering.
3. Outline processes used to create genetically altered organisms.
4. Describe the use of genetic transformation with plants.
5. Explain the role of genetic transformation with animals.

TERMS

ballistic gene transfer
callus
competent cell
conjugation
donor cell
electroporation
gene transfer
GMO
GMO-free food
in vitro
ligation
marker gene
microinjection
pharmacology
plasmid
recipient cell
restriction endonuclease
totipotent
transformation
transformed cell
transgene integration
vector

MAKING NATURE BETTER

Organisms have certain characteristics. These are the traits that nature gave them. The traits are sometimes changed to make organisms more desirable and productive. How they are changed is important. Changes need to be beneficial and not result in damage to people nor the environment. Some of the changes involve molecular biotechnology.

Molecular biotechnology deals with the genetic information in cells and how it can be changed. It includes the study of the structure and parts of cells. The goal is to have organisms that are more useful. The organisms might produce new products or perform important functions.

11-2. Molecular biologists are analyzing electrophoretic data to confirm the DNA sequences. (Courtesy, Agricultural Research Service, USDA)

GENETIC ENGINEERING

Genetic engineering is the process of changing the genetic information in a cell. Genes in the cells are manipulated. The functions of the genes may be changed. New genes may be added. Genes may be removed. Organisms with different characteristics result.

A major tool in genetic engineering is recombinant DNA. Recombinant DNA is DNA that has been joined to other unrelated or foreign DNA. The new DNA has been altered from its natural form. It is sometimes called gene splicing—tiny segments of a gene are taken out and replaced.

TRANSGENIC ORGANISMS

Organisms that are altered by genetic engineering are transgenic. Their genetic material has been changed by other than random natural breeding. The goal is to create a desired quality in the "new" organism.

11-3. Distinguishing between GMO and GMO-free soybeans, oil, and meal products is difficult. The beans must be segregated from the time of harvest. (Courtesy, American Soybean Association)

Transgenic organisms are created by a process that is sometimes called gene transfer. ***Gene transfer*** is moving a gene from one organism to another. Recombinant DNA methods are used. The methods involved require skill and knowledge to carry out.

In the food industry, products from transgenic organisms are known as GMO products. A ***GMO*** is a genetically modified organism. Food products are labeled as GMO products by the marketing sector, including international markets. Some buyers shy away from GMO foods and pay a premium for GMO-free foods. A ***GMO-free food*** is a product in which no transgenic materials were used in its manufacture, such as soybeans used in making cooking oil. People who grow and market soybeans find it difficult to segregate GMO from non-GMO soybeans. (Question: Do you think you would know the difference in your foods if some were from GMO products?) The term GEO (genetically enhanced organism) is being used by some people.

PRACTICAL APPLICATIONS OF GENETIC ENGINEERING

Genetic engineering has many important applications. The technique goes far beyond traditional selective breeding. Genetic material can be shared across scientific kingdoms to get desired products. An example is bacteria. Bacteria can be engineered to produce human proteins. Another example is plants. Plants can be engineered to use genes from insects for their benefit. The potential of genetic engineering is virtually endless.

AGRICULTURE

The world's population is increasing. This increases the demand for food. A decrease is occurring in the world's natural resources used to produce food. Some scientists feel that meeting the demand for food in the years ahead will be a major challenge.

11-4. **Will biotechnology be the answer to blackspot disease in roses? Researchers are studying how to best solve the problem.**

Food processors may also be affected by genetic engineering. Shelf-life, storage, and food-handling may be extended and simplified. This will help foods resist spoilage and provide improved consumer appeal.

Genetic engineering is changing agricultural production. Plants are being transformed to be insect, disease, and herbicide resistant. Animals are being treated with engineered hormones to produce more milk and leaner meat. These applications are discussed in detail elsewhere in this book.

HEALTH AND MEDICINE

Genetic engineering is affecting the health care and medical industry. Altering growth with hormones and replacing organs are both relatively com-

11-5. **A beef animal is being examined for parasites. Biotechnology may be used to control parasites in the future. (Courtesy, Agricultural Research Service, USDA)**

mon. Some use of animal tissues in humans is now underway. Bone and joint replacement has become common. The materials used may be rejected by the human or animal body unless appropriate hormones are offered.

Much research is being carried out to investigate the production of skin and organ tissue for use in surgery. Some of this focuses on synthetic biology–making cells in the chemical lab. Most of it deals with using animals similar to humans, such as pigs.

PHARMACEUTICAL PRODUCTS

Pharmacology is the preparation, use, and affect of drugs. It is closely tied to health and medicine. The potential use of genetic engineering in the production of drugs is great.

Hormone production occurs naturally in the endocrine system of the mammalian body. Hormone therapy to stimulate or replace natural production has been common for the past fifty years. However, treatment was sometimes limited to the availability of the hormone. Availability was limited because the hormones often had to be extracted from a living entity.

Genetic engineering has changed hormone therapy. Bacteria can now be engineered to produce a hormone. Fermentation allows for the production of an enormous amount of hormone, thus increasing the availability to the consumer. This procedure could be applied to the production of insulin, human growth hormone, and bovine (cattle) and porcine (pig) somatotropin.

Fermentation is the process of growing bacteria to produce various compounds. The microbes are incubated under specific conditions. The incubation is in a large tank known as a fermenter. Fermentation is also known as

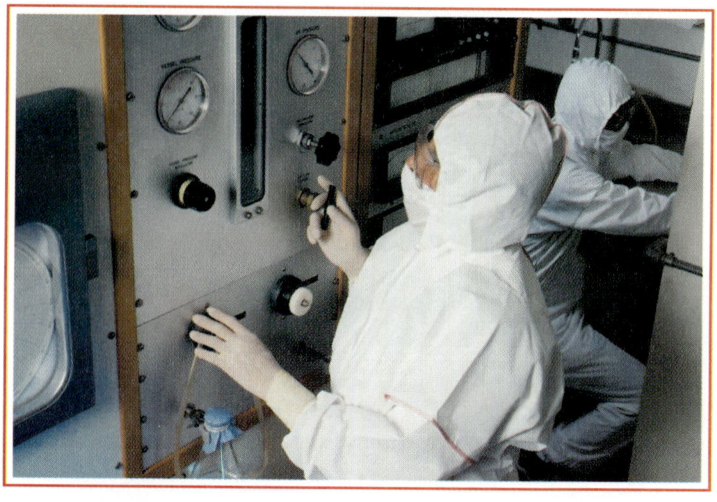

11-6. A pharmaceutical research lab.

bioprocessing. It is widely used for chemical and pharmaceutical production. The process of producing hormones with bacteria and fermentation is shown in Figure 11-7.

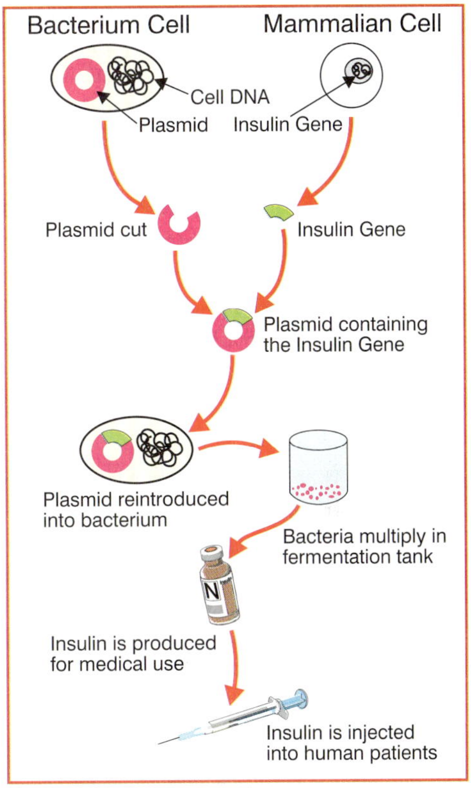

11-7. The process of producing insulin using mammal cells in the presence of bacteria followed by fermentation.

INDUSTRY AND ENVIRONMENT

Genetic engineering has the potential to substantially impact industry and manufacturing. New processes are being developed. More efficient use of scarce minerals can be made with transgenic organisms, such as greater efficiency in mining.

An area of major concern has been waste management. How can wastes from manufacturing be disposed of without damaging the environment? Bacteria have already been engineered to effectively biodegrade a number of waste products, including sewage and petroleum products.

PROCESSES IN CREATING TRANSGENIC ORGANISMS

Transgenic organisms are created using the process of recombinant DNA. Simply, and as explained in Chapter 2, recombinant DNA is taking the DNA from one organism and moving it to another. It sounds simple. But, the process is not simple. The rate of success is often low. The work involves moving very small materials into very small places!

Recombinant DNA is also known as gene splicing, gene cloning, and molecular cloning. It is useful because of the actions of enzymes. The enzymes recognize specific DNA sequences and hold (cleave) the double-stranded DNA in both strands.

Career Profile

GENETIC ENGINEER

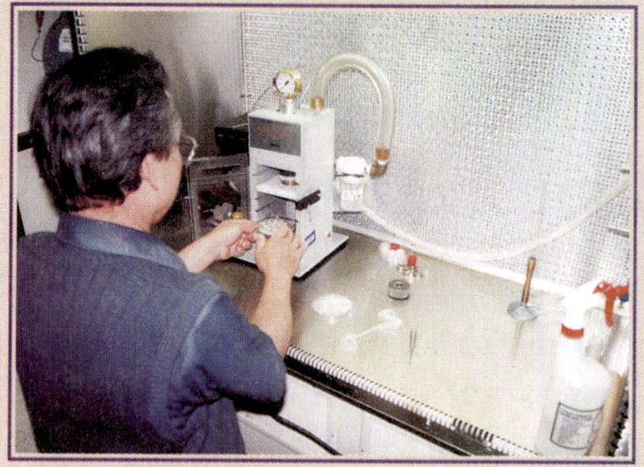

A genetic engineer is responsible for mapping genes, designing and building gene constructs, and developing transformation systems. They may carry out protocols to develop transgenic organisms. This photo shows a gene gun being used to transfer genetic material. (Courtesy, University of Illinois)

Genetic engineers usually have doctoral degrees in molecular biology, cellular physiology, or a related field. Some have postdoctoral education in research. Undergraduate education may be in agriculture, genetics, biology, or another science-based area. High school courses in science, as well as agriculture and horticulture, are useful.

Jobs are with agricultural, pharmaceutical, and genetic engineering companies. Some are with universities, research and development agencies, medical schools, and the like.

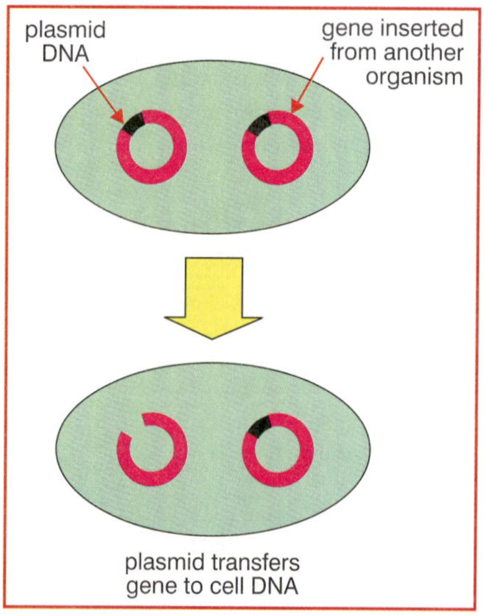

11-8. A simplified illustration of recombinant DNA. (Figure 11-11 provides more details.)

Recombinant DNA includes a number of protocols. These are all intended to transfer genetic information. No single approach always works. In simple terms, the process is cutting a gene out of a DNA strand and inserting the gene into another DNA strand. The DNA is combined to have a desired quality in the transgenic organism.

This section covers some fundamentals associated with recombinant DNA protocols. Remember, no one protocol is used in all recombinant DNA work.

ENDONUCLEASES

An endonuclease is a type of enzyme in a DNA strand. It produces nucleic acid strand breaks in the interior of the nucleic acid strand.

Restriction endonuclease is an enzyme produced by bacteria that is used in recombinant DNA. It cleaves DNA at a location defined by the specific sequence of the nucleotide bases. Restriction endonuclease recognizes a specific DNA sequence. It cleaves phosphodiester bonds on both strands between nucleotides. One of the first restriction endonucleases used was from *Escherichia coli,* and known as *Eco*RI.

A phosphodiester bond is the linkage of a phosphate group to the 3′ carbon of one nucleotide to the 5′ carbon of another nucleotide. It forms a linkage between nucleotides of the same nucleic acid strand. (Refer to Figure 10-17 in Chapter 10 for the location of 3′ and 5′ carbons.)

Restriction endonucleases recognize certain DNA sequences and digest DNA at those recognition sites. This process "cuts-open" the bacterial plasmid. The gene construct is engineered into the plasmid with ligasees. Engineered plasmids are then put back into a bacterium.

The engineered plasmid is documented with a plasmid map that may look something like the one labeled below.

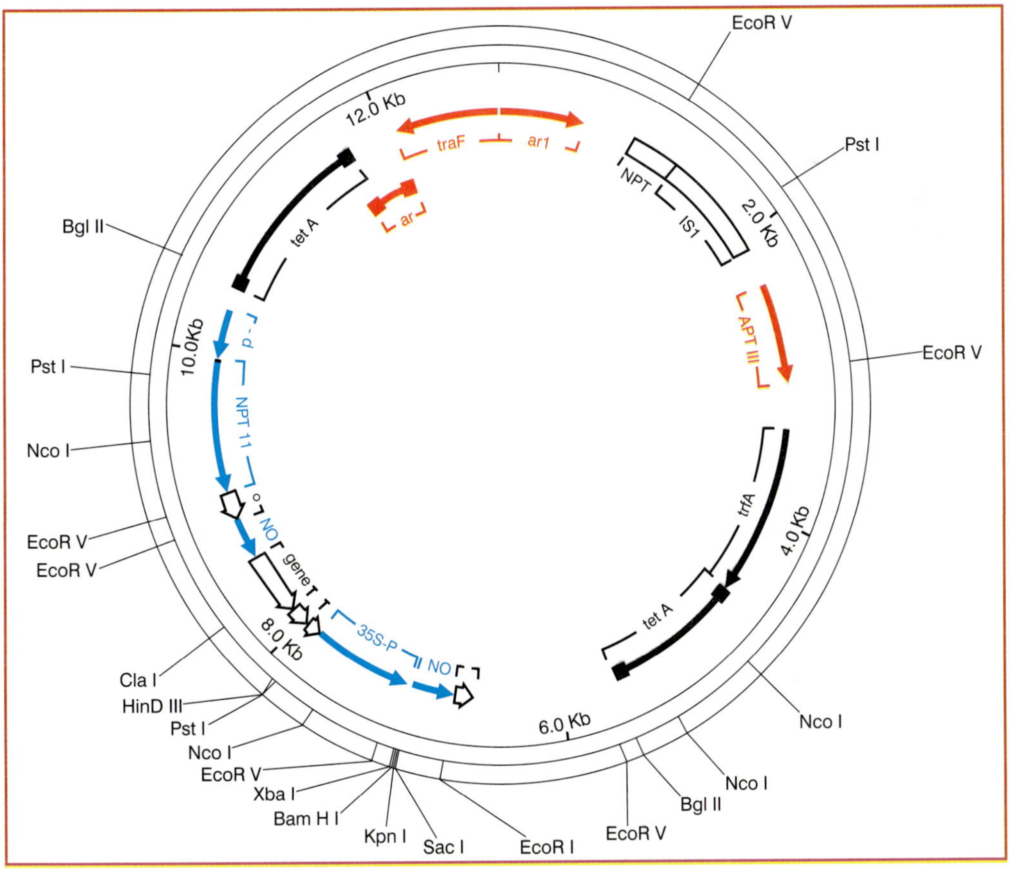

11-9. **A sample plasmid map documents an engineered plasmid.**

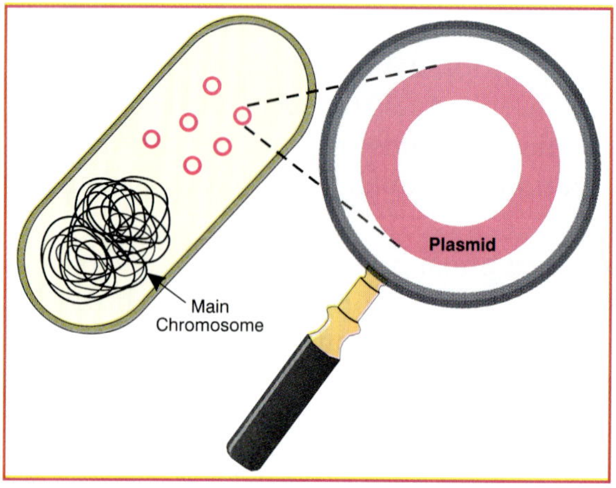

11-10. A simple illustration of a plasmid in *E. coli*.

CLONING VECTORS

A ***vector*** is a carrier for DNA during the recombinant DNA process. Yeast cells and *E. coli* bacteria have been most commonly used as vectors. Vectors are primarily from organisms that are easy to work with.

A ***plasmid*** is a piece of free-floating DNA in the cytoplasm of bacteria. Plasmids are double-stranded, circular DNA molecules that replicate independently of the chromosome. Nearly all genera of bacteria have plasmids. They are the simplest vectors. In addition to being good vectors, some natural plasmids carry genes that have resistance to antibiotic drugs.

Certain bacteria naturally transfer their DNA to other organisms. This is by replicating and passing the plasmid across the cellular membrane. This is the phenomenon that makes most genetic engineering possible.

Plasmid cloning vectors are used in some genetic engineering protocols. They are useful because they do not replicate in most host cells. Some plasmid cloning vectors have specific names or numbers, such as *p*BR322. This is probably the most widely studied vector. The p stands for plasmid and the BR stands for the last names of the two researchers who created the plasmid (F. Bolivar and R. Rodriguez). The number, 322, is a numerical designation that was important to its creators.

A genetically engineered plasmid will contain many components, including, but not limited to:

- Promoter gene—A promoter gene is sequence of bases in a nucleic acid strand that serves as a signal for the start of transcription of a given gene. It aligns the RNA polymerase so transcription will initiate at a specific nucleotide. A promoter gene is a 150- to 300-base pair unit that contains binding sites for RNA polymerase and proteins that control DNA transcription.
- Chromosomal DNA construct—A chromosomal DNA construct is the gene(s) of interest. It is obtained from the chromosomal DNA of some organism.
- Antibiotic resistant gene—Antibiotic resistant genes are used as a marker system for transformed cells. Cells that have been successfully engineered will not die when placed in media containing antibiotics.

- "Gus" (beta-glucorodinase) gene—The "Gus" gene is also used as a marker system for transformed cells. In the presence of certain chemicals in cell culture media, engineered cells will express a blue color.

TRANSFORMATION

Transformation is the process of introducing free DNA into bacteria. This is a step necessary for bacteria to serve as vectors. The DNA is foreign to the bacteria. *E. coli* is the main host cell for recombinant DNA.

The host cells are usually treated with ice cold $CaCl_2$ and then exposed to 42°C for about 90 seconds to promote uptake of the plasmid DNA. The rate of transformation is 1 per 1,000 cells treated. This depends on cell competence.

A ***competent cell*** is one that is capable of taking up DNA. With many, including *E. coli*, competence must be induced. If competence is not natural or cannot be induced, another host cell must be selected. Electroporation is used with some bacteria to promote transformation.

Electroporation is the use of an electric shock to momentarily open or disrupt cell walls so genetic information can enter the cell. The electric field promotes the uptake of free DNA. The amount of electric current varies with the species of bacteria.

LIGATION AND CONJUGATION

Ligation is uniting or attaching two DNA fragments. The DNA in the recipient cell must be cut and joined with new DNA. The ***recipient cell*** is the cell that is receiving the "new" DNA. The ***donor cell*** is the cell that provided the DNA fragment for the recipient cell.

Ligation is often in test tubes or other lab glassware. Because it is done in glass, it is

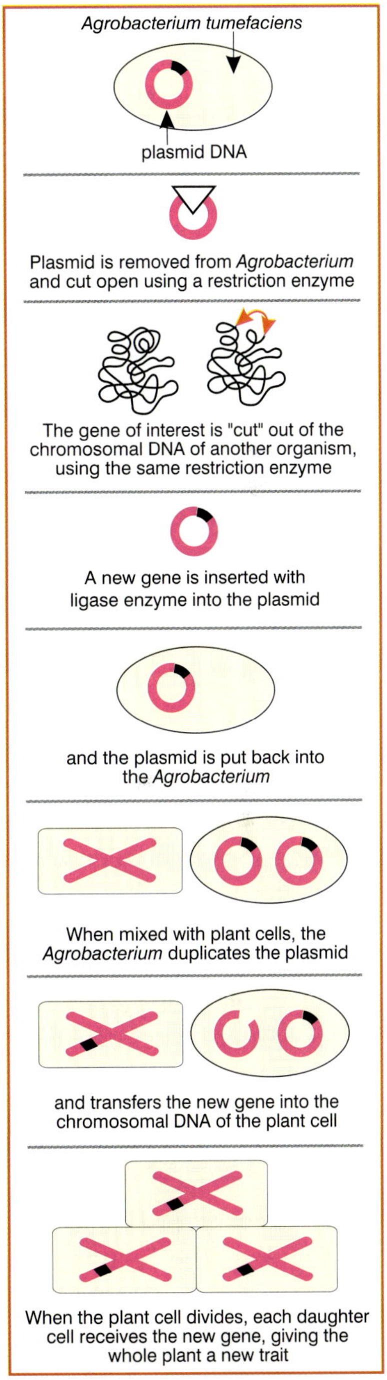

11-11. The general process of creating transgenic plants using *Agrobacterium tumefaciens*. A plasmid serves as the vector.

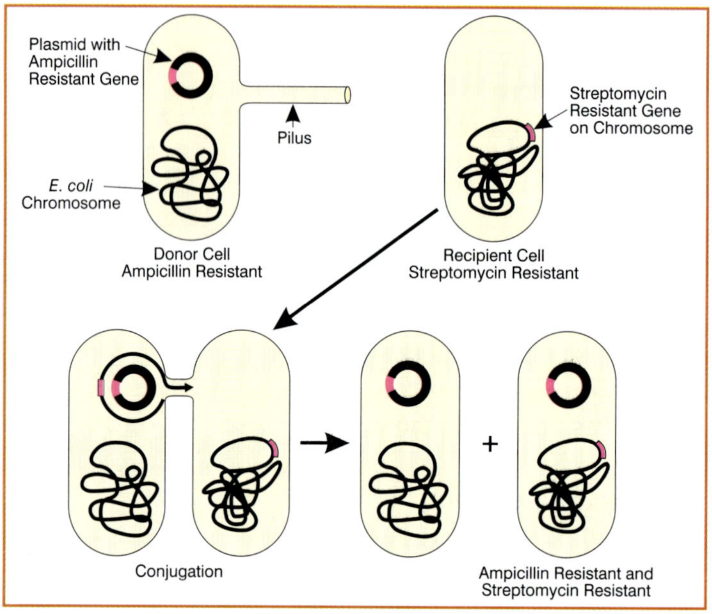

11-12. A simple illustration of conjugation by *E. coli*. This shows the transfer of a plasmid that is resistant to ampicillin into a cell with streptomycin resistance. The transgenic cell has resistance to both ampicillin and streptomycin.

known as ***in vitro*** (or "in glass"). Genes are spliced together using ligases, enzymes that "glue" DNA together. The sum total is known as the gene construct.

Some bacteria exchange DNA by conjugation. ***Conjugation*** is the contact of bacteria that involves the exchange of DNA with a mating tube. The conjugation is between a bacterium that contains the small DNA segment and a bacterium without the segment.

The cell with the new DNA is a ***transformed cell***. It becomes the basis for an organism being transgenic or genetically altered.

A ***marker gene*** is a gene that identifies which organisms have been successfully transformed. They produce an easily identified phenotype. Marker genes are often antibiotic resistant. Sometimes they create a color change.

OTHER PROCESSES

Several protocols are used in genetic engineering. The goal is to alter the DNA of the recipient cell to achieve a desired quality.

GENETIC ENGINEERING OF PLANTS

Genetic alteration begins in cells. To genetically modify an entire organism, the initial transfer of genetic information occurs at the single cell level.

New genes cannot be "shot" into a whole, growing plant. Miraculously, plants are totipotent.

Totipotent means that an organism has the ability to grow from a single cell. Cells differentiate and specialize. A functioning organism develops from a single cell. Totipotency is especially important with plants. Some people call this process regeneration–one cell may form an entire new plant.

The primary difficulty in transforming plants has been getting genetic information past a cell wall. A number of techniques have been developed to move genes through the plant cell wall. Some include physical manipulation and the use of a naturally occurring bacterium, *Agrobacterium tumefacians.*

AGROBACTERIUM TRANSFORMATION

Agrobacterium tumefacians is a bacterium that causes a disease known as crown gall in plants. *Agrobacterium* infects the plant by transferring its genetic material into the plant cell. The bacterium does this by moving its plasmid DNA into an injured site on the plant. This could be compared to an infection that you might get when germs enter a scratch on your body.

Biotech Connection

HUMAN GROWTH HORMONE

Most young people grow and develop normally. A few have a condition known as dwarfism. Their bodies fail to produce a human growth hormone as children. In the past, the costs of treating children for this condition were exorbitant. Costs were high because a very limited supply of the growth hormone was available. (It had to be extracted from cadavers.)

Now, thanks to genetic engineering, human growth hormone can be produced in very large quantities. Bacteria are engineered to produce the hormone and then multiplied in huge fermentation tanks. It has made the treatment affordable for most families.

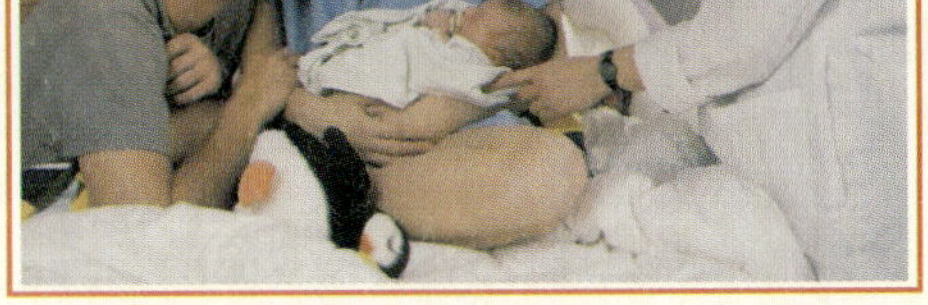

All parents of newborn babies wonder if their offspring will grow normally. New human growth hormone products help assure normal growth.

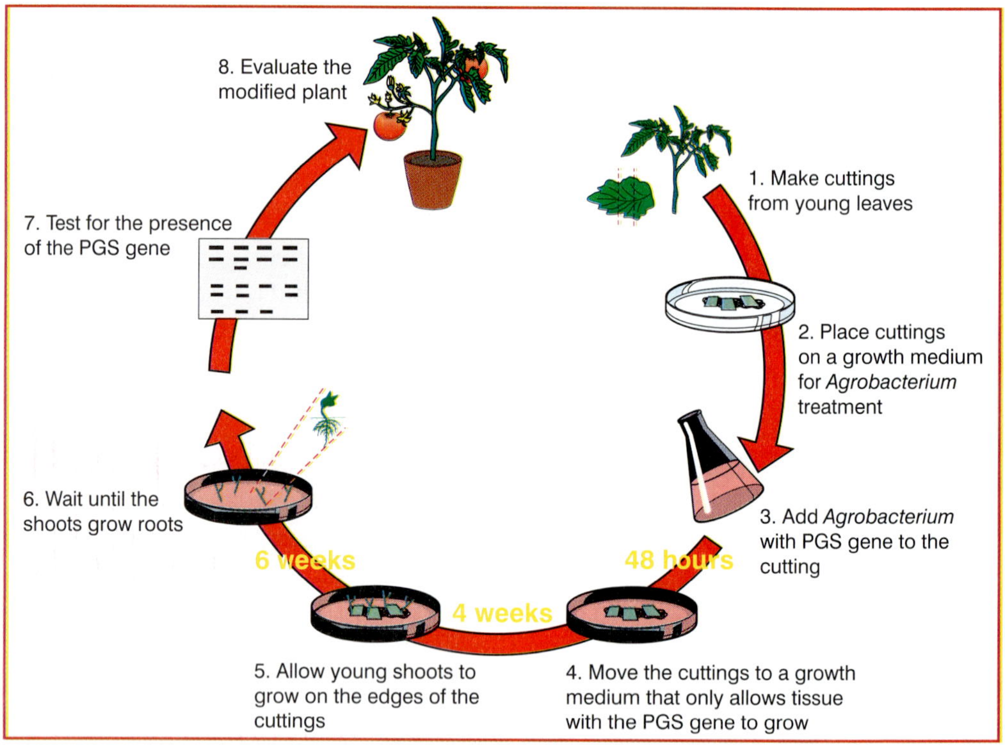

11-13. The procedure in using *Agrobacterium tumefacians* in creating a transgenic tomato.

Because of these natural abilities, *Agrobacterium* transformation is the most common technique for genetically engineering dicotyledonous plants. Dicotyledonous plants are those with two seed leaves, known as cotyledons. Examples of dicot plants include soybeans, peas, and cotton.

BALLISTIC GENE TRANSFER

Some cells do not respond well to the use of bacteria as vectors. For example, *Agrobacterium* cannot infect monocot plants, such as corn and rye grass. Other methods of gene transfer are used with them.

Ballistic gene transfer is the use of tiny DNA-coated projectiles as carriers. It is used to transport DNA through the walls of intended recipient cells. The projectiles are often known as microprojectiles. They are tiny gold particles. The projectiles are released at a velocity that penetrates the walls of cells. The success rate varies from less than 1 percent to about 10 percent. Ballistic gene transfer is also known as biolistics.

"Ballistic" transformation is accomplished with a device known as a "gene gun" or microprojectile unit. A gene gun typically uses an enclosed,

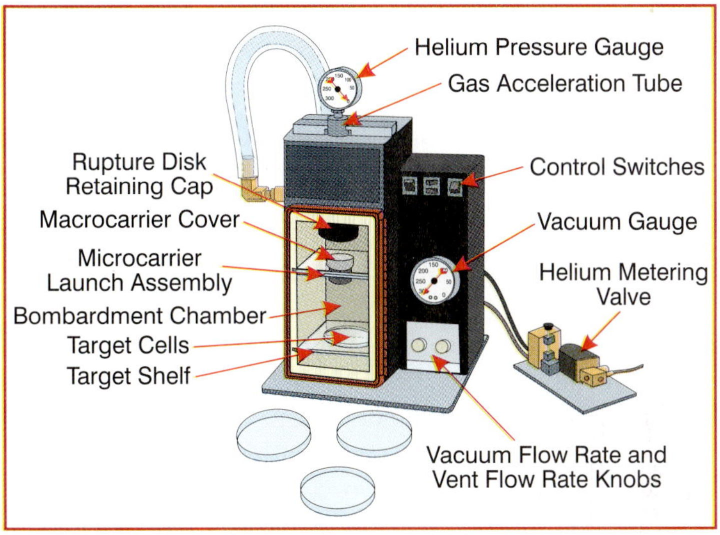

11-14. The major parts of a gene gun.

vacuum system with a high-powered ammunition charge to "shoot" DNA into plant cells.

Creating transgenic organisms with a "gene gun" has been most useful with important agricultural crops. It has been used to create herbicide and insect resistant corn.

THE ROLE OF TISSUE CULTURE

Once genetically engineered cells have been identified, they must be cultured on growth-specific media in an aseptic environment.

Cells begin on a callus medium. A ***callus*** is a mass of undifferentiated plant cells. By growing engineered cells into a callus, the number of transformed plants from each transformed cell is increased. The callus is then transferred to a shooting/multiplication medium where plant cells begin to form differentiated leaf and shoot tissue. This is manipulated by controlling levels of plant growth hormones in the culture medium. Shoots are then placed on a rooting media, where roots form in the presence of low auxin/cytokinin concentrations.

11-15. Tissue culture has resulted in small plants from callus.

GENETIC ENGINEERING OF ANIMALS

Genetic engineering of animals is a controversial and much talked about subject. It is not a common research practice because, unlike plants and bacteria, DNA transfer cannot be closely controlled.

Due to early cell differentiation, genetic transformation in animals must occur in the early embryo stage following the union of sex cells (fertilization). It must be done before the cells begin to divide (mitosis). This is accomplished by using microinjection, to randomly inject DNA into the fertilized cell.

Most of the research with transgenic animals has been with mammals, primarily mice, pigs, sheep, goats, and cattle. The only proven method of producing transgenic livestock is by microinjection of DNA into one-celled embryos or the nuclei of two-celled embryos. ***Microinjection*** is the injection of materials directly into cells using a small glass pipette.

The process of creating transgenic animals involves three major steps:

1. Obtaining embryos.
2. Microinjecting embryos.
3. Culturing and transferring zygotes.

Embryo survival is a problem. The percentage has ranged from about 15 percent survival of genetically altered mice embryos to 1 percent or less with cattle. Of those that are born, the transgene integration rate is low. ***Transgene integration*** is expression by the animal of the desired trait. In research, transgene integration has been lowest with cattle and highest with mice.

More information on animal biotechnology is presented in Chapter 14.

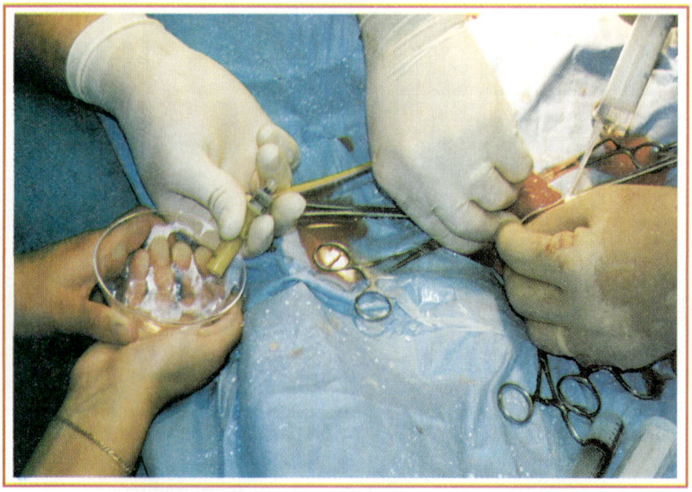

11-16. Embryos are being surgically harvested from a donor ewe (female sheep). The embryos are removed through a tube and collected in a glass dish. (Courtesy, Elite Genetics, Waukon, Iowa)

REVIEWING

MAIN IDEAS

Genetic engineering is the incorporation or deletion of genetic material in an organism. It is now being used in agriculture, health and medicine, pharmaceuticals, and industry and the environment. Currently, bacteria are being genetically transformed to produce pharmaceutical products and perform other functions, and plants are being engineered to have a number of beneficial traits.

Genetic engineering techniques differ in procedure, but the most common use a bacterial plasmid to carry genetic information. Plasmids are free-floating rings of DNA that can be altered to contain desired genetic constructs with the use of specialized enzymes.

Plant cells can be engineered with the use of: *Agrobacterium tumefacians*, ballistic gene transfer, or electroporation. Plants are grown into mature organisms using tissue culture.

Animal genetic engineering has been received with less enthusiasm. People have concerns about altering animals. Issues surround altering the genetic material in animals.

QUESTIONS

Answer the following questions using correct spelling and complete sentences.

1. What is genetic engineering? Why is it useful?
2. What is GMO food? GMO-free food?
3. List four areas in which practical application of genetic engineering is made. Briefly describe each.
4. What is recombinant DNA?
5. What is restriction endonuclease? Why is it important in recombinant DNA?
6. Why are *Escherichia coli* used in some genetic engineering work?
7. What is a plasmid? Why are they important in genetic engineering?
8. Distinguish between a recipient cell and a donor cell.
9. Briefly describe the overall general protocol that may be used in genetic engineering.
10. List and describe three methods used to genetically engineer plants.
11. Explain why researchers do not commonly try to genetically engineer animals.

EVALUATING

Match the term with the correct definition. Write the letter of the term in the blank that is provided.

a. electroporation
b. plasmid
c. restriction endonuclease
d. callus
e. transformation
f. *in vitro*
g. ligation
h. totipotency
i. transgene integration
j. conjugation

_____ 1. An enzyme produced by bacteria and used in recombinant DNA.
_____ 2. Free-floating DNA in the cytoplasm of bacteria.
_____ 3. The process of introducing free DNA into bacteria.
_____ 4. The use of electric shock to disrupt cell walls.
_____ 5. The ability of an organism to grow from a single cell.
_____ 6. A mass of undifferentiated plant cells.
_____ 7. The expression by an organism of a desired trait.
_____ 8. The contact of bacteria so that DNA is exchanged.
_____ 9. Uniting two DNA fragments.
_____ 10. A process done in a test tube or other container normally made of glass.

EXPLORING

1. Use your imagination and create an idea for a genetically engineered product. Collect information from scientists, by reading, or in other ways. Develop a detailed product "fact sheet," stating all of the information about your product and design a full-page, full-color advertisement for your product that could be used in magazines, newspapers, or the Internet. Set up a table top display about your idea in your classroom.

2. Write to a major chemical company, such as Monsanto, Dow-Elanco, and American Cyanamid, and ask for promotional information on genetically engineered products they have developed. Design a class bulletin board to display your information.

3. Make a field trip to a lab where transgenic organisms are being created. Have a scientist explain the purpose of the work and the processes that are being used. Prepare a written report on your observations.

4. Job shadow a scientist involved with biotechnology. Carefully observe the protocol being used. Prepare a report on your observations.

12

MICROBIAL BIOTECHNOLOGY

Most people know about the living things they see each day. We identify pets, farm animals, trees, and other humans. However, there are many living things that we cannot readily see. Our eyes need help to see them.

These tiny organisms are microbes. Microbes are found all over Earth. Since microbes are so small, they can fit into many different places. Microbes are found in your stomach, in soil, in food, and on your desk. They do many useful things for us, but they sometimes cause problems, such as disease.

Biotechnology scientists often use microbes in doing research. Their size and fast reproduction help make them very useful. Many thousands of microbes can be grown in a single dish. Some will divide to form more organisms in only a couple of hours. In one day, a scientist can produce hundreds of thousands of microbes. Microbes have made it possible for much of our biotechnology work to be successful.

12-1. **A cultured sample of the bacterium, *Brucella abortus*, is being examined in a laboratory. The disease caused by *Brucella abortus* is known as brucellosis. Cows with the disease abort in the first six months of pregnancy. *Brucella abortus* is a nonmotile, non-sporing, and Gram-negative coccobacilli. (Courtesy, Agricultural Research Service, USDA)**

OBJECTIVES

This chapter provides details on the use of microbes in biotechnology, including the kinds of microorganisms used, habitats required, tools of expression, and importance in science and health. It has the following objectives:

1. Explain microorganisms as related to biotechnology.
2. List and explain four advantages of microbes in biotechnology.
3. Name and describe the major types of microbes and their habitats.
4. Describe the observation of microbes, including slide preparation, staining, and culture.
5. Identify and describe examples of microbial biotechnology.

TERMS

aerobe
agar
anaerobe
bacteria culture
biochemical process
biodegradation
bioremediation
biotransformation
cilia
diversity
flagellum
gram stain
medium
microbial biotechnology
microbiology
parasite
penicillin
spore
strain
symbiosis
thallus

MICROORGANISMS AND BIOTECHNOLOGY

Microbiology is the study of microbes (microorganisms). A microbe is a free-living organism that usually exists as a single cell. Some are in clusters of cells. Unlike microbes, the cells found in animals and plants are unable to live alone. They only exist in multicellular relationships.

Microbes have been studied for a few hundred years. A great deal is known about the different types and their biology. With the invention of the microscope, Robert Hook described the first microorganisms in 1664. The first person to study microorganisms in detail was the Dutch microscope builder Anton van Leeuwenhoek in 1684. The invention of better microscopes has allowed scientists to study deeper into the mysteries of these one-celled organisms.

Microbes carry out many processes that are vital for the life of higher organisms. Without microbes, higher organisms would quickly disappear from the earth.

12-2. The effects of some microbes can be easily seen, such as this unattractive mildew growing on the painted wood trim of a house. (Mildew is caused by a type of fungus. It can be removed by washing the area with a bleach solution.)

MICROBIAL BIOTECHNOLOGY

Microbiology helped lay the foundation for biotechnology. Single-cell organisms carry out life processes in one cell. Researchers often study microbes to learn about these processes in all living organisms. More is known about the biological and chemical processes in microbes than in complex plants and animals.

12-3. Oysters are being prepared for study. The goal is to see if substances are present that make them unfit to eat. (Courtesy, Agricultural Research Service, USDA)

12-4. A sample of aflatoxin is being prepared for residue testing. (Aflatoxin is produced by some species of the fungus, *Aspergillus flavus*. It attacks corn and other grain making them unfit for food.) (Courtesy, Agricultural Research Service, USDA)

Microbial biotechnology uses microbes for biotechnology research and to produce materials. Chemicals, polymers, pharmaceuticals, and antibiotics are produced using microbes. Industry uses them because they reduce pollution and are less expensive than chemical methods.

Microbes are used in many areas of food production, medicine, and the environment. Foods include cheese, yogurt, and bread preparation. Specific species of yeast called bakers' yeast have been isolated for the preparation of bread. Different species of bacteria are used to prepare cheese and yogurt. In modern times, microbiology has been used to produce antibiotics. Antibiotics are compounds used to treat bacterial infections in animals. Discovered by Alexander Fleming in 1929, ***penicillin*** was one of the first effective antibiotics. The development of antibiotics was a major advance in health care.

Microbes naturally perform many useful roles in the environment. They help to keep it clean. They decompose many materials and recycle nutrients. A good example is the fallen leaves from trees. Microbial activity breaks down the leaves and returns nutrients to the soil. Scientists often use the term geochemical cycling to describe these processes.

Microbes are a part of many life processes on Earth. They help animals in obtaining food nutrients. One example is ruminant animals, such as cattle. Bacteria in the rumen help make nutrients available to the animal from low-quality grasses and other plant materials.

PROGRESS WITH MICROBES

The DNA sequence of some microbes has been determined. This is a useful tool in research. The first recombinant DNA work was with microbes. Combining DNA from two different sources and introducing it into a host cell creates recombinant DNA. Microbes commonly serve as hosts and vectors in recombinant DNA research. Viruses are often used to incorporate DNA into microbes or the cells of higher organisms. New DNA might be inserted so the microbe produces a specific compound or enzyme.

Many of the major enzymes used to manipulate DNA, DNA polymerase, restriction nucleases, and DNA ligase, were isolated from bacteria. These enzymes are important laboratory tools in biotechnology. The basis of biotechnology has been worked out using microbes. Scientists are beginning to apply these techniques to higher organisms.

USEFULNESS IN BIOTECH

Many of the techniques used in biotechnology were developed with microbes. Microbes are the most common organisms used in biotechnology research. Microbes have advantages over higher organisms, such as plants and animals, in research.

The four main advantages of microbes are:

- Size—Because microbes are small, hundreds of thousands can be grown in a very small area, such as a petri dish. Their size also lessens the amount of nutrients needed to grow a large number. Imagine having to grow 100 mice and how much space and food you would need!
- Rate of Reproduction—Many microbes will reproduce rapidly. In some cases, the number of organisms will double in only four hours. Scientists can produce many organisms in a short time. Higher organisms can take much longer to reproduce. For example, to produce 100 mice from two mice might take a couple months.

12-5. Many thousands of microbes can be grown in a small petri dish with a good medium.

Biotech Connection

GEOTHERMAL POOLS

Some areas have natural pools of heated water. These are known as geothermal pools. Many geothermal pools can be found in Yellowstone National Park. Even though they are warm, some microbes grow in them.

Some microbiologists are interested in the microbes that grow in and around hot thermal pools. Microbiologists hope to find microorganisms that produce new proteins and enzymes that are stable at hot temperatures. The enzymes and proteins in ordinary living organisms break down at high temperatures and over time. These newly identified compounds will hopefully be more stable under different conditions and over time.

This photo shows a geothermal pool area in Yellowstone National Park. The color is due to minerals and microbes growing in the warm water as it slowly trickles down the terraces at Mammoth Hot Springs.

- Diversity—***Diversity*** means that many different forms and types exist. There are more types of microbes than there are types of higher organisms. Many microbes have not been identified yet. If one organism does not seem to work well for the researcher, another that has been well studied can be used.
- Single Celled—Because each organism consists of one cell, it is possible to modify the organism by changing its cell. In higher organisms, all of the cells might have to be modified. This is impossible in a fully developed organism.

Because of these advantages, scientists often prefer to work with microbes. Of course, many biotechnology advances will be made on higher organisms. However, the details of the science often will have been developed using microbes.

TYPES OF MICROBES

Microbes are found in three of the five kingdoms: Monera, Protista, and Fungi. None are in the Plantae and Animalia kingdoms.

GROUPS OF MICROBES

Microbiologists often classify microbes into one of five major groups: protozoa, algae, fungi, bacteria, and viruses. These groups can be further classified.

Cell Structure

Fungi, protozoa, and algae are eukaryotic organisms. Bacteria are known as prokaryotic organisms. As you remember from earlier in the book, eukaryotic organisms (also known as eukaryotes) have a nucleus in the cell that contains the genetic material (DNA). Prokaryotic organisms (also known as prokaryotes) do not have a nucleus and their single DNA strand rests in the cell. Viruses do not have cell structures that live independently.

Most microbes are quite small. Some can be relatively large, such as a mushroom that may grow several inches high. Algae may grow long filaments that are easy to see without a microscope.

Need for Oxygen

Microbes can also be classified by their need for oxygen. An ***anaerobe*** is a microbe that does not require free oxygen for growth. Free oxygen is gaseous oxygen as found in the air. An ***aerobe*** is a microbe that requires free oxygen for growth. For example, some microbes may live in water where little or no free oxygen is present.

The actions of aerobic and anaerobic organisms in decaying materials, such as dead animals, vary. Aerobic organisms do not produce odors nor methane gas. Anaerobic organisms do produce odors and methane.

12-6. The composting of chicken litter using aerobic bacteria does not produce an odor. (Chicken litter is the manure, feathers, bedding, and other materials removed from a poultry house.)

The design of a biological process must consider odor emission. A good example is composting. Composting that uses aerobic microbes will not result in unpleasant odors. Another example is the treatment of animal wastes from a farm. Aerobic decomposition of manure will not produce odors. Anaerobic decomposition will often result in a highly objectionable odor.

Table 12-1. Size Comparisons of Selected Microbes

Microbe	Size Range*	Cell Nature
Protozoa	2-1,000 μm	Eukaryote
Algae	1 μm to a meter or more	Eukaryote
Fungi	2 μm to about 1 meter	Eukaryote
Bacteria	0.1-10 μm	Prokaryote
Viruses	0.01-0.25 μm	Acellular

*μ = micro (0.000.001 m or 10^{-6})

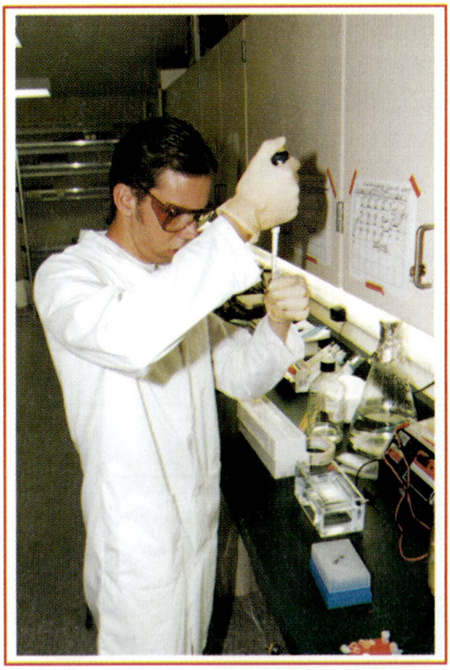

12-7. **Because microbes are all about us, care must be taken to prevent contamination of lab work.**

Found All Around

Microbes can be found just about everywhere. Each environment on Earth, no matter how slight the difference, has its own particular microbes that differ from others. Since microbes are so small, their environments are small.

Every day, new environments are being created on Earth for microbes. New volcanic islands or new lakes create new habitats. The use of pesticides, fertilizers, and detergents can change an area such that a new microbial environment is created. The release of industrial pollutants creates new environments.

Microbes are found in all oceans, lakes, and rivers. Yes, they are often found in the tap water of your home! Often these waters are kept clean by some forms of microbes. Some kinds of bacteria can be added to certain filter systems to clean the water. Some forms of algae, for example phytoplankton, add oxygen to the water and remove carbon dioxide. The oxygen in the water can be used by higher organisms, such as fish. Some microbes simply serve as food for higher organisms. For example, shrimp, oysters, and even whales feed on microbes. Harmful bacteria can also be found in water. The protozoa, *Giardia lamblia,* found in freshwater streams can make humans sick.

Microbes are found in the atmosphere. They do not reproduce in the air. Blowing through the air is a method for microbes to transport themselves from one place to another.

PROTOZOA

Protozoa are microbes in the Kingdom Protista. As unicellular organisms, protozoa have a wide range of sizes and appearances. Some protozoa have flagella; others do not. A ***flagellum*** is long, moving thread on a protozoan.

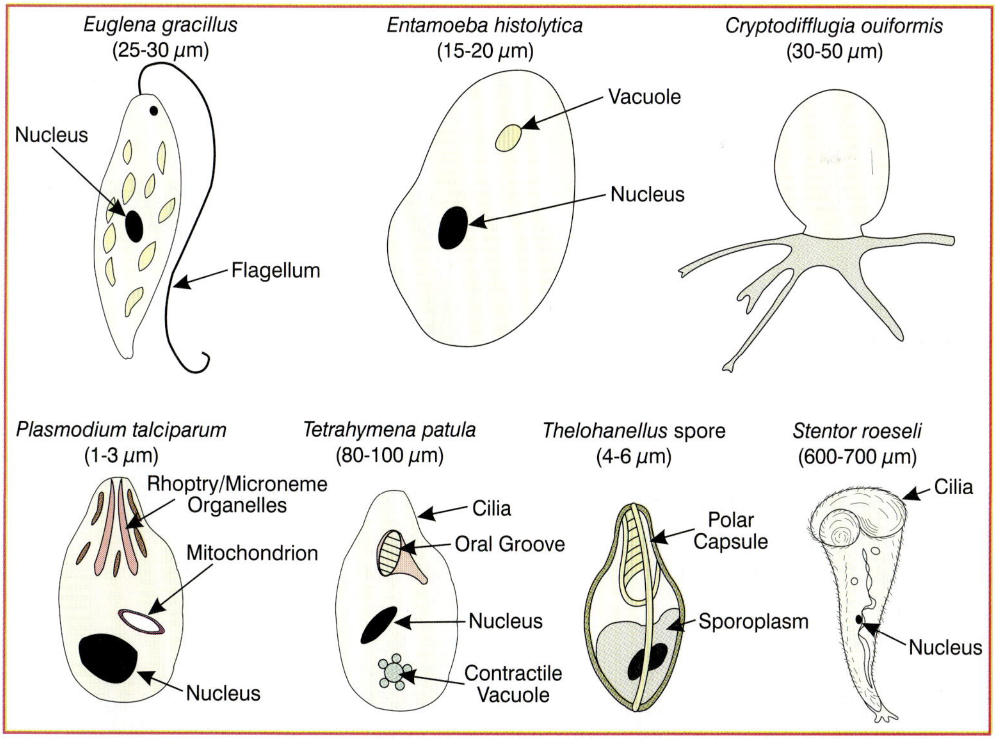

12-8. Examples of protozoa. (This figure helps compare the morphology of selected protozoa. Careful observation with a microscope may reveal the presence of protozoa in water, feces, or other specimens.)

Movement of the flagellum propels the protozoan through its liquid environment.

Some protozoans have cilia; others do not. ***Cilia*** are short threadlike projections from protozoa that move back and forth. The six phyla of protozoa are sarcomastigophora, labyrinthomorpha, apicomplexa, microspora, myxozoa, and ciliophora. Several of these are further classified into subphyla.

Protozoa are found in a wide range of environments. Temperature, moisture, and the presence of oxygen are three environmental factors that affect their presence.

Species of protozoa vary in reproduction. Some reproduce asexually; others reproduce sexually. Asexual reproduction is most common.

Protozoa are both beneficial and harmful. Some protozoa exist in symbiotic relationships with members of other kingdoms. ***Symbiosis*** is the close relationship between two different types of organisms for their mutual benefit. For example, some protozoa are found in the rumen of a cow. The protozoa help control the population of other microbes, such as bacteria and fungi,

Career Profile

MICROBIOLOGIST

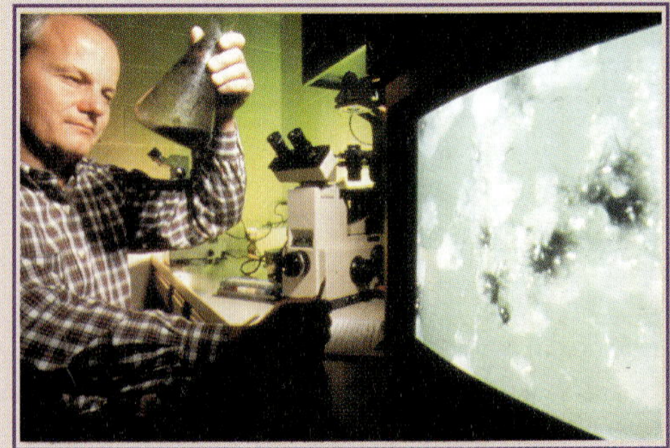

Microbiologists identify and study microbes. Their work includes identifying new microbes. They also study microbes that cause disease and environmental damage. They seek solutions to problems and strive to improve the environment. The chief tool of the microbiologist is the microscope. Microbiologists use microscopes to observe tiny microbes for research and identification. Many use light microscopes; others use electron microscopes.

Most microbiologists have college degrees in biology, biochemistry, microbiology, or related areas. Advanced positions may require a masters or doctors degree. Begin preparing to study the area in college by taking courses in science in high school.

Jobs for microbiologists are with private industries, medical facilities, government agencies, and with colleges and universities. This photo shows a microbiologist observing fungi. (Courtesy, Agricultural Research Service, USDA)

by feeding on them. In turn, the rumen provides a good environment for this particular species of protozoans.

Some protozoa cause disease. They are best known for having parasite-like effects on animals. A ***parasite*** is an organism that lives in or on another organism (host) and gains sustenance from its host. Blood and digestive diseases are often due to protozoa. In the intestines, some protozoa cause diarrhea. These protozoa are on the food that the organism eats. (Always wash uncooked food thoroughly!) Other protozoa may be spread by mosquitoes and cause diseases in the blood of humans and other animals. For example, the *Plasmodium malariae* is transferred by the *Anopheles* mosquito and causes a disease known as malaria in humans. Other diseases caused by protozoa include sleeping sickness, Chagas disease, and visceral leishmaniasis.

ALGAE

Algae are members of the Kingdom Protista. Most contain chloroplasts, which helps distinguish them from the protozoans. Green algae and brown algae are two examples.

Algae may form thalli, but they never form true tissues. A ***thallus*** (singular form of thalli) is an aggregate of filaments in plantlike form. Though

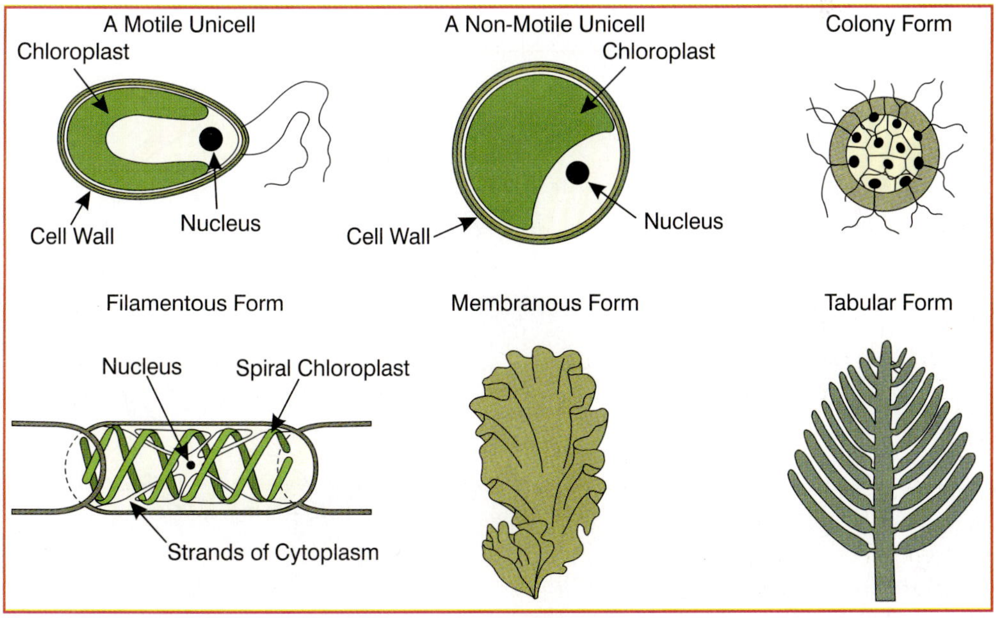

12-9. **Examples of algae forms.**

thalli may resemble plants, they are not plants. Seaweed is a good example. Some algae have flagella; others do not.

Some algae carry out photosynthesis. Most must have bright light for photosynthesis to occur; others can carry out photosynthesis in low light. We typically think that all photosynthesis requires bright light. That is not the case with some algae, such as species that exist deep in the ocean where light levels are low. Those that carry out photosynthesis under these conditions may have thalli near the surface that take in the energy of light.

Some algae reproduce sexually. Other algae reproduce asexually. For example, filamentous green algae reproduce sexually. Various stages of development are involved.

Algae have both beneficial and harmful effects. Benefits of algae include use of algae as food, manufacturing of substances from algae, and the production of oxygen by photosynthesis. Algae can be damaging to our environment. Large amounts of algae in lakes and streams may be toxic to other organisms. Water with large populations of algae and bacteria may suffer eutrophication. It usually results in water with high amounts of nutrients. An oxygen deficiency results. Other organisms cannot survive.

FUNGI

Fungi (plural form of fungus) are in their own kingdom–Fungi. Sizes of fungi vary from non-visible organisms to large organisms, such as mush-

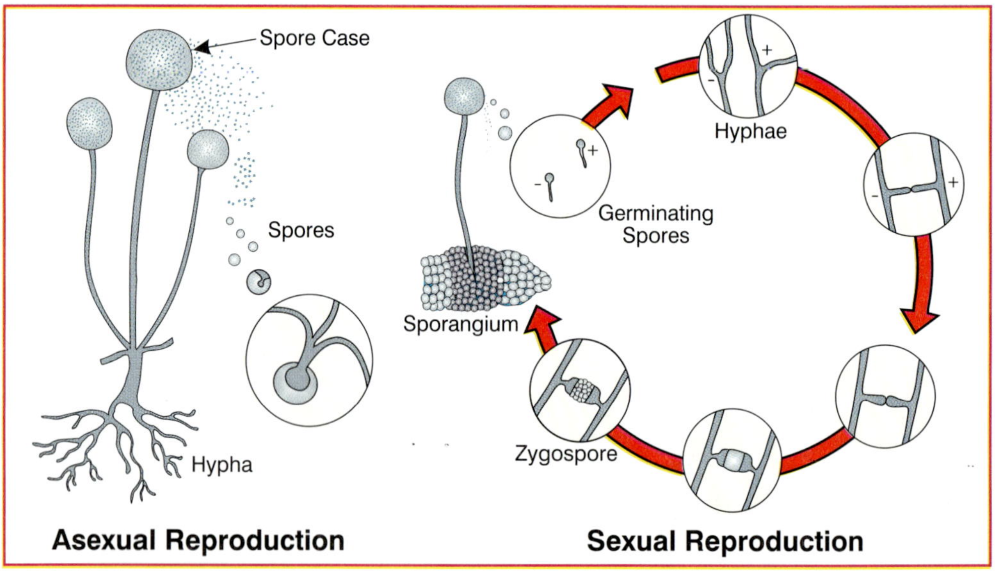

12-10. **Reproduction of the *Rhizopus* genus of fungi may be sexual and/or asexual.**

rooms. They have filamentous structures. The fine threads that form the filaments are known as hyphae. Fungi gain their food supply from other organisms or sources by absorbing it. An example is mold on bread that gains nutrients from the bread.

Fungi are both harmful and beneficial. Mold forms of fungi destroy foods. Mildew damages buildings. Parasitic fungi cause diseases, such as athlete's foot and ringworm. Useful products from fungi include edible mushrooms, yeasts that cause bread to rise or fermentation for alcohol, and lichen-forms that detect the presence of pollutants in the air. Fungi are also used in solving environmental problems.

Reproduction varies. It may be sexual or asexual, depending on the species and stage of growth. Some fungi form spores. A ***spore*** is a thick-walled dormant form of a fungus. Spores are easily moved about by air currents. They can come into contact with food and materials. Spores can cause human discomfort, such as allergies.

BACTERIA

Bacteria are important microbes in biotechnology. They are in the Kingdom Monera. Bacteria are found in large numbers in most areas of our environment. One small spoonful of soil may contain as many as 10 billion bacteria!

Bacteria are both useful and harmful. *Escherichia coli* is one of the most common species of bacteria found around humans. *E. coli* are found in the

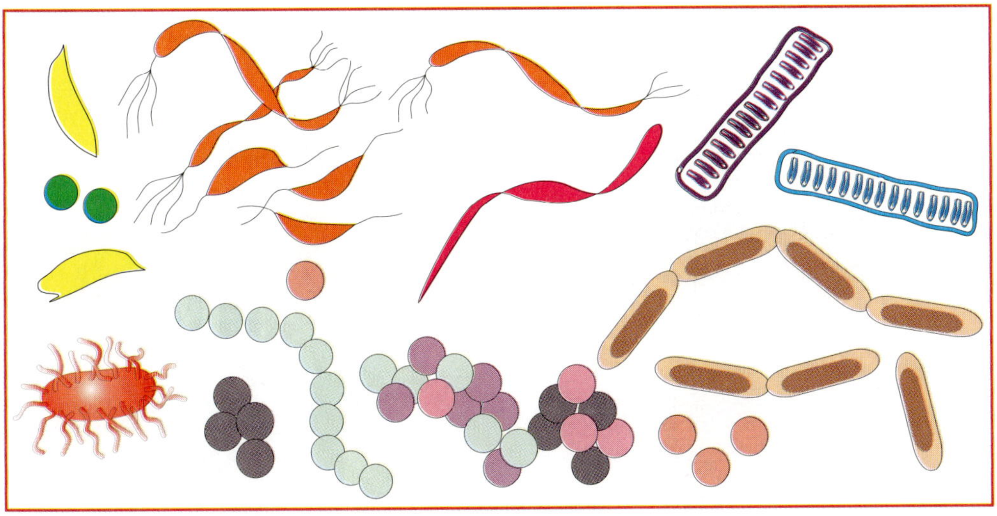

12-11. Examples of bacteria.

intestines of humans where they help digest food and provide vitamins. Unfortunately, some strains of *E. coli* can cause disease in humans also. A ***strain*** is a group within a species that has some distinguishing characteristics. The difference is slight, but it is enough to allow distinction.

Bacteria cause materials to decay and return elements to the soil. Bacteria break down dead plants and animals into forms that can be used again by new plants and animals. In the soil, some types of bacteria will convert nitrogen from the air into a form that can be used by plants. This reduces the need for fertilizer in some situations.

Other sections of this chapter cover the study and use of bacteria in more detail. Chapter 11 introduced bacteria as vectors, particularly *E. coli*.

VIRUSES

Viruses are quite small, acellular particles. Acellular means that they are not cells. Their tissue or matter is not divided into cells. They do not carry out life processes. Association with another living system is essential.

Viruses are sometimes said to be incomplete organisms. They contain genetic material and proteins. Viruses are not capable of reproducing by themselves. Viruses are useful in biotechnology to introduce DNA into cells. Electron microscopes are used in studying viruses.

Viruses contain RNA, lipids, DNA, proteins, and other structures and materials. Figure 12-12 shows line drawings of two virus shapes.

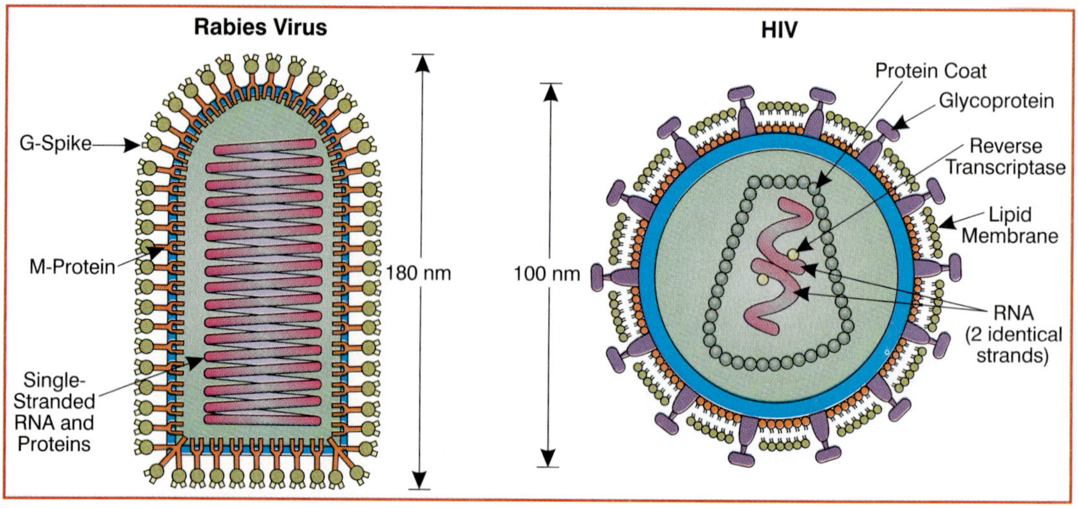

12-12. **Two virus shapes as shown by the rabies virus and the human immunodeficiency virus (HIV).**

MICROBE IDENTIFICATION AND MORPHOLOGY

The identity of microbes is often needed. Their identity is needed to treat a disease, analyze a symptom, or identify a source of contamination. Microbes are identified by shape, cell wall structure, flagella, growth habits, and reactions to specialized media and staining agents.

12-13. A microscope is the main tool used in microbe study.

BACTERIAL IDENTIFICATION

Some 5,000 different species of bacteria have been identified. The distinguishing characteristics are often based on the shape of the cell and the colonies that bacteria form. The shapes are spherical (cocci), rod (bacilli), spirochaetes (tightly coiled), vibrio (comma-shaped), spirillum (long, curved rods), appendaged (cells with stalks or tube-like extensions), and plemoric (other varying shapes). The cocci-shaped bacteria are typically *Streptococcus* and *Staphylococcus* genera. Single-rod shaped bacteria are *Pseudomonas* genus.

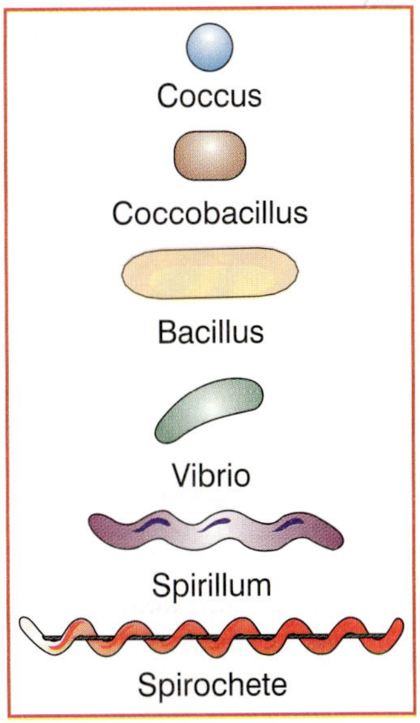

12-14. Bacteria shapes vary widely. Some form colonies and others have flagella.

Microscope Slide Preparation

Identifying bacteria requires the use of a microscope. In order to view bacteria with a microscope, a microscope slide must be prepared. In some cases, a culture of the bacteria may need to be grown. (Growing cultures is covered in the next section of this chapter.)

Here are the steps to follow in preparing a slide:

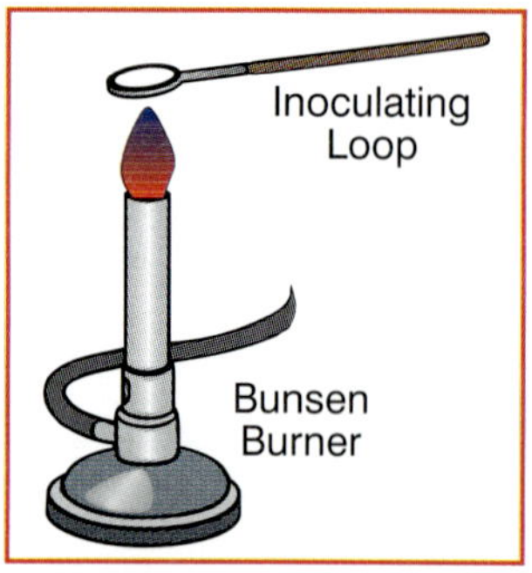

12-15. Passing an inoculating loop through a flame destroys microbes that could contaminate work.

1. In preparing a slide, use a glass or plastic slide and follow these steps. (Note: Be sure to follow all safety procedures.)
2. Prepare a microscope slide by placing a drop of water in the center of the slide.
3. Flame the inoculating loop to rid it of other microbes.
4. Remove a sample of the cultured bacteria from an isolated colony. Lift the lid on the culture dish just as little as possible to prevent contamination.
5. Smear the water with the inoculating loop exposed to the bacteria.
6. Flame the inoculating loop.
7. Allow the slide to air-dry.
8. "Fix" the slide by passing it through flame briefly.
9. Allow the slide to cool.

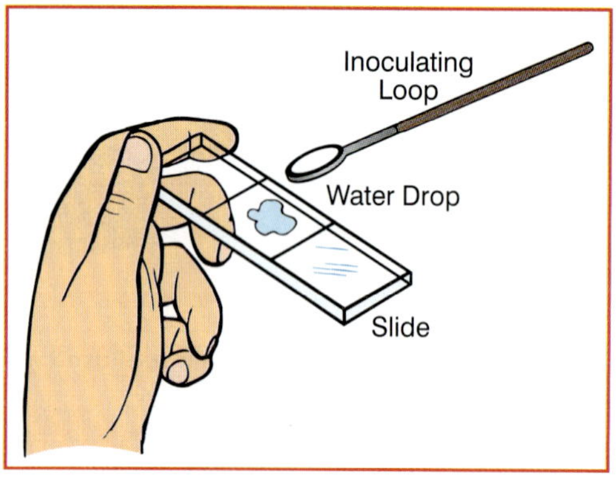

12-16. Using an inoculating loop to prepare a slide.

Gram Staining

In the presence of different chemicals, bacteria will stain a different color,

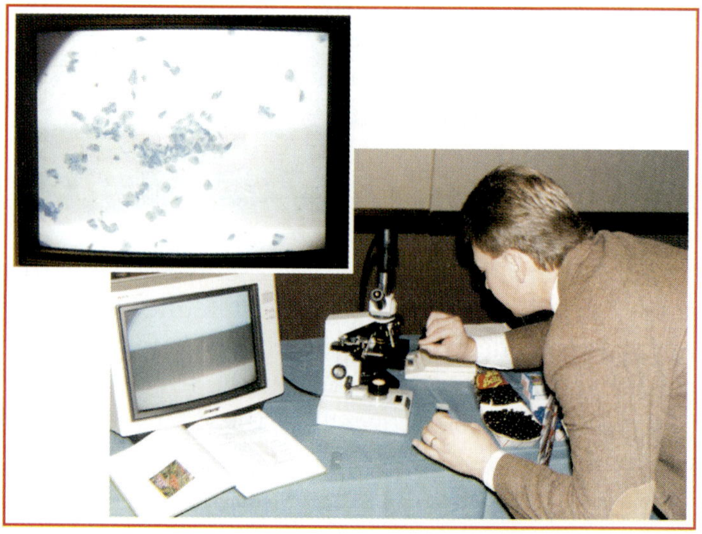

12-17. Using video microscopy to view microbes. (A video camera presents the microscope image on the screen.)

based on their identity. A common test used to identify bacteria is the ***gram stain***. Gram staining separates bacteria into two large groups: Gram-negative bacteria, which will stain pink, and Gram-positive bacteria, which will stain blue. The test also allows for ready visual analysis of the morphology of bacteria.

Here are the steps to follow in gram staining a slide: (Note: Be sure to follow appropriate safety procedure to protect the eyes and skin from exposure.)

1. Prepare the slide as described above.
2. Flood the sample with Hucker ammonium oxalate crystal violet for 60 seconds.
3. Rinse the slide with tap water.
4. Flood the slide with Gram's iodine solution for 60 seconds.
5. Rinse the slide with tap water.

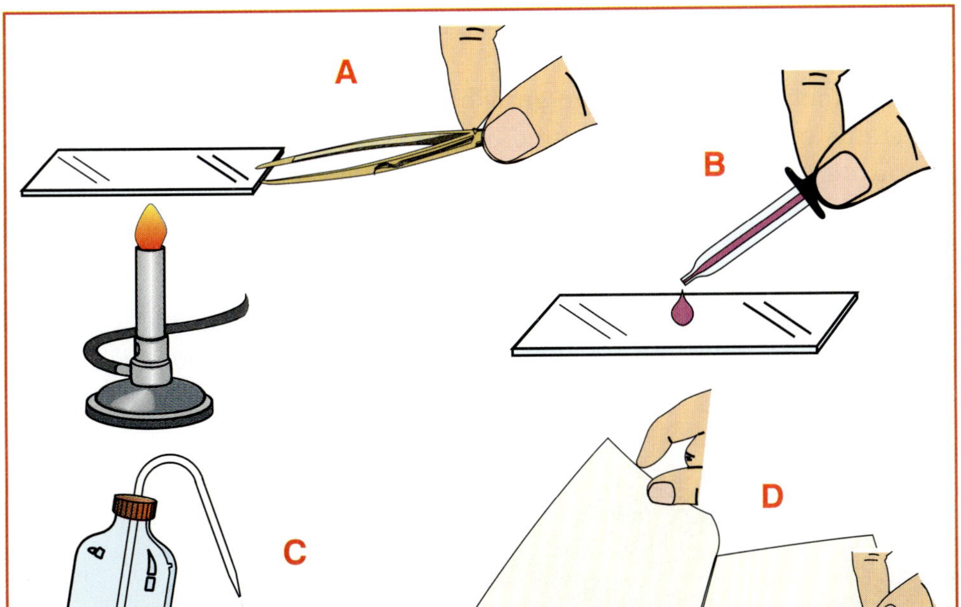

12-18. **Staining a slide with a bacterial smear involves: (A) pass a slide you have prepared (smear side up) through a flame; (B) add stain and let it stand for one minute; (C) gently rinse with water; and (D) carefully blot it between two paper towels.**

6. De-stain the sample area on the slide with 95 percent ethanol. Rinse with ethanol until the runoff is clear.
7. Rinse the slide with tap water.
8. Flood the sample area with safranin for 60 seconds.
9. Rinse with tap water
10. Blot carefully with blotting paper or a clean paper towel.
11. Observe the sample with the microscope and, if available, an oil immersion lens.
12. Compare the bacteria or other organisms with drawings or prepared specimen slides of known species. (You may wish to draw the different organisms that you see.)

Bacteria can be further analyzed by culture and growth on species-specific media.

BACTERIA CULTURE

It is sometimes necessary to grow bacteria in a lab. Preparing slides and gram stains often requires cultured bacteria. Growing bacteria is known as ***bacteria culture***.

Media

Bacteria are cultured in a medium. A ***medium*** is a substance that provides the needed environment, including nutrients, for the growth of the bac-

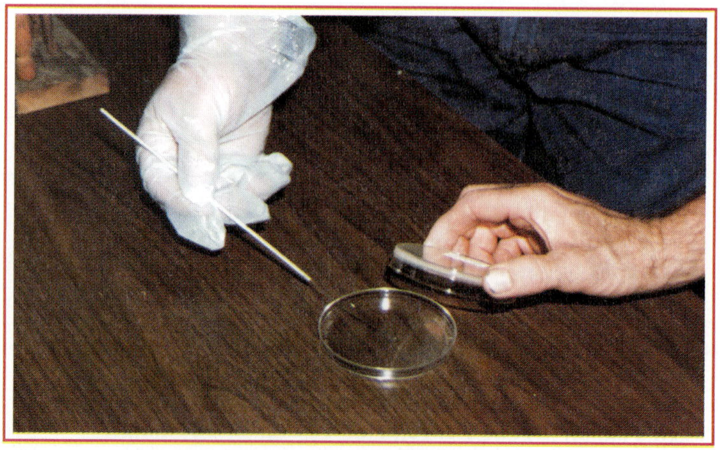

12-19. Inoculating agar in a petri dish.

teria. It is also known as culture medium. Media (the plural form of medium) may be liquid or solid. Media are selected because of ingredients that promote growth of the desired species. Liquid medium is often broth that has a high nutrient content. Solid medium includes agar. ***Agar*** is a gelatin-like medium that is prepared using hot water. As it cools, it becomes gelatin-like. Seaweed, blood, and other substances may be used to create the desired agar. Blood is often an additive when culturing *Streptococcus* bacteria.

Vessels in which liquid media are used are known as fermenters. Solid media are typically used in petri dishes. All vessels used for culture should be sterile before beginning the culture process. The culture must be kept at the proper temperature. Oxygen must be available. Depending on the species of bacteria, pH and water activity are important.

The medium should be carefully placed in the vessels to prevent contamination.

Isolation

Once a medium has been prepared, it must be inoculated with the desired bacterium or other microbe. Here is a general procedure for isolating individual bacteria colonies: (Note: Follow all appropriate safety procedures. Use eye and hand protection.)

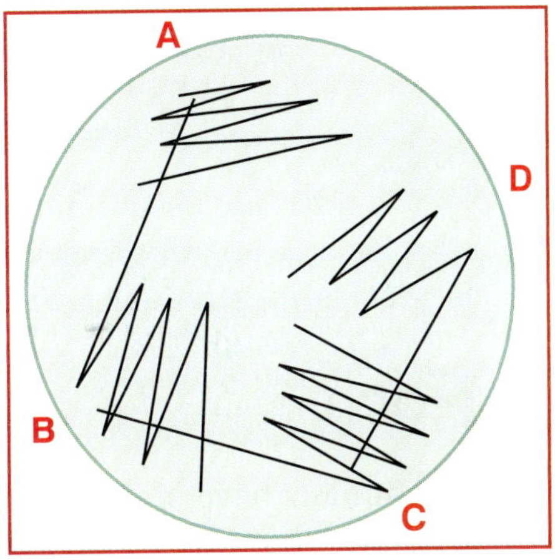

12-20. **Steps in streaking a plate.**

1. Hold the sample in one hand and flame an inoculating loop. Let it cool, and, with the other hand, take a small sample of the organism to be studied.
2. Raise the cover of the petri plate to be used for isolation just enough to allow access to the inoculating loop. Streak the plate according to the diagram in Figure 12-20.
3. At area A, streak the plate in a zigzag pattern, flame the inoculating loop, and let it cool.
4. Turn the plate and make one streak from area A to area B, using that streak to make a zigzag pattern in area B. Flame the inoculating loop and let cool.
5. Turn the plate to area C, using that streak to make a zigzag pattern. Flame the inoculating loop and let it cool, and repeat procedure from

area C to area D. Always flame the inoculating loop and cool before putting it away.

6. Leave the samples sealed, upside down, and at room temperature.

Isolated colonies should develop in 24 to 48 hours. Bacteria are then identified based on their morphology.

FUNGI AND YEASTS

Spore-producing yeasts and fungi can be cultured and isolated in much the same manner as bacteria. However, fungi and yeasts are identified by their growth on specific media. Fungi and yeasts that do not produce spores can be cultured by taking small, solid samples and pressing into culture media.

MICROBES IN BIOTECHNOLOGY

Microbes are used in many ways. Some of the uses are based on the natural roles of a microbe. Newer uses are in biotechnology where they are used as vectors in genetic engineering.

FOOD

Humans have often used microbes in producing food. Historically, common uses were with bread, yogurt, and cheese. Today, humans use microbes to produce new materials. Modern biotechnology began with the use of microbes to produce antibiotics and drugs.

Microbes were the first individual cells to be studied. We know more about their biochemical processes than in higher organisms. A ***biochemical process*** is the series of reactions occurring within an organism that enables the production of specific compounds or energy. Every living organism carries out biochemical processes.

BIOTRANSFORMATION

Biotransformation is a process that promotes the production of chemicals by microbes. Industry uses specific compounds in a fermentation process. These are added to a fermenter (large container) so the microbes produce commercially valuable compounds.

12-21. Using a bioreactor (fermentation vat) to study bacteria. (Courtesy, Agricultural Research Service, USDA)

AGRICULTURE

Agriculture uses microbes in many ways. Some uses are a natural part of the environment. Other uses involve genetic engineering.

Microbes can be used to protect plants from cold temperature and pests. Scientists have developed bacteria that will discourage the formation of ice crystals under freezing conditions. The formation of ice crystals on plants at freezing temperature will kill or damage the plant. The bacteria that normally live on plants will encourage the formation of ice crystals at freezing temperatures. But the new bacteria have been changed using biotechnology so they do not produce ice crystals until much colder temperatures.

Some microbes are natural pesticides. Over the years, chemical pesticides have received a bad reputation. Some are said to be unhealthy for some animals. The pests sometimes develop resistance. Some pesticides, such as DDT (dichlorodiphenyltrichloroethane), have been banned from use after being determined unhealthy for the environment. Microbes for the control of pests are being developed. These microbes are degradable and pest specific.

INDUSTRY

Microbes, specifically bacteria, are used by industry and research to produce enzymes for the production of certain products. Enzymes are compounds produced by organisms that can be used to create specific chemicals, foods, and pharmaceuticals. These enzymes can also be used to destroy toxic materials. The U.S. Army is using biotechnology to isolate and produce enzymes that can destroy chemical warfare agents and protect soldiers and civilians.

12-22. **Modern wastewater treatment, such as this city system, uses bioremediation processes.**

ENVIRONMENTAL CLEANUP

Biotechnology research is also being done to produce microbes that convert wastes into fuel. Materials that would otherwise be discarded and would create a waste problem might be utilized to generate fuel.

Cleaning up contaminated areas is one of the promising future uses of biotechnology. Microbes can be used to cleanup accidents and spills. Bacteria have been developed through biotechnology that can be used to degrade oil spills on water.

Some people feel that biotechnology will produce products to help remove a variety of waste materials, such as toxic metals, plastics, and pesticides. This cleaning-up process is known as ***bioremediation.*** It is a method that uses biotechnology and microbes to reclaim a contaminated site.

The slow degrading of plastics is a problem in the environment. Many plastics degrade very slowly. They remain in the environment for a long time. Some plastics produce toxic materials or kill animals by choking, suffocating, or drowning them.

Currently, scientists are studying organisms that might break down plastics into friendly compounds or produce plastics that are biodegradable within a short time. ***Biodegradation*** is the natural process by which microbes break down a product to environmentally friendly or useful compounds. The process occurs in nature all of the time. An example is the degrading of grass clippings from mowing a lawn. The clippings are attacked by microbes and converted to organic matter and other substances.

REVIEWING

MAIN IDEAS

Microbial biotechnology is the study of microbes. Microbes can be found all around our environment and all over the earth. There are more types of microorganisms on Earth than there are higher organisms. The fundamental tools and techniques of biotechnology were developed using microbes.

The five major types of microbes are protozoa, algae, fungi, bacteria, and viruses. Bacteria are the most important in biotechnology. This is because of their cell structure and large numbers available with different characteristics.

The study of microbes involves their identification. Shapes, colors, environment, and other factors are used. Some are cultured to assure that strains have been isolated for study. Microscope slides are prepared using gram staining procedures.

Microbes are used in industry to produce pharmaceuticals, antibiotics, and chemicals. Biotechnology can be used to produce microbes that help cleanup soil and water through bioremediation. Agriculture uses microbes to protect plants from cold temperatures and pests.

QUESTIONS

Answer the following questions using correct spelling and complete sentences.

1. What is microbiology?
2. What is a microorganism?
3. How is penicillin important to medicine?
4. Name and describe the four advantages of using microbes in biotechnology instead of using higher organisms.
5. Name the five major groups or types of microorganisms and list the kingdom that each is in.
6. How is a microscope slide prepared to study bacteria?
7. What is gram staining? Why is it used?
8. Describe the difference between eukaryotes and prokaryotes.
9. What is a culture medium? What two major kinds are used?
10. What is the process of streaking growing medium?
11. Name two ways that biotechnology is using microbes in agriculture.
12. How are microbes used to cleanup the environment?

EVALUATING

Match the term with the correct definition. Write the letter of the term in the blank that is provided.

a. microbiology
b. anaerobe
c. aerobe
d. symbiosis
e. thallus
f. strain
g. spore
h. gram stain
i. agar
j. medium

_____ 1. A gelatin-like medium prepared using hot water and poured into a petri dish or other container for culturing microbes.

_____ 2. A test used to identify bacteria.

_____ 3. The study of microbes.

_____ 4. A substance that provides the needed environment for growing bacteria in a laboratory.

_____ 5. An organism that can live without available free oxygen.

_____ 6. A relationship of mutual benefit between two organisms.

_____ 7. A filament-type structure that may give algae a plant-like appearance.

_____ 8. A group within a species.

_____ 9. An organism that requires free oxygen to live.

_____ 10. A thick-walled dormant stage of fungi.

EXPLORING

1. Use the World Wide Web and school library to identify microbe products that you use every day. Prepare a written report on your findings.

2. Take samples of water from different locations, such as ponds or streams, in your area. Prepare a slide and use a microscope to observe the microbes. Identify the organisms that you see.(It might be advantageous to also view prepared slides from a biological supply company. "Fixed" microbes can be purchased on the microscope slide for best viewing.) (Note: Be sure to follow all safety procedures with this activity.)

3. Prepare bread using yeast and without yeast. What do you see as the differences in the two breads? Also, visit a local bakery and ask someone to discuss how bread is prepared and what type of yeast is used.

13

PLANT BIOTECHNOLOGY

Just how important are plants? We sometimes fail to remember what we get from plants. Plants are the main sources of the food we eat, the fiber for the clothes we wear, and the pulp for the paper and wood we use.

Plants are efficient providers. They use energy from the Sun and manufacture food for themselves. This food is stored in leaves, roots, fruit, stems, seed, and other plant parts. When we harvest crops, we collect what the plants have manufactured.

Plants are now being engineered to produce more than food materials for themselves. They are producing pharmaceutical products and chemical compounds. In addition, ornamental horticulture and landscaping are enormous industries. There are millions of plant products. Every person depends on plants for existence.

13-1. **A geneticist is transferring selected alfalfa DNA clones in the process of making genetic markers. (Courtesy, Agricultural Research Service, USDA)**

OBJECTIVES

This chapter covers how plants are used in biotechnology. The techniques of biotechnology in plant research and development are included. It has the following objectives:

1. Outline the plant breeding process.
2. Explain the use of genetic engineering to create transgenic plants.
3. Describe the process of plant tissue culture.
4. List and describe examples of plant products created through genetic engineering.

TERMS

artificial selection
crossbreeding
cultivar
explant
fungicide
herbicide
hybridization
insecticide
line breeding
mutation
parent plant
plant breeding
plantlet
plant tissue culture
taxol
weed

CREATING NEW PLANTS WITH TRADITIONAL BREEDING

People have been trying to make plants better for centuries. Some people have specialized in plant breeding. Today, biotechnology is being used to develop new cultivars.

A ***cultivar*** is a cultivated variety of a crop plant. Cultivars are distinguished by minor differences, such as length of time required to reach maturity. A number of new technologies have been combined to develop plant cultivars for the 21st century.

13-2. Unwanted breeding of corn plants has been eliminated by covering the silking shoots. The tassels (at the top of the plants) produce pollen that fertilizes the silks that form on the female shoot.

PLANT BREEDING

Plant breeding is the improvement of plants by selectively breeding chosen individuals. It has been widely used with crops and ornamental plants. The goal of plant breeding is to develop plant cultivars that have new or improved traits over the cultivars that already exist.

Here are a few techniques in plant breeding:

- Line Breeding—***Line breeding*** is breeding offspring (successive generations) among themselves. This approach is used to maintain desired qualities. The generations are mated enough times to be fairly certain that the desired trait

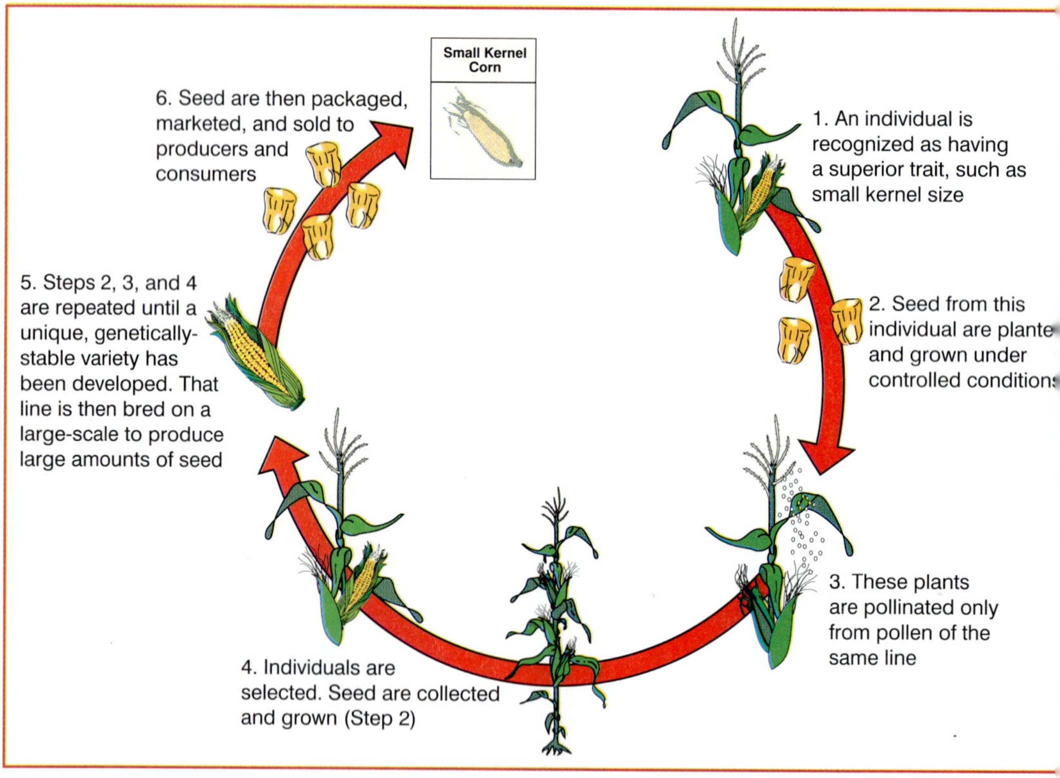

13-3. **An example of the process of line breeding to obtain a desired trait, such as small kernel siz**

will be expressed in future generations. Line breeding is sometimes said to be self-crossing.

- Crossbreeding—***Crossbreeding*** is breeding individuals of different varieties or species. The goal is to have offspring with desired qualities.

Traditional plant breeding is based upon two strategies: artificial selection and hybridization.

SELECTION

Selection occurs naturally or may be promoted by human effort. ***Artificial selection*** is the process of choosing individuals from each generation that possess a superior quality. Qualities vary. All selection is typically within a species. Selection may be to improve the size of fruit, quantity of production, sturdiness of stalk, or other characteristics. Line breeding is used

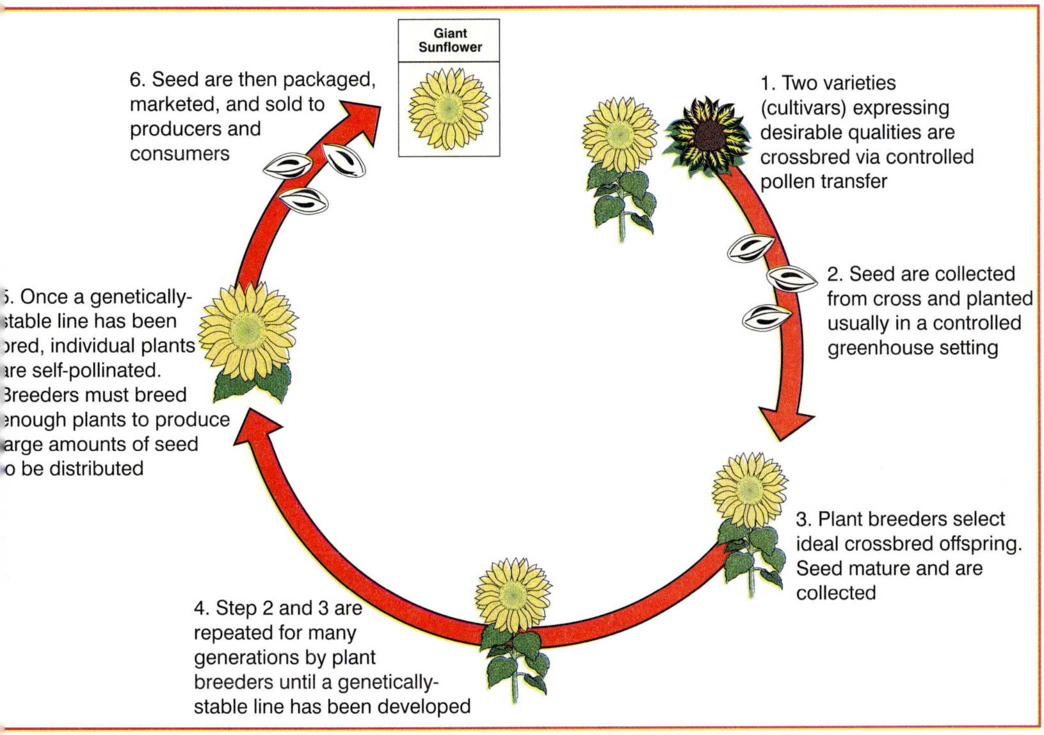

3-4. An example of the process of traditional plant breeding using crossbreeding to obtain a gi-nt sunflower.

with those that are selected. This is breeding the selected individuals to one another.

The result of selection and line breeding is multiple offspring with the desired superior qualities. This breeding process is continued until the offspring possess the desired trait. The seed is then marketed as a unique cultivar.

13-5. Petunias are popular ornamental plants that have been improved through hybridization.

HYBRIDIZATION

Hybridization is the process of breeding two individuals from distinctly different varieties. This is also known as cross-pollination. This technique results in a condition known as "hybrid vigor," meaning the offspring may exhibit greater yield, height, disease resistance, or other desirable, "healthy" traits. Again, offspring possessing the desired trait will be line bred to create a new plant variety. The offspring of hybridization are known as hybrids.

Of all hybrid crops, hybrid corn is probably the best known. Many years were required for hybrid corn varieties to be widely planted. Some people were reluctant to plant the corn because of fear or lack of awareness. Planting hybrid corn had some of the same resistance by growers as genetically modified corn has today.

CREATING NEW PLANTS WITH GENETIC ENGINEERING

While traditional plant breeding techniques remain important, technology is making an impact. No longer do plant breeders rely just on visual observations. They can study the genetics of plants using DNA profiling.

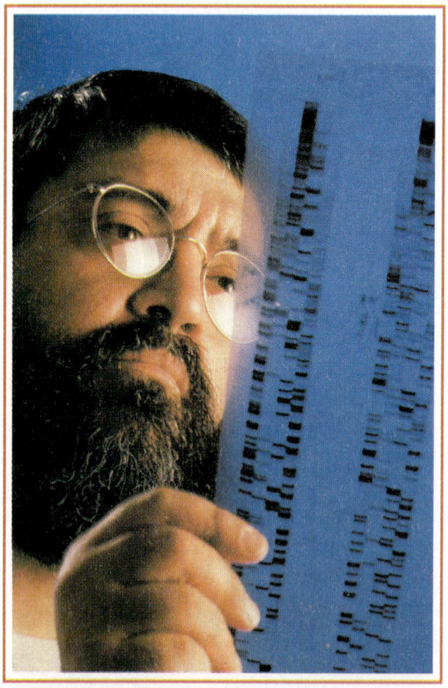

13-6. Studying the DNA profile of a plant. (Courtesy, Agricultural Research Service, USDA)

DNA PROFILING

DNA profiling (molecular markers) can be used to gain genetic information about plants that may be trait carriers. The DNA reveals genetic differences in plants with desired traits and those that do not have the traits. This information serves as the basis for selecting which plants to breed.

As breeding begins, testing is done to screen plants for the desired genetic makeup. Continued selection results in plants with similar molecular sequences being bred. After several generations, the desired trait is stable and appears in every individual offspring.

GENETIC ENGINEERING

Genetic engineering has been used to modify many plant species. DNA profiling,

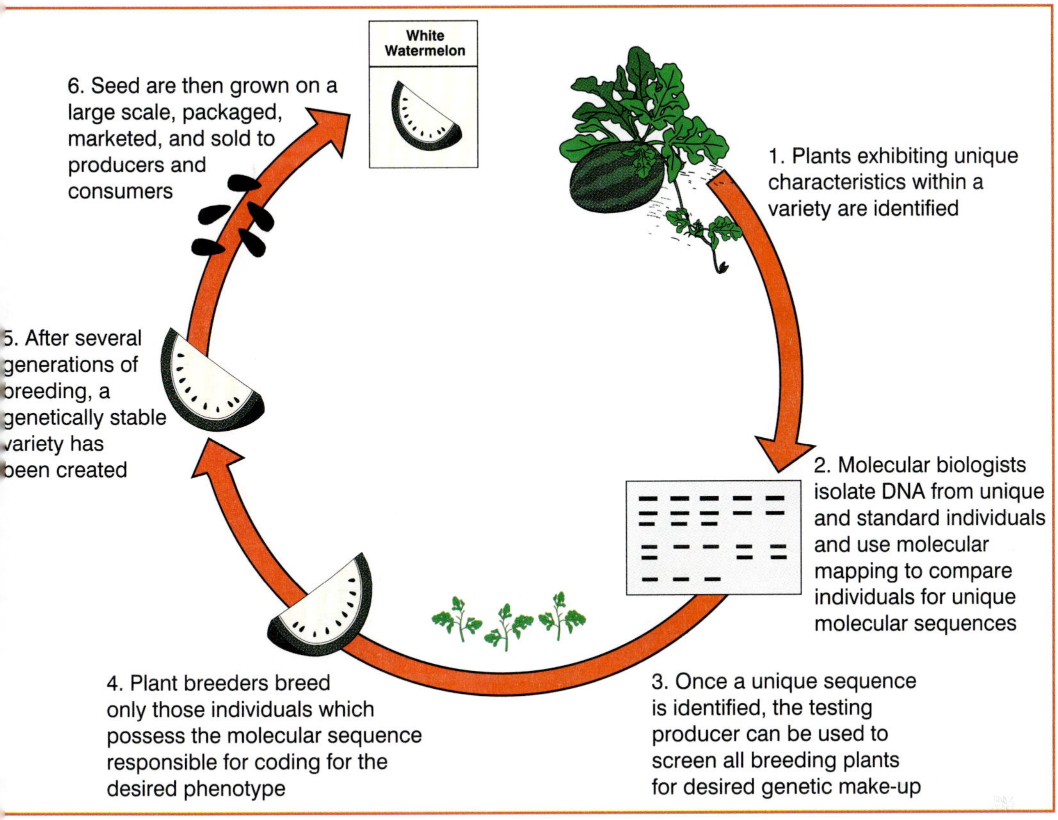

13-7. The process of plant breeding using a DNA profile. (In this example, the goal is to produce a white watermelon.)

selection, and similar traditional approaches do not always result in cultivars with the desired traits. Genetic engineering can be used to achieve the desired cultivar traits. Molecular technologies are used to "find" genes that can be incorporated into plants via genetic engineering.

Three approaches are used in plant genetic engineering. Several other approaches hold future potential. The three most widely used are:

- *Agrobacterium*-mediated gene transfer—This technique involves using bacteria as vectors in the change of genetic material. The *Agrobacterium tumefacians* bacterium naturally readily transfers its genetic material into some plant cells as a disease known as crown gall. Biotechnologists have captured this natural trait and have used it in genetic engineering. *A. rhizogenes* is also used in *Agrobacterium*-mediated transfer. Chapter 11 presents this approach in detail.

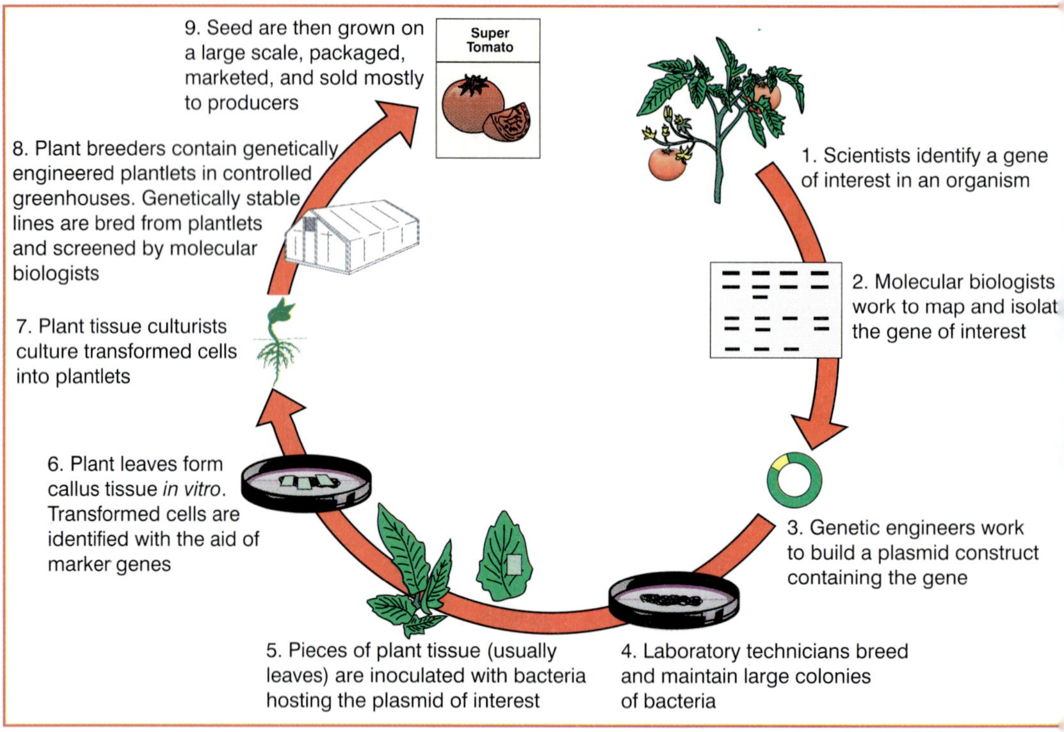

13-8. An example of the process of developing a cultivar using the *Agrobacterium* transformation approach in genetic engineering.

- Ballistic gene transfer—This technique uses a gene gun that uses projectiles as carriers of DNA into recipient cells. Chapter 11 presents this approach in more detail.

13-9. Ballistic gene transfer has been used to create transgenic barley.

13-10. **Direct gene, *Agrobacterium*-mediated, and ballistic gene transfers have all been used to create transgenic tobacco.**

- Direct gene transfer to protoplasts—Enzyme action is used to remove the walls of the intended recipient cells. Once the walls are removed, the remaining materials are protoplasts. These protoplasts can be incubated with nucleic acids having the desired traits. The nucleic acid permeates the membranes surrounding the protoplasts. Gene transfer occurs between the nucleic acid carrying isolated plasmid DNA and the DNA in the protoplasts. This technique occurs when polyethylene glycol is present.

Biotech Connection

PIZZA TOMATOES

Have you had a good pizza recently? What made it good? You might have had a variety of toppings. The tomato sauce was probably one of the ingredients that was unnoticed. If it had been missing, you would have known it!

Pizzas have a thick tomato sauce and spices. Tomatoes from the field must be processed to remove skin, seeds, and most of the water content. Like people, water makes up more than two-thirds of a normal tomato's composition.

Scientists have worked to genetically engineer a tomato that has a higher content of solids (the fleshy part of the tomato). This results in the need to use fewer tomatoes to make every can of tomato or pizza sauce. You benefit in the form of lower costs for everything that uses tomato sauce: ketchup, pizza sauce, salsa, spaghetti sauce, and many others. This photo shows a tomato breeder inspecting a new crop.

Table 13-1. Examples of Genetically Modified Plants*

Common Name	Scientific Name	Method Used**
Alfalfa	*Medicago sativa*	AM
Apple	*Malus pumila*	AM
Barley	*Hordeum vulgare*	BM
Canola	*Brassica napus*	BM
Carnation	*Dianthus sp. Caryophyllus*	AM
Carrot	*Daucus carota*	AM
Cotton	*Gossypium hirsutum*	AM
Corn	*Zea mays*	DP
Cucumber	*Cucumis sativus*	AM
Kiwi fruit	*Actinidia chinensis*	AM
Lettuce	*Lactuca sativa*	AM
Oat	*Avena sativa*	BM
Peanut	*Arachis hypogaea*	AM
Petunia	*Petunia hybrida*	DP
Potato	*Solanum tuberosum*	AM
Rice	*Oryza sativa*	AM
Soybean	*Glycine max*	AM and BM
Strawberry	*Fragaria X Anannassa*	AM
Sugar beet	*Beta vulgaris*	AM
Sugarcane	*Saccharum officinarum*	BM
Tobacco	*Nicotiana tabacum*	AM, BM, and DP
Watermelon	*Citrullus vulgaris*	AM
Wheat	*Triticum aestivum*	BM and DP

*These are but a few examples. More complete lists can be found in references listed in the bibliography of this book.

**Method used is as follows: AM=*Agrobacterium*-mediated transformation; BM=ballistics-mediated transformation; and DP=direct gene transfer to protoplasts.

Once a plant has been genetically engineered and propagated, a new variety has been created. Plant tissue culture is used as a tool in genetic engineering. It can be used to asexually propagate large numbers of unique plants.

13-11. **Both *Agrobacterium*-mediated and ballistic gene transfers have been used to create transgenic peanuts.**

PLANT TISSUE CULTURE

Plant tissue culture is used to clone new plants. It is also used in some approaches in producing transgenic plants. It does not create transgenic plants, but is a tool to reproduce plants once they are transgenic. Plant tissue culture is also known as micropropagation.

THE PROCESS

The process of tissue culture involves taking one or a few cells and producing another organ or organism. With plants, small tissues can be grown into plants with the characteristics of the species. Cell differentiation occurs as the tissue develops into a new plant. The plant that develops will be identical in genotype to its parent.

The general procedure is plant tissue culture is:

1. Prepare an aseptic environment for the work.
2. Small pieces of plant material known as ***explants*** are removed from the parent plant. A ***parent plant*** is the plant that is the source of explants. Carefully use a sharp scalpel in preparing the explants.
3. The explants are cleaned and disinfected with a mild bleach solution. This removes bacteria and fungi that could cause disease and interfere with the culturing process.

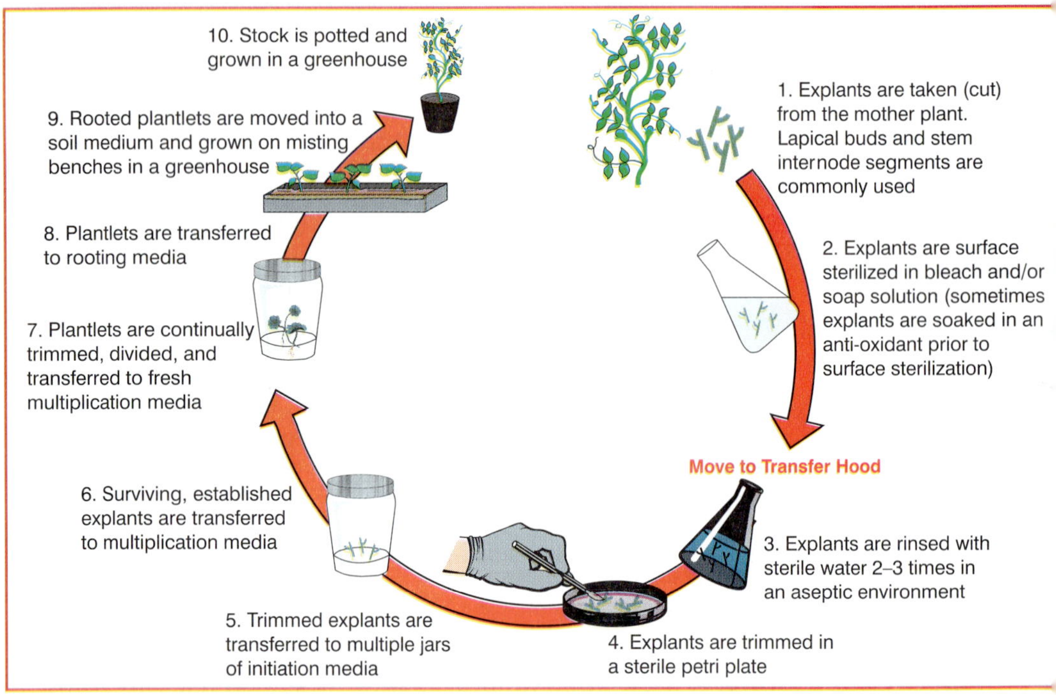

13-12. **The general process of making plant tissue cultures in mass micropropagation.**

4. Explants are trimmed to the desired size in a sterile petri dish.
5. The trimmed explants are placed in jars or test tubes with an appropriate medium. The jars or tubes are sealed with covers that allow some gas exchange.
6. The jars or tubes with explants are placed in an area for growing. The temperature and light should be ideal for plant growth.
7. The first cell that forms from an explant is a callus. A callus is an undifferentiated mass of cells. Some kinds of media promote cell differentiation. (In some cases, only genetically engineered plants go through a callus stage. Plants that are mass-micropropagated do not always go through a callus stage. Inducting the callus on plant tissue dramatically increases the chances of a mutation occurring. A ***mutation*** is a genetic variation that occurs naturally and without prediction that it will occur.)
8. As the callus grows (in some cases the explant), a plantlet is formed. A ***plantlet*** is a plant with tiny leaves, stems, and roots that have not developed into the normal-size parts of the parent plant.
9. Plantlets are transferred to a rooting medium.

13-13. Explants are being trimmed from a cutting taken from a parent plant.

13-14. An explant is being placed in a sterile jar with agar for growth.

13-15. Plantlets have formed in these test tubes. (Courtesy, Agricultural Research Service, USDA)

Career Profile

PLANT BREEDER

A plant breeder is responsible for developing new and improved plant varieties. Most plant breeders work with horticultural and agricultural crops. Although, plant breeding incorporates a number of strategies, most work is completed in a greenhouse or field setting. Some plant breeders travel searching for new species to incorporate into a breeding program and/or overseeing operations in other countries.

There are many plant breeders working today who entered the field with an undergraduate degree in plant breeding, plant science, or other related area. However, most people now entering this career have completed a related graduate degree program.

Jobs are with seed companies, government agencies, horticultural breeders, universities and colleges, and research stations. This photo shows a grain sorghum breeder inspecting grain (seed) heads that have been covered to protect from unwanted pollination. (Courtesy, Agricultural Research Service, USDA)

10. Plantlets with roots are transferred to a soil or other medium in a greenhouse.

11. Once the plantlets have grown roots and normal-size leaves and stems, they are transferred to a nursery or field for additional growth.

12. When adequate size has been reached, the plants are marketed or used in production.

THE MEDIA

The media used with different species of plants varies. Media may also vary for every stage of growth. Most media are synthesized growing substrates that promote tissue growth. Although there are hundreds of proven types of plant tissue culture media, most are composed of seven basic ingredients:

- Water—clean, purified
- Macronutrients—nitrogen, phosphorus, potassium
- Micronutrients—manganese, iron, boron, calcium, copper, zinc, etc.
- Vitamins—thiamine, folic acid, and others
- Organic Carbohydrate Source—sucrose, fructose
- Plant Growth Hormones—auxins, cytokinins, and gibberellins
- Gelling agent—Most media contain a gelling agent, such as gelatin.

Media and culture vessels are autoclaved. Explants must be surface sterilized prior to transfer onto media. All micropropagated plants must be cultured using appropriate aseptic technique.

WHY PLANTS ARE GENETICALLY ENGINEERED

Genetic engineering is used to improve the natural capacity of a plant. In some cases, it is used to create plants that are not damaged by the use of herbicides and other chemicals during crop growth.

INSECT AND DISEASE RESISTANCE

Losses from insects and diseases can be devastating to crop producers. Due to concern for the environment, the use of chemicals to control pests is being reduced.

A chemical used to control insects is an ***insecticide***. Most insecticides target specific insect pests. Some insecticides are highly poisonous. They can kill other living things, including beneficial insects, when sprayed onto a crop.

The chemicals used to control plant diseases are chosen based on the kind of disease. For example, diseases caused by fungi

13-16. Genetic engineering is being used with corn to control damage by certain insects. It is likely that the loss caused by the worm on this ear of corn could have been prevented by using insecticides.

13-17. **Genetic engineering with potatoes is being used to control insects, such as the Colorado potato beetle (*Leptinotarsa decemlineata*).**

are controlled with fungicides. A ***fungicide*** is a chemical compound used to control fungal diseases.

Considerable research has been done to find solutions to insect and disease problems without using chemicals. Some chemical companies have invested millions into the genetic engineering of crops that are resistant to insects and/or disease. Many of these resistant cultivars are currently being grown and marketed with more being added every year.

HERBICIDE RESISTANCE

Weeds cause large losses in some crops. A ***weed*** is a plant that is growing where it is not wanted. It is a plant pest. A chemical used to control weeds is

13-18. **A field of soybeans with no visible weeds.**

a ***herbicide***. Some herbicides are selective; others are not. A selective herbicide kills only certain weeds, such as grass in a crop with broad leaves.

A few herbicides are available. These are approved for use in reducing weed populations, thus reducing competition for resources with crop plants. Prior to genetic engineering, many herbicides could not be used in agronomic situations. If the chemical was applied over a large area, it not only killed the weeds, it killed the crop as well. Corn, cotton, canola, and soybeans have been engineered to have "herbicide resistance." Herbicides can now be applied to these crops and have no detrimental effect on the crop itself.

NUTRITIONAL QUALITY, TASTE, AND PROCESSABILITY

People want food that tastes good and has the nutrients they need to have good health. Genetic engineering is being used in an attempt to meet these demands of consumers.

There are many traits being engineered into plants that are of direct benefit to the consumer. Soybeans and canola are being engineered to produce modified oils and to have increased oil production. Tomatoes have been created to ripen more slowly and to have an increased solids content. Currently, researchers are working to improve the nutritional qualities of a variety of crop cultivars.

Many food products are processed in some way. Special steps may be taken to prevent spoilage. Canning, freezing, and pickling are three examples. Food crops are being engineered to reduce the rate of spoilage and help the fresh product last in storage and while being transported.

13-19. **Consumers want vegetables without deterioration. These asparagus spears show no signs of deterioration.**

PHARMACEUTICAL PRODUCTION

Plants are sources of substances to promote good health. These substances are extracted from the leaves, bark, stems, seeds, roots, or other parts of plants. One example is the Pacific yew. The bark of this tree is used to make ***taxol***, which holds promise in treating human cancer.

Plants can be engineered to produce pharmaceuticals, enzymes, and vaccines. Plants can also be engineered to produce some proteins, however, the processes used to extract and purify them are costly and inefficient. In the future, people may "take their medicine" by eating these plants. The plants may be cooked and served as part of a meal. Some crop producers may find these crops profitable to grow.

Farming crops containing pharmaceuticals is sometimes known as "pharming." The term is also applied to animals, such as "pharm animals."

13-20. The bark is being removed from the small log of a Pacific yew to extract taxol. (Courtesy, U.S. Forest Service)

INDUSTRIAL APPLICATIONS

Once the entire genome for a given plant species has been decoded, industrial uses increase. The biosynthetic pathways of the plant can be genetically altered to produce many different lipids and proteins. An example is the potato. A precursor is present in potatoes that may allow the production of plastic from that crop.

Commercial oils now produced chemically may be extracted from crops in the future. The crops will be genetically transformed to produce those oils.

REVIEWING

MAIN IDEAS

Plant breeding is used to improve the qualities of crop and ornamental plants. It is a form of biotechnology that uses the process of artificial selection and breeding. In the mid-1900s, hybridization became important in improving plant quality.

New technologies are being used to improve plants. These include DNA profiling, genetic engineering, and plant tissue culture. The technology makes it possible to create and propagate new plant cultivars.

Tissue culture is a biotechnology process that may or may not involve genetic engineering. In effect, plant tissue culture clones a plant. The practice of plant tissue culture is fairly easy to use provided the basic procedures are followed.

Genetic engineering is used with plants to improve them or obtain desired products. Some plants are engineered to resist insects and disease. Others are engineered to facilitate weed control. Genetic engineering is also used to make plants that are better suited to mass food processing and marketing. Others are engineered to obtain desired pharmaceutical benefits.

QUESTIONS

Answer the following questions using spelling grammar and complete sentences.

1. What is plant breeding?
2. Why is plant breeding used?
3. Distinguish between line breeding and crossbreeding.
4. What is hybridization? Name an example of a plant that is widely planted as a hybrid.
5. How does DNA profiling improve on traditional plant breeding?
6. What three approaches are used in genetically engineering plants? Briefly describe each and name plants where each has been used.
7. Describe the general process of plant tissue culture.
8. Why is an aseptic environment needed for plant tissue culture work?
9. What seven basic components are in most media?
10. What five areas are reasons for genetically engineering plants? Briefly describe each area.

EVALUATING

Match the term with the correct definition. Write the letter by the term in the blank that is provided.

a. weed
b. herbicide
c. insecticide
d. cultivar
e. line breeding
f. parent plant
g. mutation
h. plantlet
i. explant
j. taxol

_____ 1. A cancer-fighting drug made from the bark of the Pacific yew tree.

_____ 2. A chemical compound used to control insects.

_____ 3. A plant that is out of place.

_____ 4. A chemical compound used to control weeds.

_____ 5. A cultivated variety of a crop plant.

_____ 6. Breeding offspring of successive generations among themselves.

_____ 7. A genetic variation that occurs without prediction.

_____ 8. The small plant that forms in plant tissue culture.

_____ 9. The plant that is the source of explants.

_____ 10. The tiny tissue taken from a parent plant that is used in tissue culture.

EXPLORING

1. Assume that you are a plant breeder for a rose grower. You have an educational background in biotechnology and you believe that you could implement innovations to help breed new, exciting types of roses. Prepare a proposal to present to the owner of the company. State the types of innovations you would like the company to consider, outline how those innovations would help to breed new types of roses, give ideas for new rose cultivars, and provide justification for your ideas.

2. Change the world! If you could genetically engineer one important agronomic crop, what would it be? What trait(s) would you try to enhance? Why? Prepare an oral report and deliver it in class.

3. Make a field trip to local horticultural or agronomic crop breeder. Observe how plants are being improved. Interview the plant breeders about their work. Prepare a report on your findings.

14

ANIMAL BIOTECHNOLOGY

Animals are a big part of our lives. Not only do animals provide companionship and love, they also provide a number of products that we use every day. For most people, meat is a major part of their food diet. Animals are also sources of leather, wool, milk, eggs, and other products. If you think about it, you can probably name a way that an animal has impacted your life already today.

People who work with animals are concerned about the animals' well-being. They want to make sure the animals' needs are met. They do not want the animals to suffer.

We live better today because of animals. Good practices will help us gain even more benefits from animals. You might think of some benefits you would like to see. Further, animals bring joy to our lives as companions and guardians.

14-1. People enjoy having top animals, such as this American Saddlebred Horse that is used for showing in the country pleasure class. (Courtesy, American Saddlebred Horse Association)

OBJECTIVES

This chapter provides information on advances in animal biotechnology. It uses approaches that respect animal well-being as well as improve animal productivity. It has the following objectives:

1. Explain the use of animals in biotechnology.
2. Describe processes of animal reproduction.
3. Relate advancements in animal reproduction technology to the livestock and other animal industries.
4. Explain the use of biotechnology in production efficiency.
5. Describe the use of genetic engineering with animals.

TERMS

anestrous
artificial insemination
copulation
embryo stage
embryo transfer
estrous cycle
estrus
fertilization
fetal stage
gender preselection
gestation
***in vitro* fertilization**
laboratory animal
lactation
models for animal research
nonhuman primate
ovum
parturition
primate
pronucleus
puberty
reproduction
semen
sexual reproduction
sperm
xenograft
zygote

ROLES OF ANIMALS IN BIOTECHNOLOGY

Animals have important roles in biotechnology. Most of the biotechnology work begins in a lab. Researchers attempt to answer questions about animals through lab experiments. Without the use of animals, humans might be in danger.

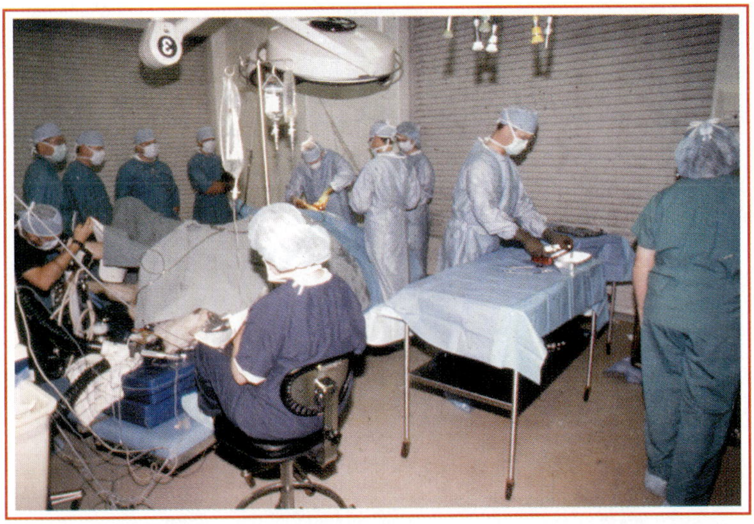

14-2. People strive to improve animals. This shows surgery being performed on a bull in a veterinary hospital. (Courtesy, College of Veterinary Medicine, Texas A&M University)

ANIMAL MODELS

It is not always necessary to use living animals in research. Alternatives can be used for some kinds of biotechnology work. Scientists have developed ***models for animal research***. One model uses live animals; the other models do not. These help alleviate problems that might arise about the well-being of animals.

Overall, four models apply for the use of animals:

Model 1: Living animals—These animals are living and usually have no threat to their well-being. Such animals may be known as laboratory or scientific animals. Agricultural research often uses experimental groups of animals.

14-3. **Knowing how to care for the well-being of animals begins with studying their basic needs.**

Model 2: Living animal tissues or systems—Animal tissues can be cultured in a lab. This saves the use of animals as well as the expense of feeding, housing, and cleaning-up after animals.

Model 3: Nonliving systems—This involves using nonliving mechanical models that reflect animal activity. These often relate to skeletal movement and locomotion. Artificial replacement parts, such as hip joints, can be studied using nonliving systems.

Model 4: Computer and mathematical approaches—Computer simulations with virtual reality and other uses help in biotechnology. Computer modeling may be done with a proposed biotechnology practice before it is tested with animals.

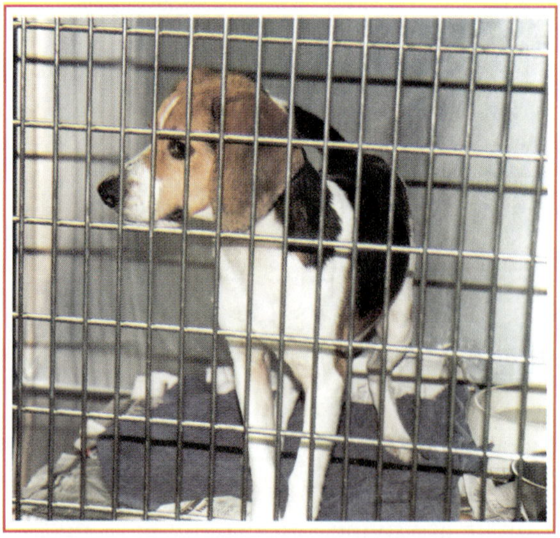

14-4. **Beagle dogs are widely used in some research. This puppy is being kept in excellent surroundings.**

LABORATORY ANIMALS

A ***laboratory animal*** is an animal used for laboratory or research purposes. Laboratories may be cages, pastures, or other facilities where the animals can be controlled and observed. They are usually carefully tended to assure the animals' well-being. Good tending is needed to assure that the research is properly carried out.

Without research involving animals, scientists would not have a base of information about animals. People would not know the nutritional needs of animals nor how to properly care for them. Most lab research focuses on helping animals to be more productive.

Small animals often work quite well in labs. Small animals eat less food and produce less wastes than larger animals. Just compare an elephant with a mouse! What are the obvious differences?

ANIMAL SPECIES

Most animals are subjects of study at one time or another. Humans, for example, may be in experiments to test new drugs. Cattle may be in experiments to determine the feeds that best meet their needs. In many cases, animals are carefully selected for study. Some are small; others are large.

The most common species used in research laboratories are mice, rats, hamsters, guinea pigs, rabbits, cats, dogs, and nonhuman primates. A ***primate*** is an animal with thumb and forefinger opposition. Look at your hand. Do your thumb and forefinger work together to make it easy for you to grasp things?

A ***nonhuman primate*** is an animal similar to humans, but is not a human. It has thumb and forefinger opposition. Common examples of nonhuman primates are monkeys and chimpanzees. Studying these animals often provides insight into human behavior.

In agriculture, when information is needed from research, cattle, hogs, sheep, horses, fish, and others are used. These animals are normally maintained in their adapted habitat. Various feeds, medicines, and other conditions are used. The results are carefully studied to determine future recommendations.

14-5. Chimpanzees are popular nonhuman primates for research purposes.

14-6. These beef animals are involved in nutrition research. The transponders hanging on the chains under their necks use radio waves to identify the animal.

ANIMAL REPRODUCTION

Reproduction is the most important aspect of successful animal production. Much of the effort in biotechnology has focused on reproduction.

Reproduction is the process by which offspring are produced. Offspring naturally have many of the same traits as their parents. They are of the same species. They often have the same color, conformation (body shape), and other features. In humans, children often have the same eye color, hair color, body features, and other characteristics as their parents.

SEXUAL REPRODUCTION

Sexual reproduction is the union of a sperm and an egg to ultimately produce a new individual. The parents are the sources of the sex cells that unite in the process of fertilization. ***Sperm*** are the sex cells from the male parent. The ***ovum*** or egg is the sex cell from the female parent. Only one sperm and one egg are needed for a new individual to be formed. Processes with birds, bony fish, and mammals vary, but all involve the union of a sperm and an egg.

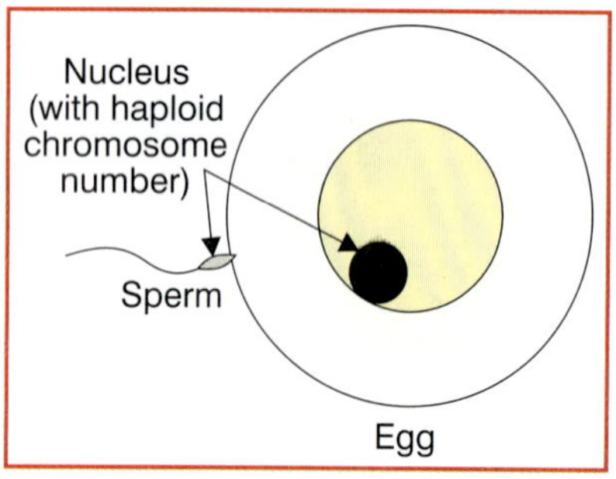

14-7. Sexual reproduction involves the union of a sperm and an ovum. Though visible with a microscope, this shows the relative size comparison of a bovine sperm and ovum. The nucleus of the sperm and ovum contain haploid chromosomes.

Ova and sperm sizes vary—though sperm are always smaller. Some eggs are quite large, such as that of an ostrich. The ostrich egg would be fertilized internally by quite small sperm. Other eggs are very small and visible only with a microscope. Since the eggs of mammals are fertilized internally, they are small, with size depending on the size of the animal. In most livestock, eggs are smaller than average garden peas. An individual sperm is only visible with a microscope.

The union of the sperm and egg is ***fertilization***. The process may be inside the reproductive tract of a female or outside the tract, depending on the species of the animal. For example, the reproductive tract of a female sheep (ewes) receives sperm from a male sheep (rams) during mating. The penis of the male ejaculates semen in the reproductive tract of the female near the egg(s). With fish, the eggs are released in a mass or other arrangement in the water. The male fish releases sperm that swim around in the water. Some sperm eventually fertilize many of the eggs.

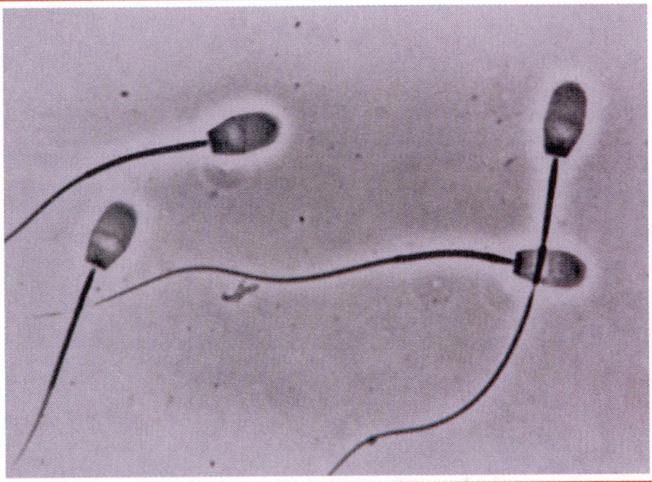

14-8. Enlarged sheep sperm viewed through a microscope. (Courtesy, Elite Genetics, Waukon, Iowa)

Mammals and some other species copulate. ***Copulation*** is the sexual uniting of animals so the male can deposit (ejaculate) semen near the eggs in the reproductive tract of the female. ***Semen,*** the fluid produced by the male reproductive organs, contains sperm. Sperm are quite active. They move

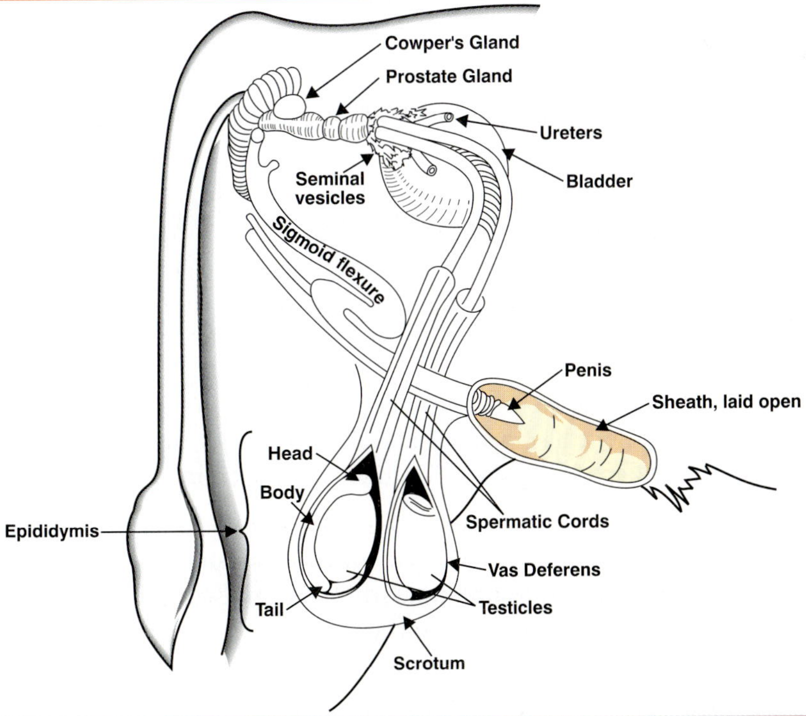

14-9. Major parts of the reproductive system of a male bovine (bull).

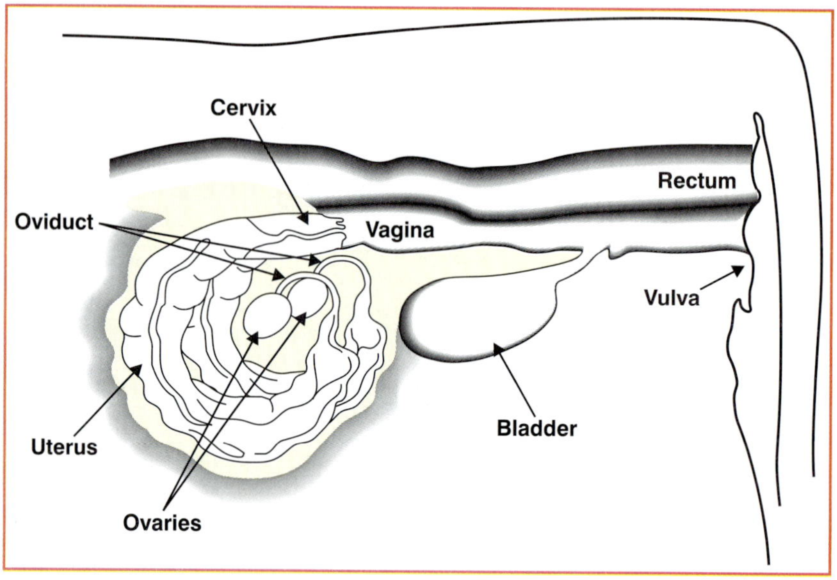

14-10. Major parts of the reproductive system of a female bovine (cow).

about seeking an ovum with which to unite. An ovum needs only one sperm to be fertilized. One mating or ejaculation may release thousands of sperm.

The reproductive systems of animals are designed to make reproduction possible. Females have ovaries that produce eggs. Males have testicles that produce sperm. The processes must work properly. If not, no new animals will be produced.

REPRODUCTIVE DEVELOPMENT

Many animals go through phases of reproductive development. The phases that are common are:

- Puberty—***Puberty*** is the time at which an animal is capable of reproduction. Sexual development has reached sufficient maturity that either sperm or ova are produced, depending on the gender of the animal.
- Fertilization—Union of the sperm and egg results in the formation of a zygote. In some cases, blockages occur in the reproductive tracts that prevent successful fertilization. With some animals, *in vitro* (in glass, such as test tubes) is used to overcome the difficulties.
- Gestation—***Gestation*** is the period of pregnancy in mammals. The first stage of development is known as the ***embryo stage***. The later stage of development is known as the ***fetal stage***. Gestation varies by species. With horses,

gestation is 337 days. This means that a colt is born an average of 337 days from the date of fertilization.

- Parturition—Once gestation has been completed, parturition occurs. ***Parturition*** is the birthing process. Uterine contractions result in the baby animal being born through the vaginal canal.
- Lactation—Female mammals produce milk as food for their babies. ***Lactation*** is the secretion of milk by the mammary glands. Milk provides the nutrients the baby needs to grow and properly develop.
- Dry time—Female mammals typically have a time when they are not lactating. The reproductive and mammary systems are preparing for the birth of the next baby animal. Dairy producers very carefully control milking so cows can replenish their systems before calving.

14-11. A Dorset ewe that has recently give birth to a lamb. What stages of reproductive development have been involved?

THE ESTROUS CYCLE

Many female mammals have an estrous cycle. ***Estrus*** is one of the periods in the cycle. It is the time when a female is fertile and receptive to a male. It is sometimes known as "heat." The ***estrous cycle*** is the time between estrus.

The periods in the reproductive system of a female mammal are cyclical or recurring. Pregnancy interrupts the estrous cycle. ***Anestrous*** is the absence of cycling. In some animals, anestrous is related to the number of hours of daylight in a day and other conditions.

The three periods in the estrous cycle that follow estrus are metestrus, diestrus, and proestrus. Metestrus immediately follows estrus. Diestrus follows metestrus. It is the period when the uterus is prepared for pregnancy. Proestrus follows diestrus. Proestrus is the period of preparation for estrus and ovulation. For pregnancy to occur, mating must occur during estrus at a time when the ovum (or ova) is ready for fertilization. Sperm must unite for the process to be complete.

Breeding times vary. Some animals are seasonal breeders and mate only at certain times of the year. Other animals regularly cycle through breeding

Biotech Connection

LEGENDS LIVE ON

How do we benefit from good horses? There have been many famous Thoroughbred race horses. After being retired from the track, these horses lived lives of luxury as prized breeding animals. The average life expectancy for a horse is about 20 years.

As a result of artificial insemination technologies, semen can be collected and frozen for an unlimited amount of time. This means that famous race horses can produce offspring long after their death. With A.I., a mare could be bred with the semen from Secretariat 100 years from now. This would insure that his racing lineage will live almost forever.

Though the technology is available, horse officials and breed associations are only now beginning to accept it. The ability to register offspring that have come from an artificial breeding has the potential to alter the animal industry. The associations are now beginning to determine the rules that will apply. This photo shows a Thoroughbred mare and her foal.

times without regard to the time of the year. For example, cows have estrous cycles throughout the year, or until pregnant or a failure occurs in the reproductive system. Most sheep are seasonal breeders. Ewes naturally come into heat in the late summer. This assures lambing (birth of lambs) in the spring. Hours of daylight, temperature, and other environmental conditions influence reproductive cycling in seasonal breeders.

ANIMAL REPRODUCTIVE TECHNOLOGIES

As mentioned earlier, reproduction is important in the production of animals. The goal is to produce offspring with the qualities that demand the highest price in the marketplace.

14-12. **These cows are being observed for reproductive development. Some of these cows appear pregnant. With natural breeding and following calving, a bull will be placed in the pasture about 10 months prior to the desired time of the next calving. The bull is pastured separately other than during the breeding season.**

Reproduction processes in humans are quite similar to those of other mammals. Many of the innovations that exist in the field of human fertility and reproduction were first created (designed) for animal production. Dairy producers have used reproductive technologies, such as *in vitro* fertilization, artificial insemination, and embryo transfer, since the late 1980s. Many livestock producers and specialty animal breeders are trying to use reproductive advancements to increase efficiency.

NATURAL BREEDING SYSTEMS

Natural breeding systems or "live cover" involves actual live pairing of a male and female. This natural process can be controlled by breeders to the extent that bloodlines can be upgraded by selective mating. Breeding involves insemination of the female by the male.

Natural breeding systems are limited by the availability of live breeding males in the area. In dairy farming, most farms rarely have a bull (male). Typically, one male can mate with a number of females. For example, one healthy bull can successfully breed 25 or more cows. For most animal producers, (cattle, swine, sheep, etc.) male offspring are castrated unless they possess the desirable characteristics that can be passed on to future generations.

Breeding males are expensive and difficult to maintain and generally producers do not keep many of them on hand. Albeit economically efficient, this does reduce the genetic diversity for a herd. Natural breeding is the simplest and the most often successful method of reproduction. Live cover can be physically stressful for some females. This is especially true if the male is

14-13. **Bulls are often kept in separate pastures from cows until the breeding season.**

physically much larger than the female. It can also be difficult to selectively breed or improve a line if there are a limited number of sires available.

ARTIFICIAL INSEMINATION

Artificial insemination is the transfer of collected semen to a recipient female. It is used with dairy cattle, beef cattle, swine, sheep, horses, turkeys, and other animals.

Generally, female estrous cycles are regulated with hormone injections. This is to assure that mature ova are available for fertilization at the time of

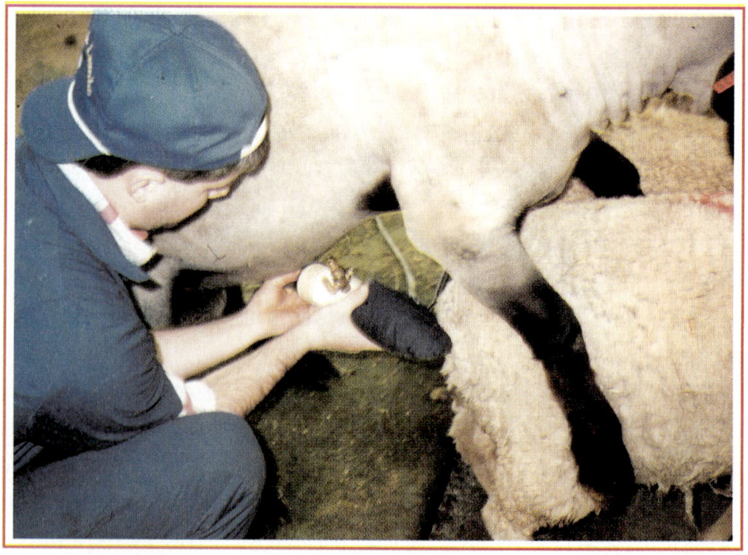

14-14. **Semen is being collected from a ram with an artificial vagina. A teaser ewe promotes sexual stimulation for the ram.**

insemination. Owners of several females will likely synchronize the estrous cycles of the animals that are to be inseminated. This assures that several animals are in heat when semen is obtained and insemination is planned. Synchronization is obtained with hormone injections to insure ovulation at the time of semen transfer. Insemination at the time of peak ovulation provides optimal conditions for successful fertilization and embryo development. Inseminating when the animal is not in "heat" will not result in conception (union of sperm and egg). The egg(s) has not been released by the ovaries.

Sperm (in semen) are collected from males by artificially promoting ejaculation. Semen is graded and carefully stored. The goal is to have quality semen with living, active sperm. Semen is typically frozen in nitrogen tanks until ready for insemination.

14-15. Collected swine semen will be filtered before microscopic examination on this Pennsylvania farm.

14-16. Semen can be stored frozen in nitrogen tanks. (Courtesy, Elite Genetics, Waukon, Iowa)

14-17. Semen in "straws" is being removed from a tank. (Courtesy, National FFA Organization)

With the use of artificial insemination, semen can be shipped long distances. This practice dramatically increases the number of available males that can be used to improve animal quality, such as a herd of dairy cattle. Depending on the species of the animal, semen may be shipped fresh or frozen in liquid nitrogen tanks.

Artificial insemination has also become common practice for the initiation of human embryo development. A.I. is also being rapidly adopted by the equine industry. Warmbloods, Arabians, and Quarter Horse registries will currently register offspring that have been produced as a result of artificial insemination. A.I. breeding systems are also being developed for exotic and companion animals.

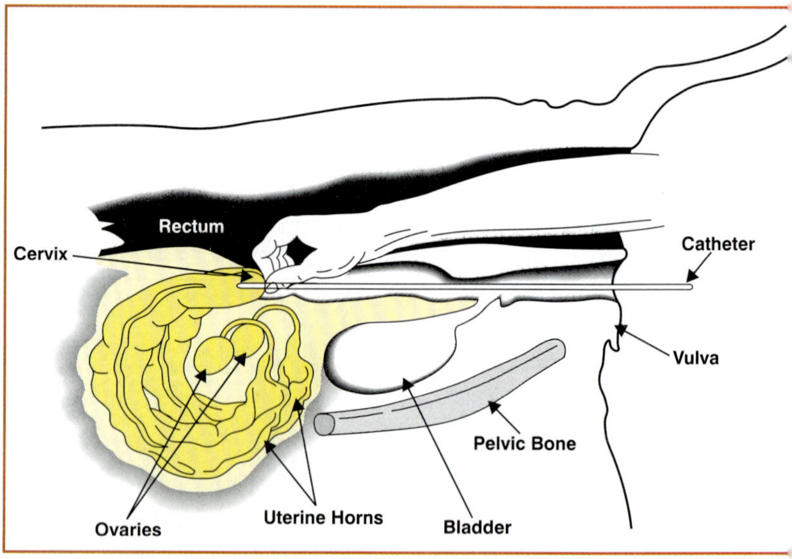

14-18. This shows the recto-vaginal method of artificially inseminating a cow. Note, the gloved hand goes in the rectum and the semen catheter (straw) is inserted into the reproductive tract. The semen will be released in the cervix near the horns of the uterus. Motile sperm will swim about seeking an ovum to fertilize.

IN VITRO FERTILIZATION

In vitro fertilization is the process of collecting ova from a donor female and fertilizing them, outside the body, with collected semen. *In vitro* is Latin for "in glass." This process is usually performed in a petri dish. It is also commonly known as producing "test-tube babies." Once fertilized, ova are implanted into a recipient female, where the embryo will mature during a regular gestation period.

In vitro fertilization is a commonly practiced reproductive therapy for humans. It is a necessary step in the embryo transfer process used in cattle.

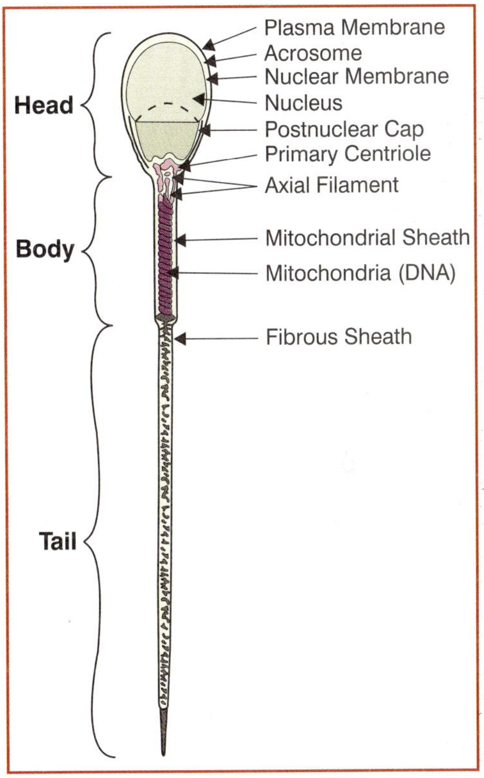

14-19. **Parts of a typical sperm cell with emphasis on the location of the DNA.**

GENDER PRESELECTION

The gender of offspring is randomly determined in natural breeding. ***Gender preselection*** is choosing the gender of a potential offspring. The sex of the offspring is determined before sperm and egg unite to form an embryo. The practice is not widely used. Considerable interest exists in some areas, such as preselecting all-female offspring of dairy cattle.

Gender preselection may be done by sorting sperm. Sorting is difficult because sperm are quite small. Some sperm create females; other sperm create males. Various approaches have been used in sorting sperm. Some have used the physical sorting of sperm. Others are based on chromosome differences. Only methods based on chromosome differences have had significant results in gender preselection. Not well proven is the notion that sperm differences result from the testicle in which they are produced.

The DNA content of sperm varies slightly based on the gender of the offspring that would result from conception. Chromosome difference is the X and Y. Some sperm have X; others have Y. Sperm with the X-chromosome produce females. Those with the Y-chromosome produce males. The Y-chromosome is smaller and has less DNA than the X-chromosome. Some scientists have achieved 100 percent success rates in cattle with gender preselection.

The processes often damage the sperm. It is difficult to sort and examine sperm without causing injury, loss of motility (ability to move about), or

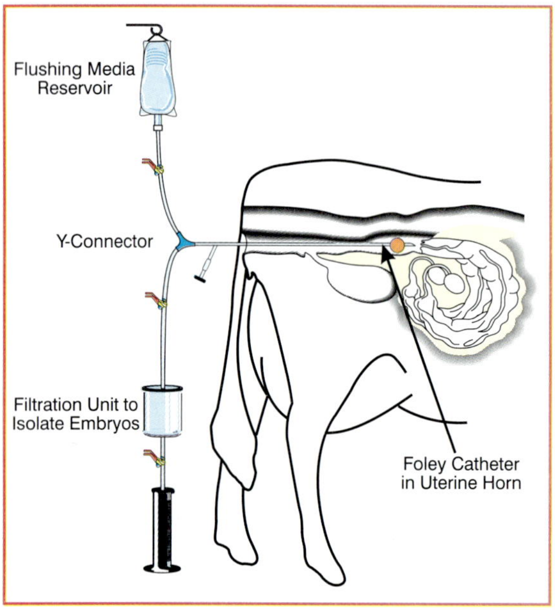

14-20. An example of procedures with nonsurgical embryo removal.

death. Once sorted, sperm may be artificially inseminated or used with *in vitro* fertilization.

EMBRYO TRANSFER

Embryo transfer is the harvesting of fertilized ova from a donor and implanting them into a recipient. The process is performed either surgically and nonsurgically. It is costly and requires specially trained personnel.

Nonsurgical transfer involves flushing the embryos from the uterine horn. The embryos are filtered from the flushing solution. The harvested embryos are transferred to a recipient. With surgical transfer, an incision is made into the uterus to remove the embryos.

Dairy farmers, who are trying to produce large numbers of high-quality female calves, are the most common enlisters of the technology. Producers of beef cattle and sheep use it to speed the process of reproduction of superior

14-21. A microscope is used to search for harvested embryos. (Courtesy, Elite Genetics, Waukon, Iowa)

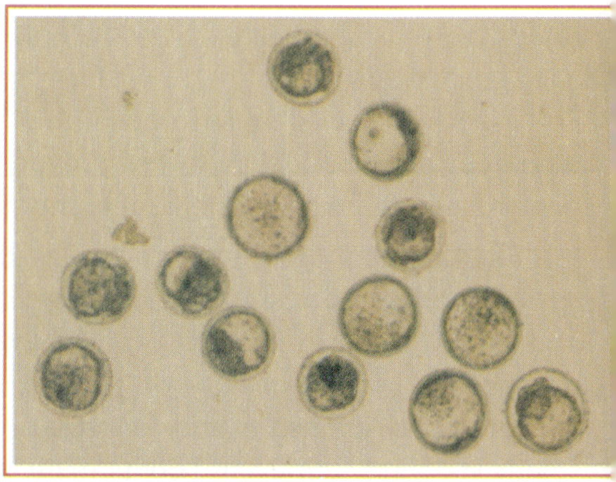

14-22. How harvested embryos appear under a microscope. Fertilized ova show cell development and are darker. Unfertilized ova show no cell division. (Courtesy, Elite Genetics, Waukon, Iowa)

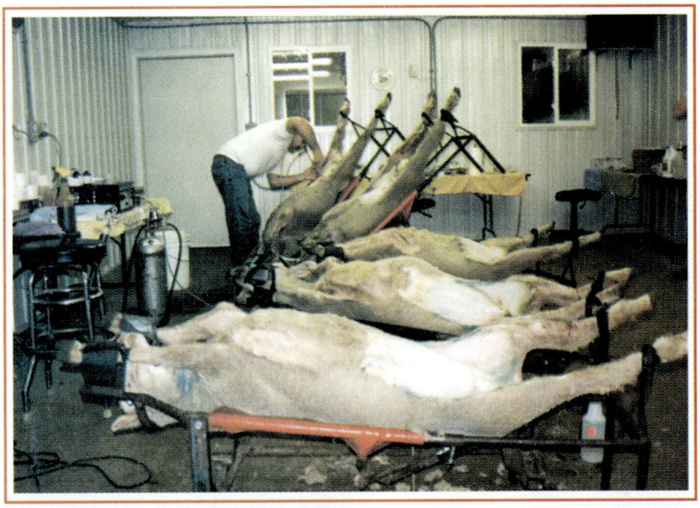

14-23. **Sedated recipient ewes are ready for the transfer procedures. Their well-being has been carefully considered. (Courtesy, Elite Genetics, Waukon, Iowa)**

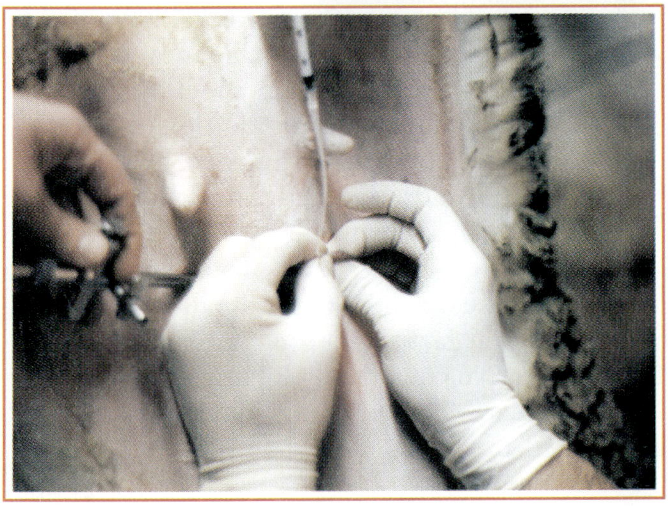

14-24. **Embryos are being transferred to the uterine horns of a recipient ewe. (Courtesy, Elite Genetics, Waukon, Iowa)**

animals. The donor is usually a top-quality, superior animal. The recipient is usually of lesser quality but quite capable of producing good offspring.

INCREASING PRODUCTION EFFICIENCY: GROWTH HORMONES

Animal producers must operate efficiently to make a profit. Livestock producers are constantly trying to improve production efficiency. This may mean increasing yield, decreasing time to maturity, or developing (produc-

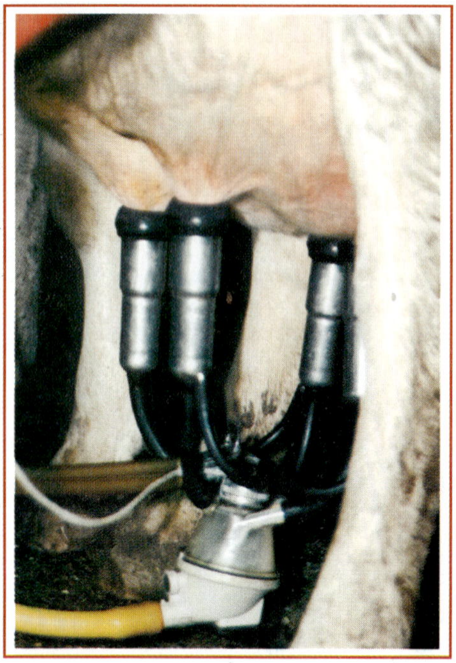

14-25. **Automated milking systems are used to keep milk from cows treated with bST safe for consumers.**

ing) a product that is more desirable to consumers. Biotechnology has resulted in several unique tools to help increase efficiency and make new products.

BOVINE SOMATOTROPIN

Bovine somatotropin (bST) is a naturally occurring growth hormone produced in the pituitary gland in the endocrine system of cattle. By treating dairy cattle with the hormone, milk production is increased approximately 15 percent. However, there is some related research that shows that cattle treated with additional bST may be "overworked." Animals show signs of stress and the overall production life of the cow may be reduced. Additionally, it is expensive to treat cattle with bST. Trained workers are needed to properly administer the drug on a regular basis.

There has been negative publicity about using bST. Some small dairy producers do not use bST and incorporate that fact in their advertising. Milk cartons will be printed with "No bST" or "No Genetically Engineered Hormones Used in the Production of This Product." These actions may promote public fear of using genetically engineered products.

PORCINE SOMATOTROPIN

Porcine somatotropin (pST) is a naturally occurring growth hormone produced in swine. Treating hogs with pST causes more efficient feed conversion. Leaner pork with less fat is produced. The major disadvantage associated with pST is the cost and the labor involved in constantly treating a large herd.

ANIMAL GENETIC ENGINEERING

Creating transgenic animals has largely been experimental. A few transgenic animals are now being used to produce specific products, such as drugs

Career Profile

A.I. TECHNICIAN

An artificial insemination (A.I.) technician is responsible for synchronizing estrous cycles and artificially inseminating live animals. Most individuals in this field work for dairy producers as a private contractor or a dairy employee. Artificial insemination is also being used extensively by the beef cattle, swine, horse, and turkey production industries.

A.I. technicians may differ in terms of educational preparation. Most, if not all, have completed an intensive, "hands-on" program offered by specialty tech schools or breeding companies. Some technicians may have completed an undergraduate program in animal science or other related field. However, a college degree is generally not a prerequisite to enter the field.

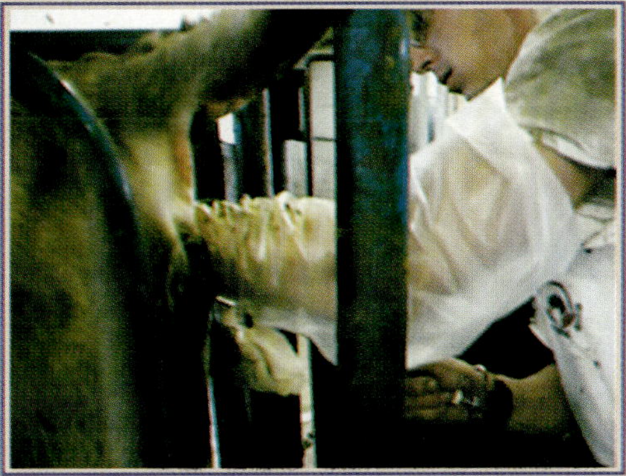

Most work is done on farms or in labs. An animal scientist or a veterinarian may oversee the work. A.I. technicians may be employed by labs, farms or ranches, or work as entrepreneurs. This photo shows an A.I. technician using the recto-vaginal method of artificial insemination with a cow. (Courtesy, Lyle Westrom, University of Minnesota–Crookston)

for human health. Drugs and other products are also produced for veterinary medical use. More efforts have been made with plant transgenic forms than with animals. Some individuals are concerned about the practice of genetically altering animals.

TRANSGENIC ANIMALS AND PRODUCTS

Several uses of animals to produce drugs have been investigated. A few have been developed to the level of satisfactory use.

- Pigs—Pigs have been used to manufacture human hemoglobin. The product from pigs is substituted for human blood. Such a practice helps alleviate the transmission of human disease through donated blood. Pigs are also being used as sources of organs in humans in xenografts. A ***xenograft*** is the practice of grafting an organ or a tissue from one species into another.

14-26. A transgenic sow with a litter of pigs produces a human blood factor in her milk. (Courtesy, Virginia Tech and Rick Rudd, University of Florida)

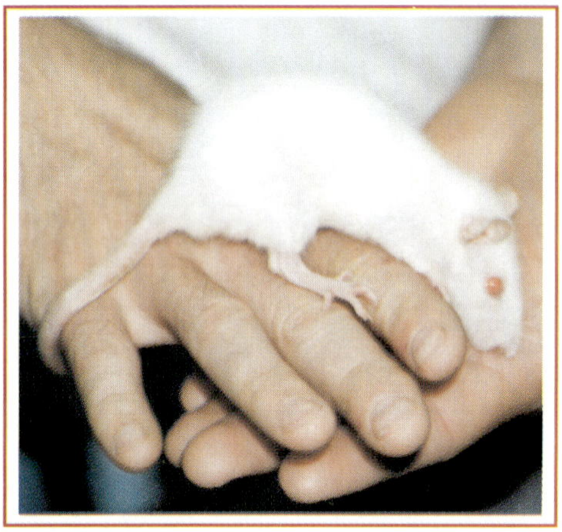

14-27. Mice are successfully used in some kinds of genetic engineering.

- Mice—Transgenic mice have been used in several ways. One of the best known is to produce human antibodies. Antibodies are needed to fight some kinds of diseases. Transgenic mice have also been used to model diseases that occur in humans. Examples of diseases studied with mice models include sickle cell anemia, atherosclerosis, and prostate cancer.
- Cattle—With cattle, much of the genetic engineering effort has focused on the mammary glands of dairy cows. Ways to alter cow milk for people who have lactose intolerance have been investigated. Another area is using biotechnology to control disease, such as mastitis in dairy cows. (Mastitis is a disease of the mammary glands that interferes with milk production.)

METHODS OF CREATING TRANSGENIC ANIMALS

Several ways to produce transgenic animals have been studied. The only proven method is by the microinjection of DNA into embryos. The embryos are in the one-cell or two-cell stages. In general, the steps in producing transgenic animals have been fairly well developed. The most common animals that these steps apply to are sheep, goats, cattle, and pigs. Three general steps follow.

Step 1: Collect embryos—Timing is important. Superovulation is promoted in donor females. With proper stimulation, far more embryos can be obtained than would be the natural result of the reproductive process. In a few cases, *in vitro* methods have been used with ova collected from animal slaughter houses.

Step 2: Inject embryos—Microscopy is used to locate embryonic cells. The pronuclei within the cells are identified. A ***pronucleus*** is the haploid (single) nucleus of the sperm or ovum that have united in fertilization to form a zygote. A ***zygote*** is the one-cell embryo formed by the union of a sperm and ovum. Some zygotes are centrifuged to increase visibility. Membranes and other structures are often not easily seen at this early stage of development. Injection needles are quite small and are guided using a microscope. The appropriate DNA must be available and used in the injection process.

Step 3: Zygote culture—Following microinjection, the zygotes are placed in the oviduct of a recipient female. This means that the female must be at the proper stage in the estrous cycle. The transfer may involve surgery or nonsurgical methods. The embryos develop much as a normal embryo. Some may be carried to full-term; others may be removed in various stages of development.

All of the steps in this process involve hazards in survival of the zygotes. Moving about, inspecting, injecting, and other procedures injure and kill zygotes. The success rate for the birth of live transgenic animals is often low. It ranges from less than 1 percent in cattle to more than 1 percent in goats. Mice have a nearly 3 percent success rate.

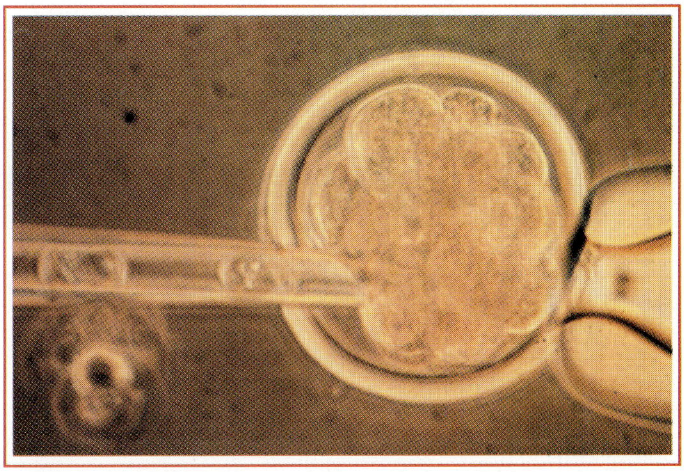

14-28. A greatly enlarged multiple-cell embryo showing the removal of selected cells. (Courtesy, Lyle Westrom, University of Minnesota-Crookston)

An Example: Growth Hormone

Genetic engineering has allowed researchers to produce animal growth hormones on a large scale. The gene for hormone production is taken from the chromosomal DNA of the related animal type, ligated into a plasmid, and put into some bacteria. Bacteria are then fermented on a large scale where they yield enormous amounts of the growth hormone. The hormone is then administered through injections to the appropriate animal.

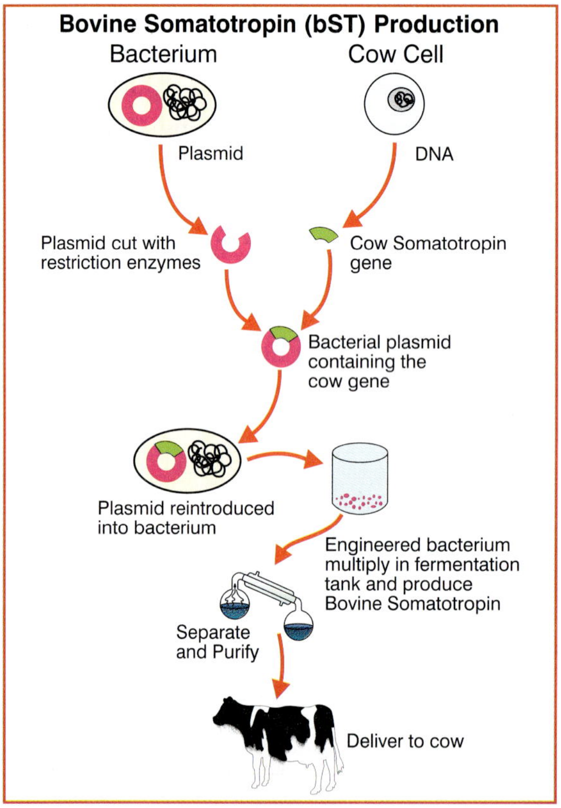

14-29. **The process of producing an "engineered" growth hormone to promote milk production.**

REVIEWING

MAIN IDEAS

Animals are important in our lives. They provide companionship, love, and many of the products we use every day. Much has been done to improve animals using non-genetic engineering methods.

Biotechnology requires the use of animals. Scientists have developed models for animal research. One model uses living animals; the other models do not. The well-being of live animals is always considered.

Biotechnology is impacting the production of animals for food and other products. Animal reproduction is being affected by several evolving technologies—artificial insemination, embryo transfer, and *in vitro* fertilization. These innovations give animal breeders an opportunity to improve a breeding line, thus incorporating beneficial traits into their stock.

Genetically engineered animal growth hormones have been introduced. Bovine somatotropin (bST) is administered to dairy cattle to increase milk production. Porcine somatotropin (pST) is used on pigs to increase feed efficiency. The benefit of using these hormones is limited by the costs associated with them.

Genetic engineering has been successfully used in several areas. Many of these uses are to benefit human health and well-being. Methods of creating transgenic animal forms often involve the microinjection of zygotes with cattle, pigs, sheep, and goats. This work is carried out using specially designed microscopes.

QUESTIONS

Answer the following questions using correct spelling and complete sentences.

1. What animal models are used in research? List the models and briefly describe each.
2. What is a laboratory animal?
3. What species of animals are used in biotechnology research? Name three, and indicate if they are primates.
4. What is sexual reproduction? What two sex cells are essential?
5. What are the stages of reproductive development in mammals? Briefly describe each stage.
6. What is the estrous cycle? What four periods are included? In which period might conception occur?
7. What is natural breeding?
8. What is artificial insemination? Briefly summarize the process.
9. What is *in vitro* fertilization?
10. What is a growth hormone? Give two examples.
11. What products are being obtained from transgenic animals? Give two examples.
12. What are the three general steps in creating a transgenic animal using microinjection?

EVALUATING

Match the term with the correct definition. Write the letter by the term in the blank that is provided.

a. zygote
b. primate
c. gender preselection
d. semen
e. puberty
f. gestation
g. anestrous
h. lactation
i. embryo transfer
j. xenograft

_____ 1. Using an organ or tissue from one species on another species.

_____ 2. The one-celled embryo formed by the union of a sperm and an ovum.

_____ 3. The absence of a reproductive cycle in a female mammal.

_____ 4. The period of pregnancy.

_____ 5. The secretion of milk by the mammary glands of a female mammal.

_____ 6. The stage of development when an animal is capable of reproduction.

_____ 7. The reproductive fluid produced by a male during ejaculation.

_____ 8. The harvesting and implanting of an embryo.

_____ 9. An animal with thumb and forefinger opposition.

_____ 10. The ability to choose the gender of an offspring.

EXPLORING

1. Organize a class debate. Possible subjects include the use of genetically engineered animal growth hormones, gender preselection, and other advanced reproductive technologies. In your debate arguments, be sure to consider many perspectives including that of public perception, actual public health concerns, advantages and risks for small producers versus large producers, and the issue of animal welfare.

2. Go to the grocery store and see if you can find animal products that are labeled as being created with genetic engineering. Likewise, see if you can find products that are labeled as NOT being created with genetically engineered products. Write a brief paragraph on why you believe products that have been created with the assistance of genetic engineering should or should not be labeled as such.

3. Make a field trip to a lab where transgenic animals are created. Observe the processes used. Interview scientists about the methods they use. Prepare a report on your observations.

15

HUMAN BIOTECHNOLOGY

Human biotechnology is about you and me, not about the other creatures on Earth. It is personal. To consider the use of biotechnology with humans gives us a different point of view—another perspective.

Could a perfect person be in the near future? The desired person might be a man that is 6 feet tall, weighs 185 pounds, and has an IQ of 150. Criminal urges and other unwanted traits have been "engineered" out. Yes, a perfect man, who is kind, considerate, and has a good income. Or, it could be a perfect woman–just choose what you want! (Glenn McGee described all of this in his book entitled *The Perfect Baby*.)

A perfect person isn't likely. We may not be sure that we want one, or if we do, what traits we would want. We will go about using biotechnology in other ways. We will improve the quality of the lives of ordinary folks like you and me. We will forget perfection for the time being.

15-1. People are the products of their genetic and cultural environments. This shows a Korean elder in traditional dress. What traits are genetic? Cultural?

OBJECTIVES

This chapter introduces biotechnology applications with human beings. It has the following objectives:

1. Relate human biology to biotechnology.
2. Describe the meaning and importance of genomics in humans as organisms.
3. Relate methods used by students in the Human Genome Project.
4. List and describe ways biotechnology is helping humans live better.
5. Explain diagnostics and give two examples involving biotechnology.

TERMS

autosome
biocompatibility
culture
developmental stages
diagnostics
dialysis machine
germ-line gene therapy
human development
human genome
Human Genome Project
microdissection
prosthetics
somatic cell gene therapy

THE HUMAN ORGANISM

Humans are similar to other animals. Yet, they have major differences from other animals. These differences set humans apart from the animals. The similarities provide certain useful benefits.

The scientific name of a human is *Homo sapiens*. The first *Homo sapiens* appeared on Earth 300,000 to 400,000 years ago.

Only one living species of *Homo sapiens* is found on Earth today. Several extinct species of humans are known through fossils. Humans are mammals in the Kingdom Animalia. Each body cell of a human has 46 chromosomes. Human organ systems carry out life processes.

THE ORGANISM

Humans consist of cells, tissues, organs, and organ systems much as those of other animals. Study of the human helps provide insight into how humans are unique. Biotechnology can be used to promote health and a long, quality life.

Humans are primates. There are about 180 species of animals that are primates. Humans are quite different from most other primates.

Some organ systems do a lot of work. If they fail, the human is no longer healthy. The heart is a good example. The heart of a person who has lived to age 70 has pumped approximately 46 million gallons (174 million L) of blood. Some wear has likely occurred. Also, substances may collect in the circulation system that cause the heart to fail. Heart failure is one of the leading causes of human death. Suppose we could use biotechnology to solve heart problems and help the grandparent you love live longer. That would be great!

The shape of the body is provided by a skeleton. The skeleton protects the internal organs from injury. Some 200 bones are in the human skeleton. Ligaments are tissues that hold the skeleton together.

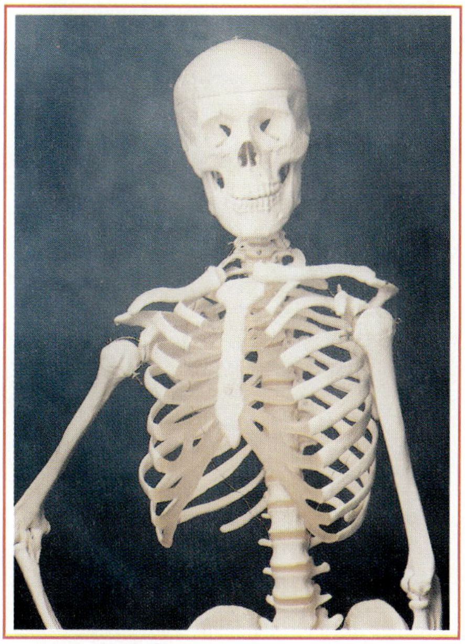

15-2. The human skeleton contains about 200 bones.

15-3. Activities, such as rock climbing, require good physical condition.

15-4. The ability to grasp tools is an important human skill. This shows a center punch being positioned for striking with a hammer in doing shop work.

Skin covers the body and performs important functions in maintaining body temperature. Nerve endings in the skin sense pain, cold, heat, and other conditions. These endings use a system of nerves to transmit the information to the brain. The brain provides a response to the conditions.

Muscles provide movement. When muscles contract, they pull tissues and ligaments causing them to move. Muscles attached to the skeleton are known as skeletal muscles. They move the appendages, such as arms and legs. Smooth muscles are in organs and control body activity, such as movement in an intestine.

To correctly function, cells must have nutrients. Proper food is needed. The digestive system is responsible for breaking the food into useful substances. The system also voids the body of certain wastes.

The respiratory system provides oxygen to the body and rids the body of carbon dioxide. The process of breathing moves oxygen and carbon dioxide. Gas exchange with the blood occurs in the lungs. The circulatory system carries oxygen from the lungs for the cells and brings wastes to the lungs and kidneys. The heart and blood vessels are the major organs in the circulatory system. Blood is the liquid material that is transported.

The urinary system removes certain wastes from the blood and flushes them from the body. The kidneys contain capillaries that filter the blood. Wastes move from the kidneys to the urinary bladder. The waste liquid (urine) is squeezed from the body through the urethra by muscular contractions.

Other systems include the reproductive and mammary systems. Reproductive systems vary with the gender of the individual. Females have ovaries that produce ova and provide a good environment for the development of an embryo and fetus in the uterus. Males have testicles that produce sperm that may fertilize released ova. The mammary system develops in females. It secretes milk for nourishing babies.

LIFE PROCESSES

Humans carry out life processes similar to those of other animals. These processes include growth and repair, obtaining food and nutrients, circulation, respiration, secretion, sensation, movement, and elimination. Reproduction, also a life process, assures that the human species continues as new generations. (The life processes were covered in detail in Chapter 7.)

Humans have a highly developed brain, which gives us many abilities. These abilities distinguish us from other animals. We can speak, reason, and create technology. The ability of humans to stand upright and walk on two legs distinguishes us from other primates. Thumb and forefinger opposition makes it possible for humans to grasp objects. Humans live longer and develop slower than most other animals.

HUMAN DEVELOPMENT

Human development is the progressive maturation of an individual from birth until death. Various stages of physical and mental growth are involved. Some stages are associated with cultural conditions. These are often related in chronological order based on the age of an individual.

Culture consists of the ways individuals have developed to go about life. The characteristics of a culture are transmitted from one generation to the next. Of course, some changes occur along the way. These are a result of the social units people form, such as families, clubs, school groups, and church organizations. Contact with each other is much a part of human cultural interaction. Culture is also reflected in government actions designed to achieve certain goals.

Developmental Stages

The combination of biological and cultural advancement has several ***developmental stages***. Sequential growth and maturity events occur during these stages. The common stages are:

15-5. **Infants need careful attention to their needs to grow and develop properly.**

15-6. **A two-year-old finds just about everything fascinating.**

- Infancy—Infancy may be further divided into newborn and the remainder of infancy. The newborn stage of development is through about 1 month of age. Infancy continues through 1 year. Infants must receive proper care from their parents.

- Preschool—The preschool stage of development is age 1 year through 5 years. Increasingly, preschool is being lowered to age 4 years with the implementation of prekindergarten. For example, laws in Georgia now provide for kindergarten for children age 4. These are very formative years. Children of this age model behavior and develop value systems that reflect the environment in which they are living.

- Child years—The child years begin following preschool and extend through age 12. In the United States, children of this age are enrolled in elementary school and are developing important social and educational skills.

- Teenage years—The teenage years are ages 13 through 19. Many changes occur in the maturation of young people. Physical growth and mental capacities are about fully achieved. Puberty is reached. Considerable emphasis is placed on education and preparation for a career and life.

15-7. **The teenage years are a time for learning and developing for adulthood.**

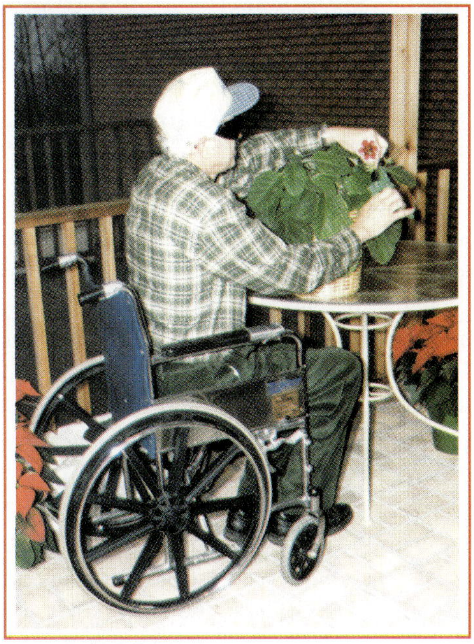

15-8. **Adults with disabilities may continue to be productive until declining health becomes a problem.**

- Adulthood—Adulthood is often said to begin at age 20. It extends through life until death. Adulthood is often divided into three stages: young adult, middle adult, and older adult. The young adult is ages 20 to 39. Culturally, this is an important stage in the career and family development of an individual. A middle adult is 40 to 65 years of age. Career maturation and completion of child rearing are a part of this stage. The older adult is above age 65. It is a stage of reflection and enjoyment of personal satisfaction. Normal age results in a slowdown of life processes. The ability to be productive continues unless declining health prevents mental or physical activity.

Biotechnology Supports Development

The use of biotechnology supports human development. It can be used to promote health and growth. These may vary with the age of the individual. Children may need hormones to promote proper growth. Older adults may need new tissues or organs or mechanical devices that support their independent living.

Ethical decisions are often involved. Which biotechnology is appropriate? Should it be used? What effect does it have on overall human well-being?

Career Profile

HUMAN BIOLOGIST

A human biologist studies the systems and processes of the human body. Areas of genetics may be investigated. How organ functions are damaged by disease are observed and recorded. Many of these scientists are involved in areas of biotechnology.

Human biologists need college degrees in biology, anatomy and physiology, or a related area. Most have advanced degrees. Some have medical degrees plus a doctorate in a specific area, such as genetics or physiology. Begin preparing in high school by taking science, mathematics, and computer courses.

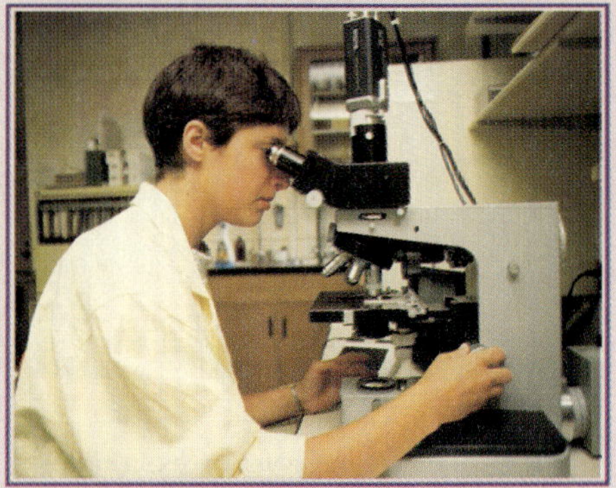

Job opportunities are found where human biology is studied. Medical and pharmaceutical companies, colleges and universities, and government agencies may have jobs for human biologists.

People vary in how they view the use of human biotechnology. Education helps people better understand what is involved. Once they understand, they can make the best decision.

HUMAN GENOMICS

The ***human genome*** is all of the genes that comprise the genetic makeup of human beings. It is the total genetic composition of humans. Efforts are now underway to identify all of the human genome. The identification of genes will allow the isolation, treatment, or replacement of those that are defective or cause disease.

HUMAN GENOME PROJECT

The ***Human Genome Project*** (HGP) is an international effort involving scientists, educators, and students in locating and identifying every human gene. The outcome will be construction of genetic and physical maps of the human genome. It is a large task that will help explain human development.

Some scientists feel that it will provide insight into the evolution of human life.

In general, this project has four complementary approaches:

1. Genetic mapping of the human genome.
2. Physical mapping of the human genome.
3. Sequencing the human genome.
4. Analyzing the genomes of other species (with comparisons to the human genome).

Background

The Human Genome Organization (HUGO) is an international group that coordinates work on the HGP among various nations. It was formed in 1988. Canada, Italy, Japan, France, the United Kingdom, and the United States have been active participants. The United Nations has promoted inclusion of developing nations in HGP work.

In the United States, work began in 1988 with formation of a committee by the National Institutes of Health (NIH) and the Department of Energy (DOE). The U.S. Congress provided funding in 1990 that officially began the HGP. A National Center for Human Genome Research has been set up at the NIH.

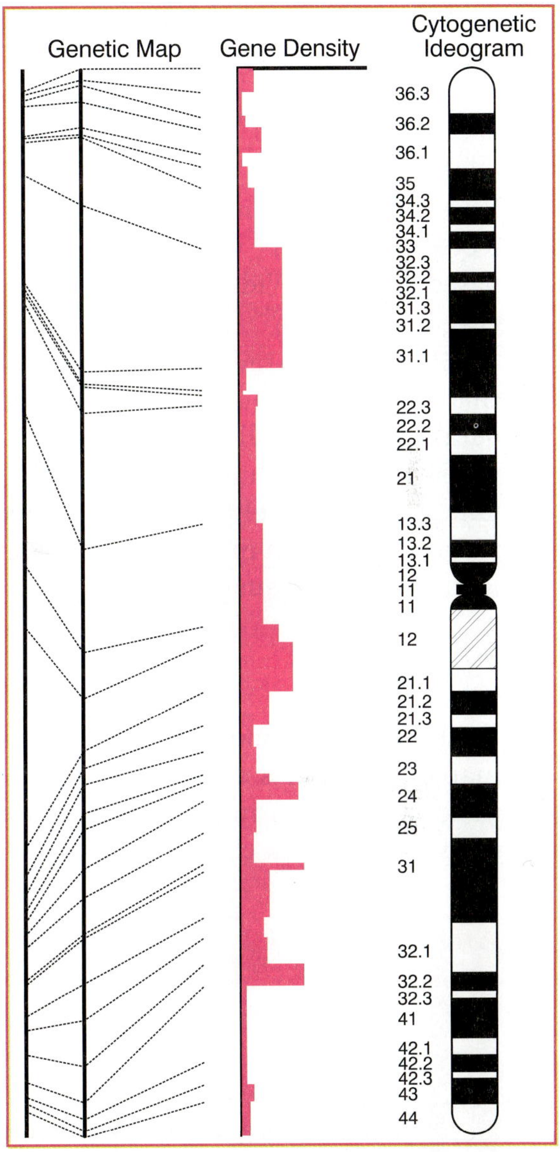

15-9. Genetic map, gene density, and cytogenetic ideogram for human chromosome 1. (A genetic map shows the location of a gene or genetic marker on a chromosome. A cytogenic ideogram relates the appearance of chromosomes to genetic phenomena.)

Efforts with mice, bacteria, and other organisms have aided the HGP. Three bacterial genomes have been completed. Plants are also being studied, with one plant (*Arabidopsis*) genome being completed. Early work focused on species with small genomes.

Initial planning was for a 15-year project to identify the human genome. Progress has been more rapid than expected. After the data have been collected, more years will be needed to identify gene expression and genetic-based diseases. Widespread use should be made of HGP information by the year 2010.

High school students are helping with the HGP. They carry out lab work that sequences a number of pieces of the genomic DNA. The work of students on very small pieces is being assembled by scientists. An example is the High School Human Genome Program at the University of Washington. One phase of this effort has high school students studying the mechanism of nicotine addiction in humans. Students in many high schools are working to identify DNA sequences that are involved.

Findings

The possible information in the human genome is tremendous. The human organism has about 100,000 genes. Just think about the amount of genetic information in 23 pairs of chromosomes, 22 autosomal chromosome pairs, and one sex chromosome pair (YY or XX). There are more than three

Biotech Connection

DNA GOES TO GREAT LENGTHS

DNA is very small stuff. Look at your skin, you can't even distinguish one cell with your unaided eyes! How could you ever see the DNA?

It is amazing what scientists have found! If you removed the DNA from one human cell, unwound it, and placed the pieces end-to-end, like a thread, it would be more than 5 feet long! One problem: You couldn't see it without a powerful microscope. The DNA would only be 50 trillionths of an inch wide!

How does one cell contain so much? Human life is truly amazing. Humans are complex organisms.

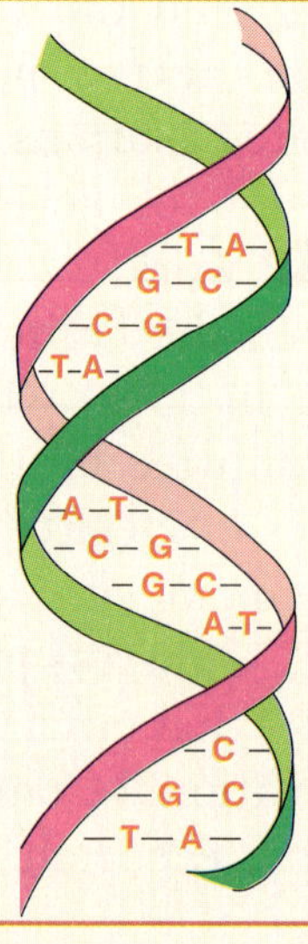

billion base pairs in the haploid genome. The human genome is far larger than that of a yeast cell or *E. coli* bacterium. (Note: An ***autosome*** is a chromosome that is not a sex chromosome. They are known as autosomal chromosomes.)

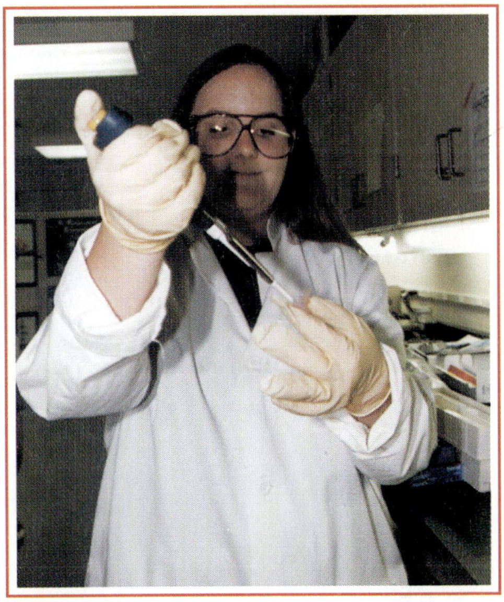

15-10. **A high school student in Arizona safely and precisely prepares materials for the High School Human Genome Program.**

Maps of the human genome will be generated by the HGP. Maps show the locations of genes and assign them to chromosomes. Two kinds of maps are being generated. Physical maps show the locations of and distances between genes and DNA fragments in specific regions of chromosomes. These maps show the complete DNA sequence of all chromosomes. The genetic linkage maps show the relative arrangement and distances between genes and markers on the same chromosomes.

Once mapped, the DNA is sequenced. This provides an orderly arrangement of the nucleotide bases of the chromosomes. The genes in the DNA are identified. These maps have powerful uses in human disease with genetic linkage.

METHODS IN THE HSGP

The High School Human Genome Program (HSGP) uses gel electrophoresis methods. As covered in a previous chapter, this involves separating genetic information based on size, electric charge, and other physical properties. An electric current moves molecules across a span of gel. Students in many schools carry out gel electrophoresis as part of the High School Human Genome Program.

Gel electrophoresis has other uses. It can help identify people who have committed crimes, are related, or have other traits.

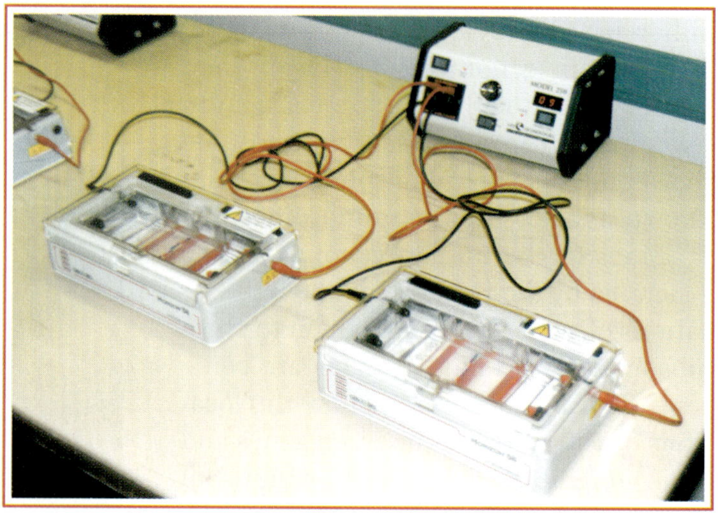

15-11. A small gel electrophoresis apparatus that is battery powered and has two tiny gel electrophoresis boxes.

A SCENARIO

Think of the gel as a race track. You are in a hot air balloon observing a horse race. When the gates are opened, you can see all four horses running at once. The race is a sprint. One horse moves ahead of the others. You can take a snapshot photo when the first horse reaches the finish line.

Assume that the first place horse was a micro-molecule. The other horses are macro-molecules. What does the photo you made of the finish line show? Relate the image to the use of gel electrophoresis.

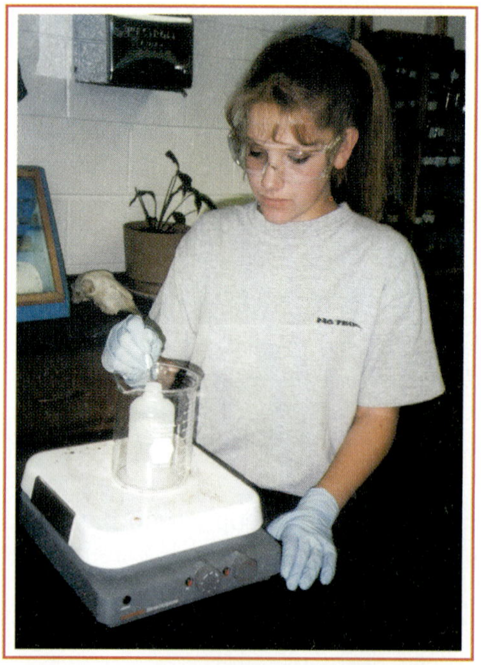

15-12. A Nevada high school student is melting gel for electrophoresis.

AN EXAMPLE

Small samples of DNA from trace amounts of blood or semen can be studied with gel electrophoresis and PCR (polymerase chain reaction). The samples are appropriately loaded into the gel and electric current is applied. The molecules in each lane will separate and form bands that are the actual genetic fingerprint of the DNA sample. Of course, the

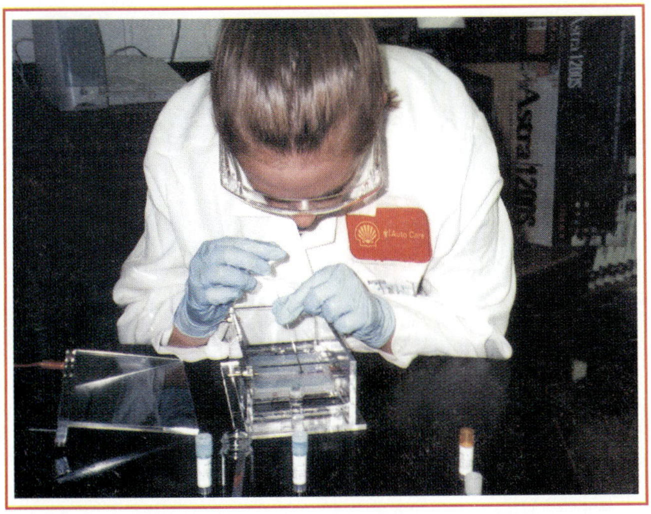

5-13. Wells are being loaded with dye for gel electrophoresis.

specific and appropriate protocol must be used. If not, the process will not work.

Here is one example of electrophoresis: (Note: Follow all safety procedures, including using eye and skin protection. Also, follow the protocol of an experienced teacher or scientist.)

1. Make an electrolyte buffer (54 g Tris base and 27 g boric acid dissolved in 1.0 L water)
2. Make an agarose gel.
3. Put the electrophoresis rig together. (Some may use a kit that has an assembled rig.)
4. Prepare the dye samples. (Red, green, blue, and yellow.)
5. Load the samples into the gel.
6. Race the dyes. (Apply electric current.)
7. Observe the results. (Explain what happened.)

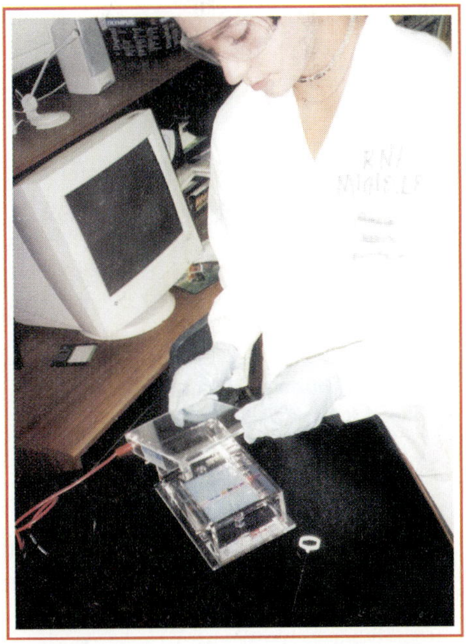

5-14. The wells have been loaded and the electrophoresis box is being closed. Electric current will be applied.

The characteristics of certain segments of DNA vary from person to person.

They form highly individual, detectable genetic fingerprints. It has been widely used in courts of law as evidence.

HELPING HUMANS LIVE BETTER

Since the 1970s, scientists have been manipulating genes at the molecular level. These efforts serve a range of attempts to improve human well-being.

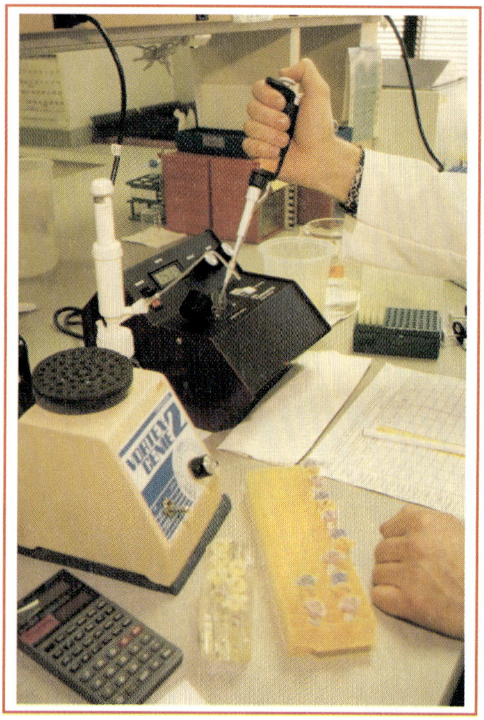

15-15. A human DNA sample is being prepared for analysis.

An example is Down's syndrome. This is a disease in which a baby is born with an extra chromosome. One in 600 to 800 babies is born with Down's syndrome. The chances of having a baby with Down's syndrome increase with the age of the mother. For example, the chance is 1 in 365 for a 35-year-old woman but only 1 in 1,205 for a 25-year-old woman. A baby's with Down syndrome has a small head, enlarged tongue, eyes that slant upward, and other traits. Mental development is slowed. Many die within a year of age. Until recently, no treatment was available. Today, genetic methods are being used to repair this defect.

GENE THERAPY

Gene therapy is the transfer of genes to a human organism. The makeup of the human is altered. The procedure is often quite controversial. Several government agencies have regulations about using transgenic technology with humans. Once therapy is done, a human is a GMO (genetically modified organism). Even though it is controversial, it has had benefits and more benefits are likely in the future.

Two general approaches are used with gene therapy: germ-line gene therapy and somatic cell gene therapy.

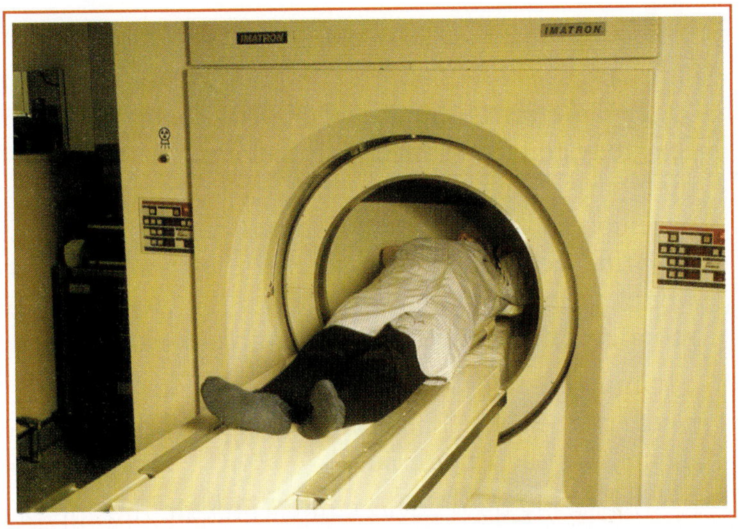

15-16. **A CT (computed tomography) scanner is used to gain a detailed image of the brain or other body part. The machine passes an x-ray beam through the body. The intensity of the beam as it emerges from the body is detected and recorded for analysis. This is an example of technology used in human health care.**

Germ-line gene therapy changes the cells that produce sperm or ova in an individual. Offspring, if any, will be artificially changed. It is considered illegal and unethical in humans.

Somatic cell gene therapy is used with cells in the body not associated with sperm or ova production. It changes the individual, but not any offspring that the individual might produce.

Gene therapy is very difficult to do successfully. One example is its use to replace defective genes in the lungs of people with cystic fibrosis. Another is the injection of genetically altered cells into cancerous prostate glands of men.

VACCINES

Vaccines help the body recognize and fight disease. Conventional vaccines use weakened or killed germs or a virus to introduce antigen proteins attached to viruses. These are used by the immune system to fight disease. The body builds resistance to the targeted disease.

Biotechnology vaccines may be used that consist of the antigen and not the virus. By isolating antigens and producing them in the lab, it is possible to make new vaccines that cannot transmit the virus itself.

A vaccine now approved for hepatitis is made by inserting the gene responsible for this antigen into yeast cells. During the formation process, each yeast cell makes a perfect copy of itself and the antigen gene. The antigen is later purified. When injected into the body, the antigen stimulates the production of antibodies that combat hepatitis.

PROSTHETICS

Prosthetics is the use of artificial parts for the human body. It includes the study and use of mechanical devices to replace or supplement natural human parts. This includes joint replacements (such as knee and hip), heart pacemakers, and intraocular lenses (those implanted into the eye). A ***dialysis machine*** is a kind of prosthesis that performs functions normal kidneys would do in removing wastes from the blood. Designing and developing appropriate prosthetic devices requires considerable research and testing.

Implanted devices must be mechanically fit for the purpose and not rejected by the body. This is known as ***biocompatibility***. Plastics are often used in making such materials.

Externally used devices must provide the appropriate mechanical action and support. Individuals with leg or arm amputations may have prosthetic devices that allow considerable freedom to lead a near-normal life. Injured joints, muscles, or nerves may be protected or strengthened with mechanical

Biotech Connection

SHARING LIFE-GIVING BLOOD

Blood is the liquid that flows in the human body. It is essential for human life. Without it, life ceases. When blood is diseased, life does not go on without problems. Fortunately, people can share blood. This is done is a special, safe way.

Blood banks across our nation collect, type, process, store, and distribute blood. Many blood donors have been recognized for their important gift to the health of others. The blood is carefully drawn from the vein of a donor using sterile equipment. The amount taken is known as a unit, which is 450 ml or nearly a pint.

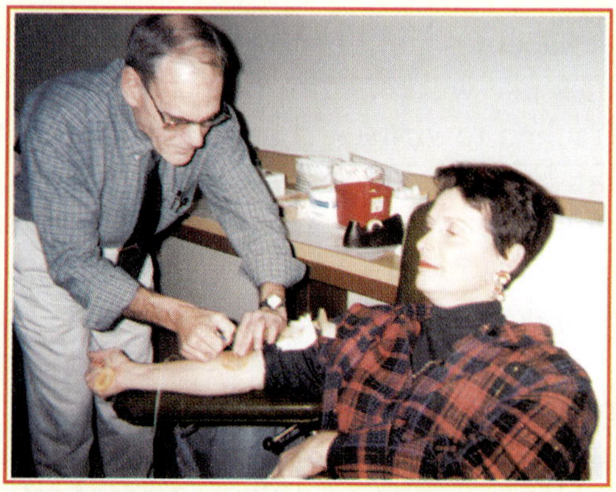

Transfusions are used to give blood to those who need it. The blood type of the donor and recipient must match. Just as it was taken from the vein of a donor, it is given into the vein of a recipient. A transfusion restores the blood lost during major surgery or in a bad accident. This photo shows a donor giving blood at a local blood bank.

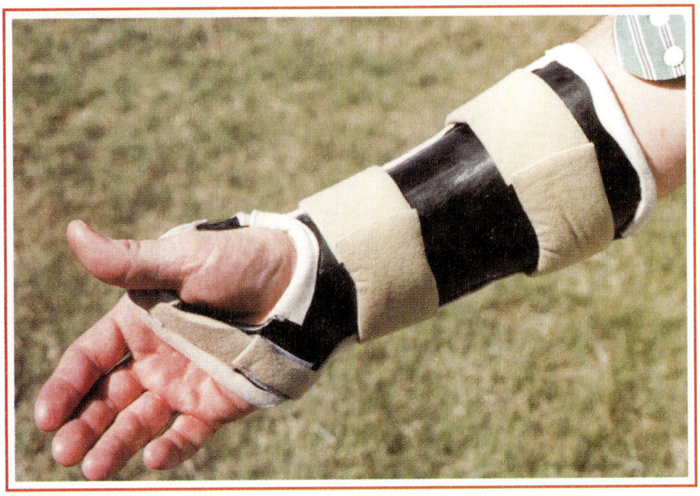

15-17. An external prosthesis that promotes healing from carpal tunnel syndrome.

devices. Carpal tunnel syndrome is a condition that develops when people do the same activities over and over. The repetitive motions injure nerves and muscles. Mechanical devices are often used to promote healing of carpal tunnel conditions.

MEDICINES

Medicine is any substance used to treat conditions that impair human health. Medicines often promote efforts of the human body in fighting off disease. They are sometimes referred to as drugs or pharmaceuticals.

Medicines are used in many ways. Large numbers of different products are available. Some are used to control blood sugar. Others are used to treat anemia, cystic fibrosis, growth deficiency, hemophilia, leukemia, genital warts, transplant rejection, and many forms of cancer.

Considerable testing is needed before a product is used to treat human disease. Large pharmaceutical companies do extensive research before a new drug is released.

DIAGNOSTICS

With humans, ***diagnostics*** is the identification of a human health problem or other condition. Knowing the approach to use in treating a disease requires accurate diagnosis of the condition. Failing to accurately diagnose a disease results in the wrong treatment. The person may die because of incorrect diagnosis.

GENOMICS AND COMPUTERS

Increasingly, diagnosis relies on human genetic information. Much more use of genetics will be possible when the Human Genome Project has been completed.

Computer databases are being used in diagnostics. Computers help match genes with health conditions. An example is the disease known as neurofibromatosis. A match between the protein made by normal copies of a newly uncovered gene was made in only a month after discovery of the genetic error involved in the disease.

MICRODISSECTION AND MOLECULAR ANALYSIS

Microdissection is a method of obtaining pure cells from diseased sections of tissues. Very small numbers of cells may be taken, such as from a small lump in a breast, blood vessel, or other organ. Microdissection is often used to test for precancerous cells or invading cancer cells.

Molecular analysis can be used to study the cells for evidence of disease. This allows comparisons with diseased DNA stored in microdissected libraries or bulk tissue libraries.

REVIEWING

MAIN IDEAS

Human beings are similar to, though set apart from, other animals. They have cells, tissues, organs, and organ systems that help them carry out life processes. Brain size and development is a major difference from other animals that allows humans to perform higher-order tasks. Disease results if an organ or system fails. Disease can also result from genetic and pathogen sources.

Humans go through stages of development. These are biological as well as culturally based. Biotechnology is sometimes used to promote appropriate development.

Human genomics deals with the genetic makeup of human beings. The Human Genome Project (HGP) is an international effort to locate and identify every human gene. Scientists, educators, and students are involved in this massive effort.

Biotechnology can help humans live better. Gene therapy can be used to treat certain genetic diseases. Genetically engineered vaccines can help the body develop antigens. Prosthetics can help provide devices to support or replace failed body parts or systems.

Diagnostics is aided by technology. Computer systems allow large data bases to be used in diagnostics. Microdissection and the use of molecular analysis also promote human well-being.

QUESTIONS

Answer the following questions using correct spelling and complete sentences.

1. What is the scientific name of a human being?
2. How are the life processes of human beings similar to other animals?
3. What major characteristics set humans apart from other animals?
4. What is human development? What stages are involved? How are the stages related to culture?
5. What is the Human Genome Project? Why is it important?
6. What is gene therapy? Should it be used?
7. What is prosthetics? Should it be used?
8. What approaches are being used in diagnostics?

EVALUATING

Match the term with the correct definition. Write the letter of the term in the blank that is provided.

a. human genome
b. dialysis machine
c. human development
d. prosthetics
e. gene therapy
f. diagnostics
g. culture
h. microdissection

_____ 1. The ways humans have developed to go about life.

_____ 2. The identification of a health problem or condition.

_____ 3. Progression of maturation of an individual from birth until death.

_____ 4. All of the genes that comprise the genetic makeup of human beings.

_____ 5. Transfer of genes to treat human disease.

_____ 6. A method of obtaining a small number of pure cells for diagnosis.

_____ 7. A kind of prosthesis.

_____ 8. Use of devices to replace or supplement human body parts.

EXPLORING

1. Use the Internet to explore the Human Genome Project. Investigate the procedures that are used and the progress that is being made. Select a particular chromosome and identify the genes that are involved. Also, explore how your school might participate in the HGP. Here is a place to begin: www.nhgri.nih.gov/HGP/

2. Use gel electrophoresis to study electric field movement of genetic molecules. Perform this activity under the supervision of a teacher or other qualified person. Follow all safety procedures.

3. Organize a debate in your class on the use of gene therapy in human medicine. If you prefer, narrow the debate to germ-line therapy or somatic cell therapy.

16

FOOD BIOTECHNOLOGY

Demand for food is up. In October 1999, the number of people on Earth reached six billion. The number will continue to increase. Each and every person needs proper food.

People want tasty and nutritious food. Food is important for several reasons. First, it provides nutrients for growth and health. Secondly, eating is a social activity. People invite friends to dinner or seek out a person with whom to eat. Thirdly, the food industry is a major economic force in the United States and world.

Biotechnology has been used in food production for a long time. It likely began with the domestication of plants and animals. The first efforts in food storage might have resulted in fermentation and other preservation simply by chance. The accidental presence of microbes likely led to yeast in bread, vinegar for pickling, and fermented beverages.

Opportunities are expanding in food biotechnology. Many people will likely work in this area in the future. Perhaps there is a career in it for you.

16-1. **Food scientists are evaluating loaves of bread made from a blend of durum and spring wheats. (Courtesy, Agricultural Research Service, USDA)**

OBJECTIVES

This chapter deals with the application of biotechnology in food production. It has the following objectives:

1. Define food biotechnology.
2. Explain the meaning of food science and describe areas of science used in food science.
3. Explain why food supply is important.
4. Describe relationships between food microbiology and biotechnology.
5. Explain relationships between catalysis and biotechnology in selected foods.

TERMS

catalysis
food biology
food-borne illness
food chemistry
food physics
food processing
food quality
food science
nutrient
per capita consumption
shelf life
spoilage
substrate
yogurt

FOOD BIOTECHNOLOGY

Food is anything that is eaten or drunk as a source of nutrients for the body. The nutrients are used for growth, maintenance, and carrying out other functions of life. Some people supplement the nutrients received from food with nutrients from nonfood sources. Biotechnology is being used more and more in providing food.

Food biotechnology is the use of technological innovations in manipulating food production and processing. As presented earlier in the book, some forms of biotechnology have been used for hundreds of years. This, however, is not the major focus of modern biotechnology.

16-2. Consumers expect to find quality food products when shopping.

Today, food biotechnology deals with cellular and subcellular manipulation of organisms. Tissue culture is sometimes included. Some scientists, however, focus beyond tissue culture. They focus on recombinant DNA (rDNA) techniques in genetic engineering. Much of this work occurs with plants and animals that are used as sources of food.

FOOD SCIENCE

Food science is the study of the substances humans eat. This includes food makeup, nutritional content, and ways of making food more available and appealing. The chemical, physical, and biological sciences are included.

FOOD CHEMISTRY

Food chemistry is the study of the substances that make up foods. The building blocks of food are the cellular structures formed by living organisms used as food sources. These structures consist of chemical elements. These may form proteins, carbohydrates, fats, water, and other food nutrients.

Food chemistry often begins with the study of water. Many food substances contain a high percentage of water–just think of the juice in an orange or a beef steak! The three states of water–liquid, solid (ice), and vapor (steam)–have a big impact on food science. These states are involved in all aspects of food production, processing, and preparation. Regulating the amount and state of water influences the ability to have good food.

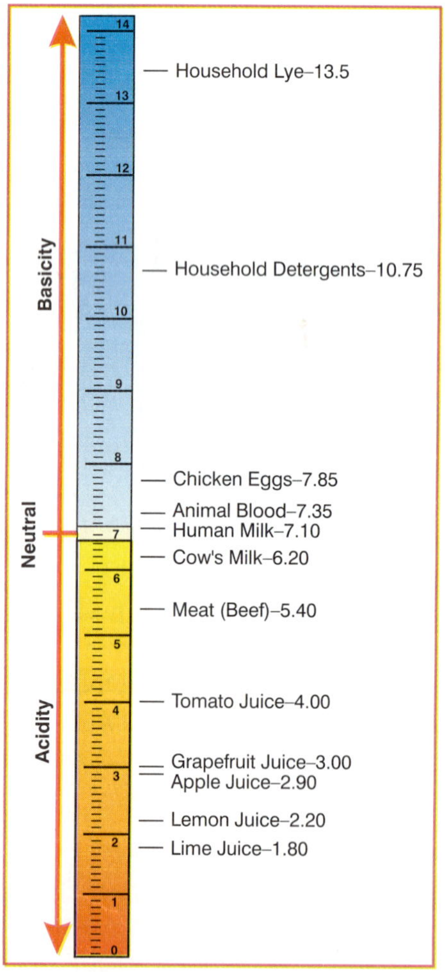

16-3. Approximate pH values of selected foods and materials used in cleaning areas where foods are handled.

Another important part of food chemistry is the pH of the food substances. pH refers to the acidity or basicity of the substances. pH is measured on a 14-point scale, with 7.0 being neutral. Values below 7.0 are acidic; those above 7.0 are basic. Knowing the pH of foods is essential in food study and manufacture. Milk has a pH range of 6.2 to 7.3, which means that it is neutral for all practical purposes. Fruit juices, on the other hand, are acidic. The pH of lemon juice is 2.2 to 2.4; the pH of tomato juice is 4.0 to 4.2; and apple juice is 2.9 to 3.8. Meat, fish, peas, corn, and many other common foods have a pH of 5.3 to 6.0.

Food chemistry also includes the study of nutrients. A ***nutrient*** is a substance in food that supports life processes. Most foods have been analyzed to determine the nutrients they contain. This helps provide a balanced ration for animals and a proper diet for humans.

The major food nutrients are proteins, fats, and carbohydrates. Proteins are formed of amino acids. The major role of protein is

Biotech Connection

PEANUTS: MIRACLE CROP

Peanuts have many uses. Mostly, they are a good source of nutrition in the human diet. They are produced by peanut plants. The nuts are more like peas than true nuts. The pods form in the ground on the end of a "peg" that grows from airborne flowers. As legumes, the plants fix nitrogen in the soil.

George Washington Carver (1864-1943) did considerable research on peanuts. He developed over 300 uses in his lab at Tuskegee Institute, Alabama. Peanuts can be boiled, roasted, ground, or pressed for oil. Dr. Carver even produced a milk substitute from peanuts! Byproducts can be used in making soap, face powder, shaving cream, shampoo, and cooking oil. Some oil is even used as an alternative fuel.

But, watch out for the allergy. A few people are very allergic to peanuts or any foods containing peanuts, peanut oil, or other products. Always alert guests to the presence of any peanut products in food. (Courtesy, Agricultural Research Service, USDA)

growth and repair of cells and tissues. Fats are known as lipids. They provide energy. Carbohydrates are typically from sugars and starches. Carbohydrates provide energy. Carbohydrate energy is less concentrated than fat energy. In fact, fats have 2.25 times more energy than carbohydrates.

16-4. **A potato appears shriveled because it has lost water. (This potato was held in cold storage for 1½ years. It has not rotted.)**

FOOD PHYSICS

Food physics is the study of the physical characteristics of food. A major focus is the relationship between food constituents

16-5. The traditional way of drying raisin-grapes in dry climates is to spread them on paper in the vineyard. (Courtesy, Agricultural Research Service, USDA)

and heat. The way most foods are affected by heat involves the water found in the food substance. Water can be added to food, removed from food, or the state of the water can be changed.

Water activity in food can be controlled to some extent and used to maintain food quality. The attraction of water molecules to themselves and to other materials creates a kind of tension in food substances. This tension relates to other physical processes that occur in foods.

Osmosis influences food characteristics. Osmosis is the flow of water from places of higher to lower concentrations. It influences the quality of many foods. Wilted vegetables, such as lettuce, have lost water. The cells in the lettuce no longer have turgor (water pressure).

Food physics includes the effects of the states of water on food. Important among these is freezing food to prevent decay. While frozen, the water molecules are not available to microbes that cause decay. Removing water from food also helps prevent decay. The raisin is a good example. A raisin is a dried grape. Drying grapes removes water and creates a condition that does not favor microbe growth.

FOOD BIOLOGY

Food biology focuses on the characteristics of living organisms and their products that are used for food. It includes how vegetables, grains, meats, eggs, milk, and other foods are kept wholesome and prepared for consumption.

Food scientists often focus on the microbiology of foods. This deals with the microbes that are on or in the food substances. It includes how these organisms may be used to improve foods or produce other foods as well as how

16-6. Temple oranges have spoiled due to the presence of microbes.

these organisms cause food to become unfit. More details on food microbiology are given later in the chapter.

FOOD SUPPLY

The quality and availability of food on the earth varies widely. Some nations have adequate food; other nations have masses of people who do not have sufficient food. Food from plant sources provides 80 percent of the protein and 88 percent of the calories (carbohydrates and fats) in the human diet. Most of this food is provided by fewer than 15 species of plants. Foods from animal sources are less important worldwide though more important in the United States.

16-7. Family gatherings during the holidays often feature an abundance of good foods.

CONSUMPTION

Consumption is the use of food substances. The average amount of food consumed by each individual in a year is known as ***per capita consumption***. The information is calculated based on all people

Career Profile

FOOD SCIENTIST

A food scientist studies foods and the systems used to provide quality food. The scientist uses chemistry, physics, and biology. Most are also involved in engineering areas of food product development and production. Food scientists must be good in communicating with technicians and others in a lab or processing plant.

Food scientists need graduate degrees in food science. Most have undergraduate degrees in agriculture, biology, chemistry, or related areas. Only a few universities have programs of study in food science. Good preparation in science and agriculture at the high school level will be useful. Practical experience is also beneficial.

Jobs for food scientists are often at research labs or processing plants. Some jobs are in new food-product development. Other jobs deal with assuring food quality in manufacturing plants. This photo shows a food scientist studying semolina pasta being extruded in a research lab. (Semolina is the part of a durum wheat kernel that is resistant to milling.) (Courtesy, Agricultural Research Service, USDA)

whether they eat a particular food or not. Here are a few examples of annual per capita consumption of foods in the United States:

- potatoes—136 pounds
- vegetables (fresh, frozen, and canned)—244 pounds
- red meats (beef, veal, lamb & mutton, and pork)—163 pounds
- chicken and turkey—61 pounds
- eggs—30 pounds
- fish and shellfish—15 pounds
- milk, cheese, and other dairy products—244 pounds
- fresh fruit and melon—124 pounds
- grains (rice, flour, and cereal products)—189 pounds

Some of these food products are more nutritionally adequate than others. Income, cultural preferences, and other factors determine the consumption patterns. In all cases, people need to have quality food.

QUALITY

Food quality is the ability of a food to meet nutritional needs. The food is wholesome. It has not spoiled nor is it contaminated with other substances. Quality food is safe to eat. Biotechnology has a role in all aspects of food quality.

16-8. Broilers are being inspected for food safety using a scanning system in a processing plant. (Courtesy, Agricultural Research Service, USDA)

Processing

Food processing is the preparation of food materials for consumption. It includes a wide range of activities in preparing food for purchase by the consumer. It typically does not include the actual preparation in homes and restaurants. Much of the role of processing is to keep food safe and make preparation easy and quick.

Food processing often includes various forms of preservation. Canning, freezing, refrigerating, drying, irradiating, pickling, and salting are common ways of preserving food. The purpose of these activities is to extend the amount of time in which food can be kept safe to eat. It prevents spoilage.

Spoilage

Spoilage is a condition that makes food unsafe to eat. The appearance, texture, odor, and flavor of food change. Spoilage is most obvious as food decays. Rotten spots on apples or mold on bread are two examples. Spoiled foods can cause food poisoning resulting in disease or death. Botulism is likely the best-known poisoning from food. Some spoiled foods have a changed appearance and odor.

16-9. A leak in a can of sweet potatoes is evidence of spoilage.

Food spoilage is caused by several conditions:

- Microorganisms—Microorganisms can grow in or on food. Some of the microorganisms are easily seen; others require magnification. We can easily see mold (fungi) on bread. We cannot easily see microorganisms, such as bacteria, in food materials.
- Insects—Insects damage food products. Eating holes in lettuce leaves or burrowing into a tomato is visible and not appealing to the consumer. Other insect damage is less obvious. In some cases, insects or insect fragments may be found in food.
- Chemical contamination—Food products are sometimes contaminated with residues of chemicals. Pesticides remaining on foods damage food quality and may cause disease when the food is eaten.
- Purity—Food products can sometimes be mixed with other foods, with weed seed or leaves, or with foreign materials. People want pure forms of food. They do not want tiny pebbles in beans nor weed seed in rice.
- Undesirable growth—Some food products can sprout. Sprouting results when these foods send out shoots to start another plant. Improperly stored seed can sprout. Onions, potatoes, and carrots can begin growing stems and leaves. These make the foods undesirable.

Storage and Transportation

Most foods are harvested and stored until needed. Many are processed in some way. Foods are transported from the point of production to a supermarket or restaurant convenient to consumers. Several weeks or months may go by. Consumers want the food to remain in good condition.

Shelf life is the length of time that a food can be stored and be safe to eat. Steps can be taken to increase shelf life. Refrigeration and canning are commonly used to increase shelf life. Freezing, drying, and salting are also used.

Biotechnology has been used to create conditions that promote longer shelf life as well as create new products with a longer shelf life. Fermentation methods increase shelf life. For example, grape juice that has fermented has longer shelf life than unfermented juice. Cucumbers that have been pickled

16-10. Eggs stored for several months will lose much of their water. This egg stored for six months has only a dried yolk and no albumen (white).

in solutions of vinegar, salt, and sugar can be kept much longer than cucumbers that have not been pickled. Pickling creates an environment in which spoilage organisms cannot readily grow.

Modern biotechnology has been used to extend storage time for foods. Genetic engineering has been used to create products with desired traits, such as longer shelf life. The first such product was the Flavr Savr Tomato in 1994. Introduction of the tomato sparked some public controversy. Some people were afraid that the product might not be wholesome food. That was false!

FOOD MICROBIOLOGY

Food microbiology deals with the microorganisms that are present in or on food. It is the study of microbes and how they affect food. Some microorganisms cause spoilage. Others cause illness or death if the food is eaten. Some microorganisms serve useful roles in fermentation and produce highly desired food products. That is how we get cheese, leavened bread, and pickled foods!

The most important microbes in foods are molds, yeasts, bacteria, and viruses. There are many different species of each. Some are useful in having desired food products; others cause spoilage.

Once natural processes and microbes are understood, biotechnology can be used to improve efficiency of the processes. Organisms can be genetically

16-11. Yeast is used in baking bread. The yeast produces a gas that causes the bread to "rise." Bubble areas or holes in bread are evidence of gas pockets that were present during baking.

modified to promote desired products. Biotechnology can be used to keep foods free of food-borne disease.

USEFUL MICROBES

The microbes that are useful in food products are those that alter the taste, texture, or other characteristics into forms that are more desirable. Food scientists must know which kinds to use and how to use them. (Chapter 12 provides background information on microbiology.)

Biotech Connection

NAMING ENZYMES: KNOWING THE ASE

How are enzymes named? Many are named by adding the suffix "ase" to a word that tells something about the enzyme. The words are usually those that indicate the nature of the reaction involved or the nature of the substrate with which the enzyme combines. For example, a lipid is a fat. An enzyme that acts on fat is a lipase.

As with most naming systems, there are exceptions. The exceptions are typically enzymes that were identified before using "ase" was started.

A few enzymes that have different names deal with food digestion. Trypsin and pepsin are two examples. Trypsin is produced by the pancreas. It is a catalyst in the breakup of paired amino acids. Pepsin is an enzyme in the acid of digestive fluids that acts on protein.

A good example is yogurt. ***Yogurt*** is made from the milk of various animals, including cows, goats, and buffaloes. Two species of bacteria (*Lactobacillus bulgaricus* and *Streptococcus thermophilus*) are added to milk to form a semisolid material known as curd. Curd is formed when the casein in milk changes to a thickened mass by a process known as coagulation. Since yogurt has been made for hundreds of years, it is likely that the first occurrence was by accident. People liked the resulting product. Some portion of the product was saved as "starter" to get other milk to form yogurt. The bacteria convert the milk sugar (lactose) into lactic acid. The acid gives yogurt its desired flavor and curtails the growth of spoilage microbes. Today, various fruits and other foods are added to yogurt to give a variety of desired flavors. Yogurt is refrigerated to increase shelf life.

Table 16-1. Examples of Useful Microbes in Food Production

Products	Microbes	Raw Materials
dill pickles	lactic acid bacteria*	cucumbers
sauerkraut	lactic acid bacteria	cabbage
vanilla flavoring	lactic acid bacteria	vanilla beans
salami (summer sausage)	lactic acid bacteria, *Scopulariopsis*, *Aspergillus*, and *Penicillium* spp.	pork, beef
country-cured hams	*Aspergillus* spp. and *Penicillium* spp.	pork (meat of swine)
cheese	lactic acid bacteria	milk
Swiss cheese	lactic acid bacteria and propionic acid bacteria	milk
sourdough bread	lactic acid bacteria and yeast	cereal grains**
leavened bread (regular bread)	yeasts	cereal grains
soy sauce	lactic acid bacteria and molds	soybeans
poi	*Geotrichum candidum* and lactic acid bacteria, and yeast	taro corms
beer	yeasts	cereal grains
wine	yeasts	fruit juices***
vinegar	acetic acid bacteria	wine
tempeh****	*Rhizopus oligosporus*	soybeans

*lactic acid bacteria include several species unique to a particular food product
**includes wheat, rye, and similar grains
***most often grape juice but can be any fruit, such as that of plums and peaches
****a food primarily produced in Indonesia

16-12. Molded snack food is unfit to eat.

16-13. Good sanitation and thorough cooking help eliminate problems with food poisoning. Wash dishes in hot water.

HARMFUL MICROBES

Harmful microbes are those that cause food to decay and become unfit for human consumption. If consumed, the food may cause disease or death. One example is the contamination of ground meat products with *Escherichia coli*. Outbreaks have occurred in various locations throughout the United States. The source of each outbreak has usually been traced to contamination at a packing plant. Most *E. coli* contamination is from feces coming into contact with the food products.

A ***food-borne illness*** is a disease that people get from the food they eat. Some people may refer to the disease as food poisoning. The nature of food-borne illness varies. Symptoms may range from mild discomfort to an upset stomach, fever, and, in severe cases, death. Food-borne illness may result from improper preparation, storing, or handling.

Food-borne illnesses are of two major types: infection and intoxication. An infection is a disease caused by ingesting a harmful microorganism.

Food-borne intoxication is caused by ingesting toxins produced by a microorganism. Botulism is the most widely known food toxin. It is caused by the *Clostridium botulinum* bacterium. These toxins are extremely dangerous. They may affect the nervous system, cause double vision, make speech and swallowing difficult, cause respiratory paralysis, and result in death. A day or more may be required for the disease to develop. If diagnosed early, a physician can administer an antitoxin to treat botulism.

Other examples of food-borne illnesses are salmonellosis, caused by *Salmonella* spp. bacteria; illness caused by *Staphylococcus aureus* bacteria; and ergotism, caused by *Claviceps purpurea* mold (fungi).

CATALYSIS

Catalysis is the alteration of the rate at which chemical reactions occur. As used here, catalysis refers to chemical reactions in foods. Some of the changes cause food to deteriorate. Other changes improve the quality of food materials. One example is the ripening of fruit. As fruit ripens, the sugar content increases. Once ripe, decay may follow if the fruit is not preserved in some way.

Processes can be used to alter natural changes in foods. These involve using substances known as catalysts. A catalyst is a chemical that increases or decreases the rate of change. The rate of natural ripening, decay, and other changes can be regulated using catalysts. Catalysts are not used up in a reaction, but participate in the reaction.

ENZYMES IN FOOD

Since most foods are of biological origin, enzymes are of major importance. An enzyme is a protein substance produced by living cells. Many kinds of enzymes are found. They comprise a group of proteins that regulate organism growth and life processes. Enzymes act as catalysts in chemical changes in cells. Each enzyme has a specific function. These functions cause changes that result in foods having desirable or undesirable qualities.

Enzymes require substrate molecules to work. A ***substrate*** is a substance on which a specific enzyme reacts. Enzymes lock onto substrate molecules.

16-14. **Millers are checking the performance of a wheat cultivar in the milling process. (Courtesy, Agricultural Research Service, USDA)**

Some fit much as a key goes into a lock; others fit by interaction of the enzyme and substrate.

Substrates are consumed by microbes when food decays. Some enzymes can be used on substrates so microbes do not readily consume the substrate. This prevents food decay. Food scientists have learned how to promote and control enzyme activity in most foods.

CHANGES BY ENZYMES

Enzymes have big impacts on foods. Some cause desirable change; others cause undesirable change.

Enzymes are classified by the types of substances they affect. Lipases affect fats (lipids). Carbohydrases affect carbohydrates. Proteases affect proteins. There are other enzymes that cause effects to occur on other substances.

Different enzymes cause different changes in foods. Some enzymes cause desirable changes, such as the use of amylase to prevent bread from becoming stale. Other enzymes produce undesirable changes–peroxidase causes unblanched vegetables to soften.

Some enzymes that create desired qualities in foods can also lead to undesirable qualities. Polygalacturonase is an example. This enzyme promotes fruit ripening but, if not stopped, it will continue ripening to the point where fruit is said to be overripe. Food biotechnology involves determining how to stop the action of polygalacturonase to assure that fruit has a good shelf-life and is properly ripened.

ENZYME ACTIVITY

Enzyme activity is influenced by several conditions. Temperature and pH are important. The presence of other substances may also influence enzyme activity.

Table 16-2. Examples of Desirable and Undesirable Changes in Foods by Enzymes

Desirable changes:	
polygalacturonase	promotes fruit to ripen
pectinase	removes juice from pulp of fruit
invertase	aids in production of soft-center chocolate-covered cherries
naringinase	removes bitterness from some fruit juices
amylase	clarifies fruit juice drinks
rennin	causes formation of curd in milk for cheese
trypsin	promotes protein digestion by pepsin
esterase	enhances cheese flavor development by lipase
amylase	promotes digestion of starches
lactase	produces low- and no-lactose dairy products (needed by people with lactose-intolerance)
Undesirable changes:	
protease	promotes food rotting
microbial lipase	promotes rancidity in fat
polygalacturonase	causes over-ripening of fruit
polyphenolase	promotes browning in some fruits and vegetables
peroxidase	promotes softening of unblanched vegetables

Warm temperatures may promote enzyme activity, while cold and hot temperatures stop enzyme activity. In general, enzyme activity is doubled for every 50°F (10°C) of increase in temperature until the point is reached where enzyme activity stops. The point where enzyme activity is at is maximum is known as the optimum temperature.

16-15. A margarine that promotes good cholesterol and discourages bad cholesterol is being used as a spread. The margarine is made from genetically engineered canola.

In processing vegetables, blanching is used prior to freezing to make catalase, peroxidase, and lipoxygenase enzymes inactive. With fresh fruit, refrigeration is used to prevent browning and rot.

Most enzymes have an optimum pH range. Changes in pH alter the enzymes that are active. A good example is in the human digestive system. In the stomach, the acid level may have a pH of 2.0. This promotes the active breakdown of protein. pH changes as food moves into the small intestine, which has a pH of about 8.0. Increasing the amount of substrate present will also increase enzyme activity to a level of excess substrate.

Once enzymes in foods are understood, biotechnology can be used to control their functions. In addition, artificial products can be produced that will replace or enhance natural enzymes in foods.

REVIEWING

MAIN IDEAS

Food provides the nutrients needed for human growth, maintenance, activity, and other purposes. The major nutrients are proteins, carbohydrates, and fats. Food biotechnology is using technology in food production and processing.

Food science includes the chemical, physical, and biological sciences. Water, pH, and protecting food from microbes are important in assuring food quality. Food from plant sources provides 80 percent or more of the nutrition in the human diet. An overall goal of food science and biotechnology is to provide wholesome, high-quality food. Food must be safe to eat. It must also be appealing and available in easy-to-prepare forms.

Food processing is preparing foods for consumption. It includes preservation. Preservation is treating food in some way so it does not spoil. Spoilage is a condition in which food is unsafe to eat. The appearance, texture, flavor, and odor of the food change. Preservation also increases the shelf life of foods. Biotechnology methods are used to increase shelf life.

Microbes are both useful and harmful in foods. Some microbes help produce desirable food products. Other microbes cause spoilage and make food unsafe to eat. Food-borne illnesses are caused by microbes or the toxins produced by microbes. Two examples of food-borne illnesses are botulism and salmonellosis.

Chemical reactions occur in food. Catalysis involves altering the rate of these reactions. Enzymes are important in these reactions. Enzyme activity can be altered by temperature, pH, and other conditions.

QUESTIONS

Answer the following questions using correct spelling and complete sentences.

1. What is food? Food biotechnology?
2. What three sciences are involved in food science? Briefly explain each.
3. Why is food important to humans?
4. What is food quality?
5. Why is food processing important?
6. What is food spoilage? What conditions cause spoilage?
7. What is shelf life? What is an example of biotechnology in extending the shelf life of a food product?
8. What is the distinction between useful and harmful microbes in food? Give an example of each.
9. Why are enzymes important in food?
10. What conditions influence enzyme activity?

EVALUATING

Match the term with the correct definition. Write the letter of the term in the blank that is provided.

a. food physics
b. food science
c. per capita consumption
d. spoilage
e. shelf life
f. yogurt
g. catalysis
h. food-borne illness
i. substrate
j. nutrient

_____ 1. Substance in food that supports life processes.

_____ 2. The study of the substances that humans eat.

_____ 3. The study of the physical characteristics of food.

_____ 4. A substance on which a specific enzyme reacts.

_____ 5. The length of time a food can be stored and be safe to eat.

_____ 6. Average amount of food consumed in a year by one person.

_____ 7. A condition that makes food unsafe to eat.

_____ 8. A product that is made by adding bacteria to milk.

_____ 9. Altering the rate of chemical reactions in food.

_____ 10. A disease people get from the food they eat.

EXPLORING

1. Use fermentation to prepare cucumber pickles. Obtain a recipe from a cookbook or other source. Be sure to follow all processes to assure a food product that is both tasty and safe to eat.

2. Investigate outbreaks of food-borne illness. Use reference materials, resource persons, the Internet, and other sources for your information. Prepare a written report that includes the source, cause, and human impact of the outbreaks.

3. Make a field trip to a food-processing facility that uses biotechnology in manufacturing. Examples include cheese plants, yogurt plants, meat packing facilities, frozen food plants, food drying plants, and canning plants.

17

ENVIRONMENTAL BIOTECHNOLOGY

People need a good environment to be healthy and happy. Maintaining a good environment is not easy. The challenge becomes greater each day with more industry and more people. Demands on resources require action to assure and sustain a good future.

The average person in the United States creates 4.3 pounds of solid waste trash and about 100 gallons of wastewater each day. In a year, this totals 1,570 pounds of solid waste trash and 36,000 gallons of water!

Wastes require proper disposal to prevent pollution. And, with only a limited amount of water, the water must be reconditioned for repeated use. These demands also cause some to wonder if we will run out of the resources needed to provide the materials that people demand. What do you think?

17-1. **A water sample is being collected from a pond to test for phosphorus. (Courtesy, Agricultural Research Service, USDA)**

OBJECTIVES

This chapter introduces the role of biotechnology in maintaining the environment. It has the following objectives:

1. Describe how biotechnology relates to environmental quality.
2. Explain the use of bioremediation.
3. Describe the use of biotechnology with solid wastes.
4. Describe the use of biotechnology with hazardous wastes.
5. Describe the use of biotechnology with wastewater treatment.
6. Explain the use of composting.

TERMS

biogeochemical processes
biological seeding
carrion
composting
crop residue
energy transfer
environmental quality
food chain
food web
hydrocarbon
incineration
landfill
leachate
recycling
solid waste
solid waste management
species diversity
wastewater

ENVIRONMENTAL QUALITY

The surroundings of an individual are important to their well-being. No one wants to breathe polluted air, drink contaminated water, or otherwise live in a poor environment.

Environmental quality is the general condition of the natural resources and other factors where we go about life. All people want to live in a place that promotes good health. Our activities, factories, homes, automobiles, farms, and other things used to provide for the needs of people, may produce substances that are harmful to the environment.

17-2. **A secluded forest area in Oregon with a crystal-clear flowing stream is a quality environment. Unfortunately, we cannot always be in such a place!**

The earth's resources constantly interact. This interaction shows the interdependence of living organisms. Many of the cycles are a part of this interdependence. A plant is a good example. The plant will grow, mature, and die. Microbes decompose the plant. The materials remaining after decomposition are nutrients for new plants that grow in the future. People have learned how to promote this cycle using biotechnology.

CONDITIONS FOR LIFE

Organisms require four conditions to live, grow, reproduce, and otherwise carry out life functions. The four conditions are:

- food—Food is the source of nutrients for organisms. How organisms get their food varies, but all must have food. Destroying the food source of organisms will result in their inability to live, grow, and reproduce.
- water—Sometimes grouped with food, water is needed for organisms to carry out life processes. The cells (and the tissues and organs that they form) have a high percentage of water. Just think about the amount of juice you can get from an orange!
- shelter—Shelter protects an organism from the harshness of the environment. All organisms need a suitable environment. For example, humans have houses and rabbits may have hollow logs in which they seek protection.
- space—Space is the area around an organism. This area is used for living. It includes air, places for protection, food sources, and areas for activity.

Biotechnology is increasingly used in these four areas. Some biological processes naturally occur in providing food and shelter and assuring good quality water.

17-3. These deer have food, water, shelter, and space in their habitat.

ENVIRONMENTAL BIOTECHNOLOGY

Environmental biotechnology is the use of living organisms or their products to manage the environment. It often involves finding ways to promote natural processes.

A good example of promoting a natural process is in crop residue management. ***Crop residue*** is the leaves, stems, and other plant materials that re-

17-4. **Chopped crop residue from the previous year is obvious in this field of young corn in Illinois. The residue protected the soil from erosion over the winter.**

main on a field after a crop is harvested. Crop producers often plow residue into the soil. This helps with decomposition and helps maintain soil fertility. The presence of these materials also helps to reduce erosion (the loss of soil). Microbe activity is promoted by this process.

Environmental biotechnology includes several areas: energy transfer, species diversity, bioremediation, and others.

ENERGY TRANSFER

Energy transfer is the exchange of energy by organisms at different levels in the environment. All energy comes from the Sun. Plants have the unique ability to capture this energy and convert it to sugar. As animals eat plants, the energy is transferred to the animal. Upon death, microbes decompose the animal and breakdown the nutrients that provide energy. These materials are again used by plants. This results in a cyclical process of energy transfer.

One flow of energy transfer is through the food chain. The ***food chain*** is the feeding sequence in which organisms get their food. Members of a food chain are interdependent. If something happens in one part of the chain, organisms in other parts of the chain are effected. The food chain can be disrupted. For example, if a drought keeps plants from growing, animals that feed on plants will not have food.

Another approach in energy transfer is with food webs. A ***food web*** is a graphical way of showing how food chains are interconnected. Food webs help people understand the importance of sustaining a good environment. Biotechnology can be used to enhance or alter food webs. Protecting certain

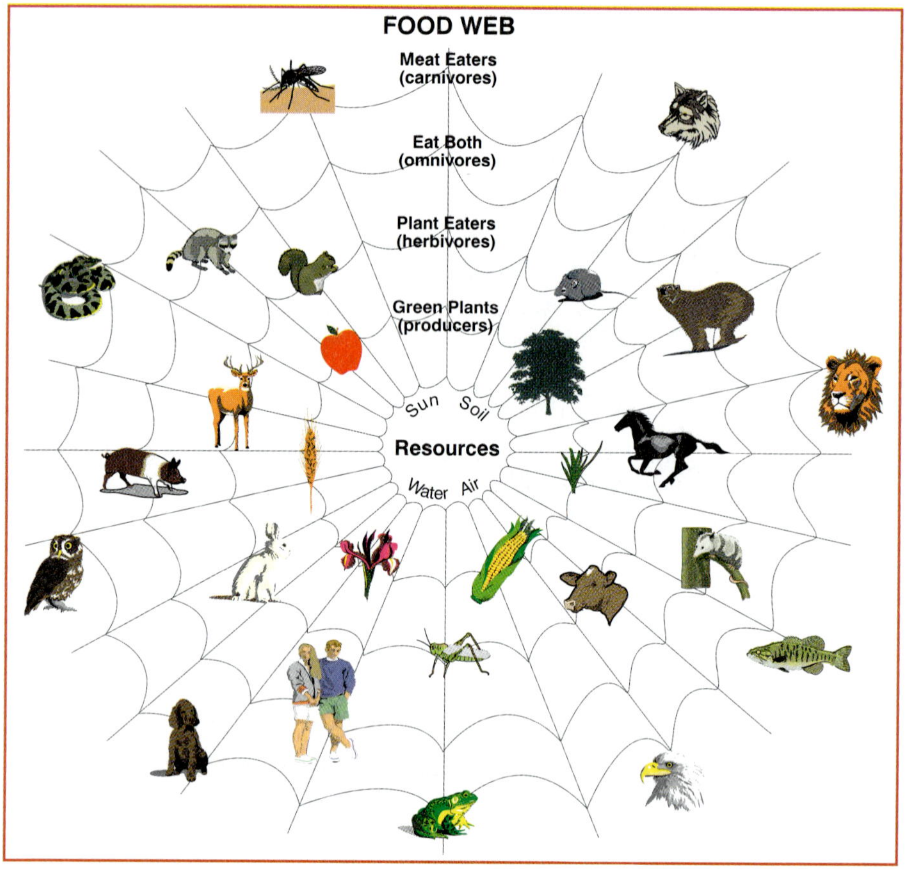

17-5. Example of a food web. Note the presence of humans.

species and promoting other species may help assure a near balanced food web.

SPECIES DIVERSITY

Every species of living organism has a unique role. Within species, organisms tend to vary yet retain their species identity. Developments in animal and plant genetics have often focused on individuals that have desired traits. Other activities have endangered or made some species extinct.

Species diversity is variation in the genetic makeup within and between species. Selecting individuals of a species for a particular trait reduces diversity. Some uses of biotechnology lead to organisms with nearly identical genetic makeups. This may result in individuals having weaknesses for other traits. A good example is with turkeys. Since consumers want meaty birds, turkey breeders have developed turkeys with larger, meatier breasts. In some cases, the turkeys have difficulty walking about and naturally breeding. Artificial insemination is used because of the inability to mate. The turkeys selected for meatiness are deficient in other attributes. The selection process creates a "narrow gene pool."

Species diversity also refers to those that are wild. They may live in tropical forests or other areas. Cutting forests destroys the homes of these species, which may lead to decline of the species and, possibly, extinction. Once extinct, the diversity of the genes of the particular organism is gone.

Scientists feel that the genetic material of wild animals may help solve disease with humans. New methods using genetic materials to cure disease may need diverse genetic information. If a species is selected for specific traits and bred for many generations, the genetic material becomes increasingly similar.

17-6. Many species are protected by law to assure species diversity. This shows a great blue heron–a water bird.

BIOREMEDIATION

Bioremediation is using biological processes to solve environmental problems. Sites that have been polluted are often cleaned up using bioremediation. Animals, plants, and microbes have a huge capacity for cleaning up wastes. Much bioremediation happens naturally. Increasingly, organisms are selected for their ability to remove harmful substances by changing them into other forms. The process is often called biodegradation.

HYDROCARBONS

A ***hydrocarbon*** is any substance that contains only carbon and hydrogen. Hydrocarbons are organic compounds. Many hydrocarbons are present in living organisms. Some may contain other substances such as oxygen, nitrogen,

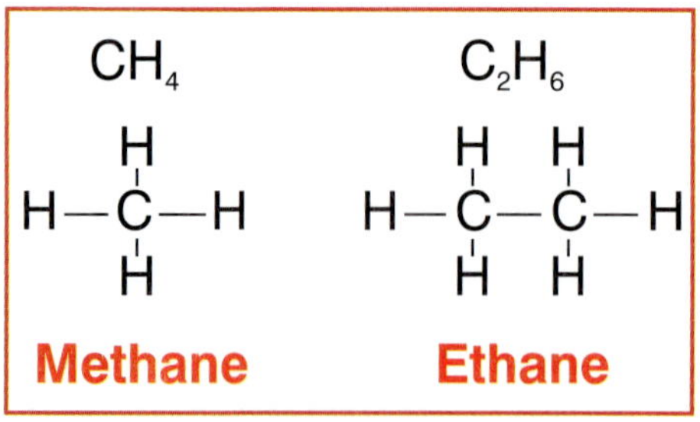

17-7. **The structure of two hydrocarbons: methane and ethane.**

phosphorus, and sulfur. Regardless, the hydrocarbon molecules are made only of the two elements.

Many of the common things that people use are hydrocarbons. Two examples are petroleum and coal. Both are fuels which are burned for energy. They originated thousands of years ago from formerly living organic material. Burning releases smoke that contains hydrocarbons into the atmosphere. Operating an engine releases exhaust. The exhaust contains substances that pollute the air, water, and land.

In chemistry, hydrocarbons are represented by rings and chains. How the rings and chains are connected by the bonding of the hydrogen and carbon elements determines the substance that is formed. For example, the structure of methane (CH_4) is similar but different from that of ethane (C_2H_6).

BIODEGRADATION

Biodegradation is the process of bacteria, fungi, and other organisms breaking material into other substances. Most biodegradation produces carbon dioxide, water, and inert biological molecules. The carbon dioxide goes into the air and is used by plants. The water evaporates, runs off, or soaks into the soil. The inert material is the solid-like material that remains. (It is sometimes known as organic matter.)

Some materials easily biodegrade; others do not. Materials from plant and animal sources readily biodegrade. Materials made of plastics, metals, glass, and similar materials are not generally biodegradable. Newer plastics have corn starch added to promote biodegrading.

Biodegradation is used to remove pollutants from the environment. Microbes, as well as animals and plants, consume wastes removing them as a pollutant in the environment. A good example is the vulture.

Vultures are large birds that feed on carrion. ***Carrion*** is the flesh of a dead animal. It includes the flesh that is rotting and often giving off a strong, unpleasant odor. Vultures (sometimes known as buzzards) eat the carrion and leave the skeletal bones. This removes the decaying flesh from the environment. Any small amounts of flesh that remain are decomposed by bacteria. In effect, the carrion has been bioremediated.

17-8. Vultures are often seen on the limbs of tall, dead trees resting and searching for carrion.

Life processes produce wastes. Plants, animals, and other living organisms produce wastes. Much of the waste is the result of carrying out life processes and cycles.

Hydrocarbons are naturally oxidized (broken down) by certain kinds of bacteria and fungi. The process provides energy for the microbes and produces carbon dioxide (CO_2), water, and remains of biological material. The remaining material varies depending on the original material that was decomposed. In some cases, the material is an organic material used as fertilizer.

Genetically-altered microbes are now being used for specific purposes. These microbes have been developed to deal with specific kinds of hydrocarbons and other materials. Oil spills, wastewater, chemicals, and other compounds are the targets of these microbes. The microbes may require fertilizer to provide nitrogen and phosphorus, which they need for their activity.

17-9. Otter have been killed by an oil spill in the ocean. Oil-eating bacteria can be used to clean up oil and prevent the death of otters. (Courtesy, U.S. Fish and Wildlife Service)

SOLID WASTES

Solid waste is garbage, refuse, sludge, and other discarded material. These are materials that do not dissolve in water. Every person produces solid waste. Just think of the napkins, cups, and forks from a fast-food restaurant or the materials from your home! Factories, businesses, and farms produce solid wastes. Junk cars, television sets, and materials from torn down houses are solid wastes. Medical wastes, including human tissues and fluids, are also solid wastes.

Solid wastes are distinguished from hazardous wastes by the level of threat they pose to the environment. Most solid wastes harbor threats but are not considered hazardous.

WASTE MANAGEMENT

Solid waste management deals with the collection and disposal of solid waste materials. Methods should be used that cause a minimum of damage to the environment. Open fires to burn wastes should not be used. Fires pol-

Biotech Connection

PRAYING FOR PREY

The praying mantis is an interesting and beneficial insect. They prey upon other insects which may damage plants and property. They may also prey on beneficial insects. Because of their beneficial work in capturing harmful insects, the "praying" is sometimes spelled "preying."

Promoting praying mantis populations helps reduce the need for chemicals that can damage the environment. A praying mantis may be 4-5 inches (10-13 cm) long. Their color allows them to blend into the greenery of plants. Unfortunately, some of the insecticides used to control pests also will kill the praying mantis. Using praying mantis to control pests is within biotechnology. (Courtesy, Gene Whitaker, U.S. Fish and Wildlife Service)

lute the air with smoke, particulate (tiny solid materials), sulfur, and gases such as nitrogen. Biodegrading-microbes are used in many approaches to waste management.

Solid waste materials are often classified into two types: those that will biodegrade and those that will not. Wood, paper, and similar materials will biodegrade. Metal, glass, and some plastics do not readily biodegrade, and should be recycled. Some of these will rust. For example, materials containing iron develop rust. Rust is formed on the surface of iron when air and moisture are present. It is a product of the oxidation process. Oxidation is a chemical degrading process that occurs without microbes being present.

Three approaches in waste management are: recycling, using landfills, and incineration.

Recycling

Recycling is reusing the materials in a product or the product itself. This reduces the demand for natural resources. For example, recycling the metal in junk cars removes solid wastes and reduces the amount of new metals that must be mined and produced. Glass, rubber, various metals, paper, and similar products can be recycled. Many communities have recycling programs for household and commercial wastes.

17-10. Recycling some materials reduces the amount of waste for disposal.

Landfilling

A ***landfill*** is an excavated area of land for the permanent disposal of wastes. Layers of solid waste are placed in the landfill and compacted. Microbes gradually decompose the materials placed in landfills. Design and location of landfills must be carefully planned. They must be located where they will not pollute streams and groundwater. Residential areas find nearby landfills objectionable. Regardless of how well they are planned, most people find them unpleasant even with the newer approaches that minimize odor and unsightly conditions.

Landfills are typically constructed in soils with high clay content. Clay tends to form a barrier that slows down the moving of liquids from the land-

17-11. A landfill with household waste.

fill through the ground. These liquids are known as ***leachate***. Leachate is produced by the decomposition or oxidation process and by the precipitation that falls on the area. Plastic liners may be used to prevent leachate from soaking into the soil. The leachate may be directed by a system of pipes to a collection area for disposal. Leachate may be treated as wastewater. It is often sprayed onto land or roadways. It should be disposed of following approved guidelines of local health and environmental agencies.

Hazardous materials should not be sent to landfills. This includes unused pesticides, paint, fuel, and wastes that do not biodegrade.

An approach similar to the activities that occur in a landfill is composting. This process is covered later in the chapter.

Incineration

Incineration is using high temperatures to destroy waste materials. The material is not used as a fuel. It is usually heated by gas or coal fires to a level where only ash remains. Carbon dioxide and other gases may be released into the atmosphere. The ash may be disposed of in landfills or otherwise as approved.

Incineration is best with certain types of materials. Animal tissues are often incinerated, such as dead chickens on a poultry farm. It may be possible to incinerate some types of hazardous materials but not those that pose potential dangers of an explosion or environmental pollution.

Incineration is more expensive than landfills. It should be used wisely and only by people who are trained.

HAZARDOUS WASTES

Hazardous wastes are handled differently from typical solid wastes. A hazardous waste is a solid or liquid that poses danger to human health or the environment. These wastes are from homes as well as factories, businesses, and farms. A few examples are unused paint, pesticides, batteries, household bleach, nail polish remover, and many petroleum products. Heavy metals such as lead, mercury, and cadmium are included.

Materials are classified as hazardous for various reasons. Some are toxic and others are flammable, reactive, or corrosive. Other wastes cause human disease. Those that cause cancer are known as carcinogens. Biodegrading is possible with most hazardous wastes. The microbes and processes must be carefully controlled to prevent accidental release into the environment. Radioactive wastes are typically placed in long-term storage areas that prevent release.

Hazardous wastes are sometimes placed in special hazardous waste landfills; others are placed in concrete barrels or buried deep in the Earth in caves. Encapsulation is sometimes used with wastes placed in a landfill. This means that the material is placed in a capsule-like container. The capsule reduces risk of release into the environment.

Biological seeding is used to assure the presence of the appropriate microbes. It involves placing the required bacteria with the waste. Biological seeding works best with materials that contain carbon and will degrade.

With genetic engineering, bacteria strains can be produced that degrade targeted wastes. The genes within the bacteria can be manipulated to enhance degrading capability.

In some cases, plants can be used to remove heavy metals and other materials from soil. The plants take-up the pollutant along with the water obtained from the soil. The pollutant is stored in the stems, leaves, and roots of the plant.

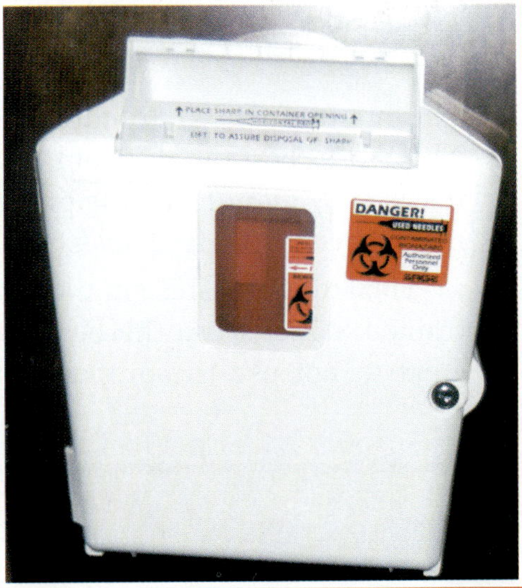

17-12. Special containers are used in health care facilities to dispose of needles and other sharps. A sharp is any instrument that can stick or cut the skin. The container is emptied by a disposal service.

Career Profile

MOLECULAR MYCOLOGIST

A molecular mycologist is a biologist who specializes in the study of fungi. The work may involve collecting samples for examination, performing lab examinations, and doing research on the useful roles of fungi, including genetically engineered fungi. The molecular mycologist in this photo is studying the use of fungi to degrade hazardous materials in liquids.

Molecular mycologists need college degrees in biology, with emphasis in mycology (study of fungi). Many have masters and doctors degrees. High school courses in biology, chemistry, environmental science, agriculture, and mathematics can provide a foundation for college studies.

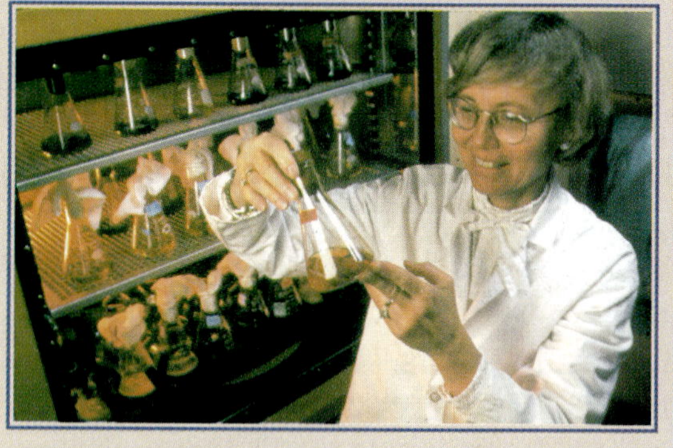

Opportunities for molecular mycologists are with government agencies, research labs, universities, and some private companies. (Courtesy, Agricultural Research Service, USDA)

The mature plants are cut or pulled and hauled away for proper disposal. This works best on areas of land where the heavy metals have been spilled or released by manufacturing processes. Such plants should not be used for human food nor to feed livestock.

Hazardous wastes should be disposed of only by people who are specially trained. Always use the best possible methods of disposal. The risk is too great to not use the proper and approved methods.

WASTEWATER

Water is important in human life, manufacturing, and many other areas. Unfortunately, use often results in the release of water that is known as wastewater.

Wastewater is used water. It often contains suspended or dissolved materials. In some cases, these materials are harmful pollutants. Regardless, wastewater must undergo a cleaning process before reuse. Nature provides a cleaning process in the water cycle. Unfortunately, most wastewater has been dirtied beyond what nature can readily clean and restore.

TREATMENT

Domestic wastewater (sewage) is from the daily living activities of people. It is the water from washing dishes and clothing, toilets, bathing, and similar uses. This wastewater contains dead skin cells, particles of food, detergents used in washing, human feces and urine, and other materials. These need to be removed before the water can be released into a stream or reused. Sanitary sewers are pipe systems used in towns and cities to carry off the water. Homes, businesses, and factories may have disposal systems. Septic tanks are used by some homes. Factories may have wastewater lagoons and other treatment facilities.

17-13. One part of a wastewater treatment facility operated by a city. Conditions are provided for microbes to act on wastes.

Most all wastewater treatment systems make use of microbes to degrade the wastes in the water. A lagoon is a lake where microbes are concentrated and conditions made ideal for their work. The microbes act on the solid materials and convert them to sludge, a semisolid material. The sludge may be further digested by microbes and dried to form biosolids. Some 300 different species of bacteria may be found in sludge. Different bacteria work together, with some species being quite selective about the materials they decompose. The oxidation process releases carbon dioxide and other substances.

Bacteria vary in need for oxygen. Bacteria that live in water where there is no free oxygen are known as anaerobic bacteria. Other bacteria require oxygen and are known as aerobic bacteria. Dense wastewater often contains little or no free oxygen. The anaerobic bacteria slowly carry out the oxidation process. People sometimes smell odors near sewage treatment facilities. This is the methane and gases that are released by the anaerobic bacteria.

Species of bacteria vary in what they oxidize. Examples are:

- *Clostridium* spp. act on sugars, amino acids, and fatty acids
- *Nitrosomonas* and *Nitrobacter* spp. convert nitrogen to ammonia and free nitrogen

Protozoa and rotifers are also found in wastewater treatment. Protozoa help in decomposition and feed on the bacteria. Rotifers are tiny animals that freely swim about in the water. They feed on bacteria and contribute to decomposition.

Treatment processes include screening out solids, using aerators to assure that water contains oxygen, using clarifiers and sediment basins, and using chemicals. A good environment must exist in wastewater for microbes to do their work. The proper microbes must be present.

BIOGEOCHEMICAL PROCESSES

Biogeochemical processes are the processes by which microbes transfer nutrients from living forms to physical forms and back to living forms. Fairly definite sequences are followed, which are sometimes known as cycles.

Three important biogeochemical cycles in wastewater are:

- nitrogen cycle—The nitrogen involves bacteria converting nitrogen compounds from harmful to less harmful forms. The first product is ammonia. Ammonia is converted to nitrite forms. The nitrite forms are oxidized into nitrate forms. Nitrate is used by plants for growth. As the leaves, seed, and other parts of plants are eaten by animals, the nitrogen is excreted by the animal and the cycle is repeated. The nitrogen cycle is important in aquatic environments to assure the well-being of fish and other species.
- phosphorus cycle—The phosphorus cycle is a series of steps in which phosphorus compounds found in wastewater are converted into other forms. The cycle is much like that of nitrogen. Excessive phosphorus may tax the ability of the cycle to naturally occur. This is frequent in wastewater with high amounts of detergents. Excess phosphorus results in the growth of too many plants and microscopic plankton. The water is often unfit for fish and other

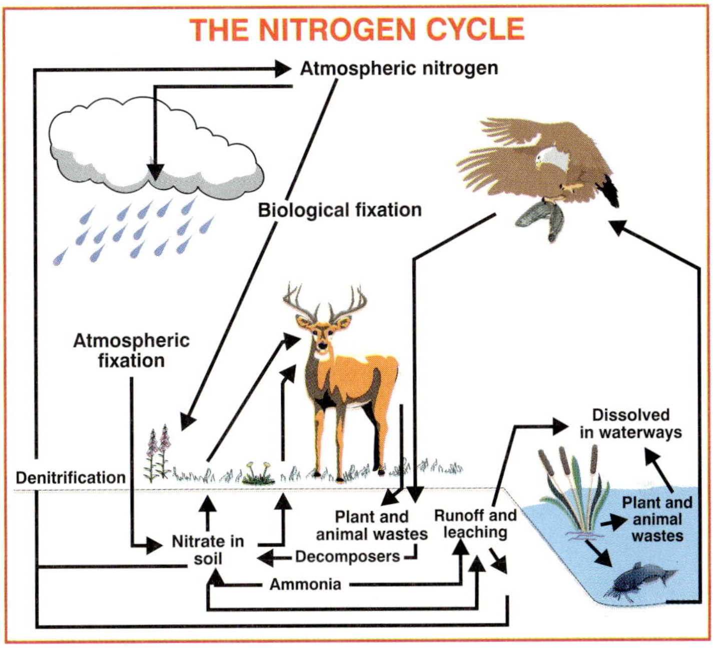

17-14. The nitrogen cycle involves various organisms and processes in converting nitrogen into various forms.

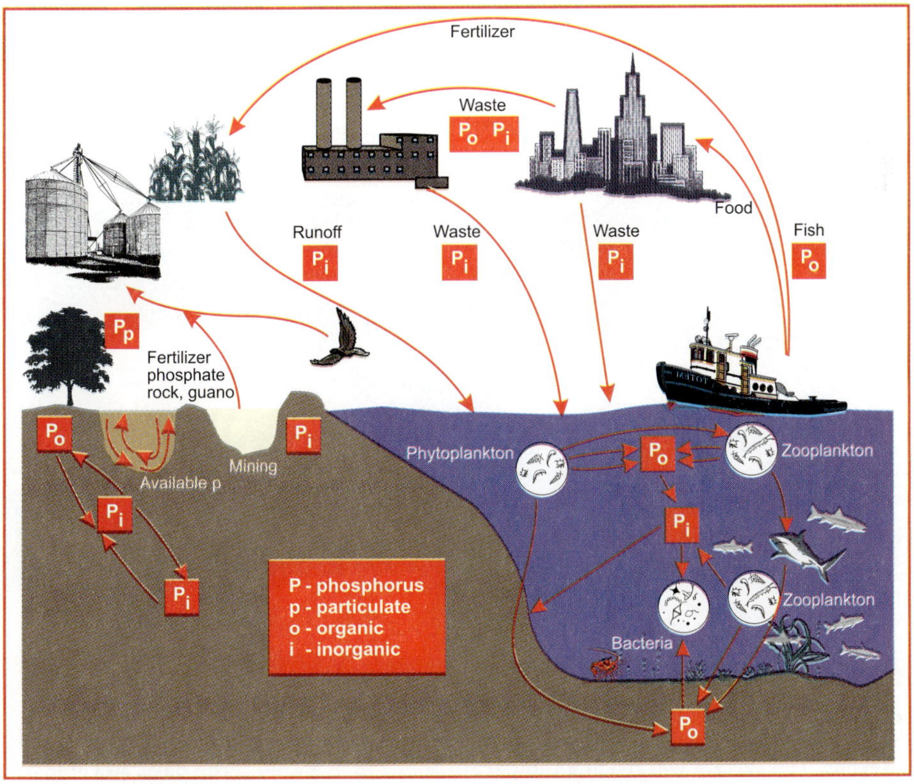

17-15. The phosphorous cycle involves processes carried out on land, in the air, and in natural bodies of water.

organisms that are naturally found in it. All of the life may be killed in a stream or lake with excess phosphorus.

- sulfur cycle—Wastewater contains some compounds with sulfur. Bacteria decompose these compounds resulting in the release of sulfur. The odor of rotten eggs near a sewage facility is evidence that the sulfur cycle is being carried out.

Scientists have learned how to promote bacterial action to speed up the cycles. Aeration is particularly important. This adds oxygen to the water so that the microbes can grow and carry out the oxidation process.

COMPOSTING

Composting is promoting the biological decomposition of organic materials. It is used at homes and commercially. Lawn clippings, leaves, and twigs are commonly composted at homes. Food scraps, paper, and other readily biodegradable wastes may also be composted. Even the solid materials from wastewater treatment plants are composted! It is based on natural processes that occur in the forests, fields, and wetlands.

17-16. An in-vessel rotating composting system is used for poultry wastes and similar materials. The system will reach a temperature of 145°F within three days. Aerobic bacteria are used. Within five days, much of the composting has been done. (Courtesy, Don Cawthon, Texas A&M University-Commerce)

Composting is used to convert wastes into forms that can be used on the land or around plants. Dead chickens can be composted. After the process has been completed, the material can be spread on pasture land. Compost from leaves and household waste can be used as mulch or worked into soil as fertilizer. Composted solids from sewage plants can be used as fertilizer.

THE PROCESS

Most composting uses aerobic bacteria. These bacteria are usually present naturally in sufficient numbers and do not create an unpleasant odor. In some cases, seeding may be used to assure that the necessary species are available. Mixing materials that have composted with new materials helps assure that bacteria are present.

A good environment is needed for the microbes. The materials being composted are placed in a pile or long row. Some may be placed in a special container known as a compost bin. The ideal temperature range is 95-140° F (35-60°C). Adequate moisture is needed. The best range of moisture is 40-50% water. A pH of 5.0-8.5 is satisfactory. Oxygen is needed for the process. Some stirring of the ingredients may be required. In some cases, agricultural fertilizer and lime are added to assure sufficient nutrients and pH for the bacteria.

HOME COMPOSTING

Home composting is useful in disposing of yard wastes and some household wastes. The material may be put in a pile or a compost bin. Alternating layers of leaves, stems, and twigs are covered with a thin layer of fertilizer. Some soil may be added at the top to assure the presence of sufficient bacteria. Once underway, new materials can be added at the top of compost piles. Regular watering may be needed to promote composting. After several months of decay, the bottom materials can be removed and used in maintaining the landscape or vegetable gardening.

Composting bins can be constructed fairly easily. Select a convenient site that is somewhat hidden from public view. The bin should be located where it will not flood nor be in the way of other activities. A five-foot square bin will hold quite a bit of material. Larger bins are needed for homes with large lawns or gardens that will produce more materials for composting. Posts may be set in the ground and net wire attached around the posts to form a box. Material is lifted over the wire and into the bin. The wire on one side should

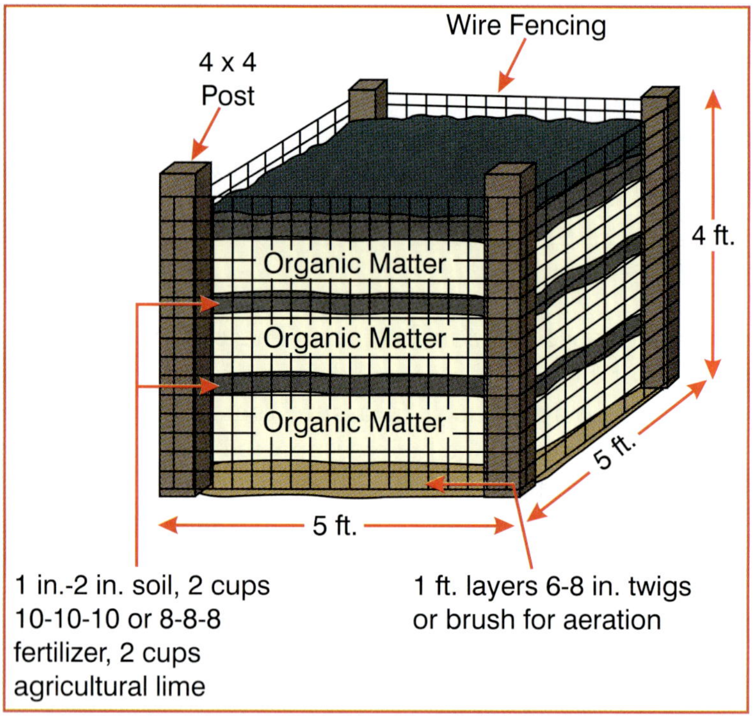

17-17. Diagram of a simple wire-enclosed compost bin.

be easily opened for removing the composted material without taking the bin apart.

The speed of composting varies with the temperature. Materials compost slowly in cold, winter weather. Composting occurs more rapidly in the summer provided adequate moisture is present. Some people place materials in a bin one growing season for use the following growing season.

REVIEWING

MAIN IDEAS

Biotechnology has many uses in helping maintain a good environment. Naturally-occurring processes clean and rid the environment of pollution. These processes sometimes need a little help. People make so much waste!

Environmental biotechnology is the use of organisms or their products in maintaining a good environment. Energy transfer, species diversity, and bioremediation are important areas.

Bioremediation is somewhat complex. Biological processes are used to solve environmental problems. These often naturally occur. Fungi, bacteria, and other organisms are active in bioremediation. Biodegradation is used by organisms to convert materials containing carbon and hydrogen into water, carbon dioxide, and inert biological molecules. However, not all materials are biodegradable, such as metals, glass, and some plastics.

Environmental biotechnology is used with solid wastes, hazardous wastes, and wastewater. Various biological processes are promoted. Genetically-engineered bacteria may be used to target specific wastes. Waste management should include recycling to conserve natural resources. Landfilling and incineration are used to dispose of domestic and industrial wastes.

Composting can be used at homes to reduce wastes. A few simple activities will yield a nutrient-rich compost material. The material is high in organic matter to improve soil. It also contains nutrients that promote plant growth.

QUESTIONS

Answer the following questions using correct spelling and complete sentences.

1. What is environmental quality? Why is it important?
2. What are the four conditions needed for life? Briefly explain the role of each.
3. What is energy transfer? How is it related to food chains and food webs?
4. Why is species diversity important?
5. What is bioremediation? Why is it useful with spills and pollution?
6. Name two organisms that are involved in biodegradation?
7. Do life processes produce wastes? Explain.
8. What are the sources of solid wastes?
9. What are three approaches used in solid waste management?
10. What is composting? How can it be used for disposing of biodegradable wastes produced at home?

EVALUATING

Match the term with the correct definition. Write the letter of the term in the blank that is provided.

a. incineration
b. leachate
c. biological seeding
d. food chain
e. species diversity
f. crop residue
g. hydrocarbon
h. environmental quality
i. solid waste
j. recycling

_____ 1. Liquids that flow from a landfill.

_____ 2. Using high temperatures to destroy wastes.

_____ 3. Reusing materials.

_____ 4. The general condition of the natural resources and other factors where we go about life.

_____ 5. The feeding sequence in which organisms get their food.

_____ 6. An activity used to assure the presence of the appropriate microbes.

_____ 7. Variation in the genetic makeup within and between species.

_____ 8. The leaves, stems, and plant materials that remain on a field after a crop is harvested.

_____ 9. Any substance that contains only carbon and hydrogen.

_____ 10. Garbage, refuse, sludge, and other discarded material.

EXPLORING

1. Construct a simple compost bin at your home or school. Operate the bin by regularly adding material and using what has been composted. Follow instructions in the chapter or from other sources to be sure the bin works properly.

2. Make a tour of the sanitary landfill near your home. Have the operator explain how it is constructed and operated. Determine how the leachate is handled. If time permits, also tour a local wastewater treatment facility. Prepare a report on your observations.

3. Make a survey of your local community or the school facilities to determine how solid wastes are handled. Develop recommendations to improve the use of wastes and reduce environmental pollution.

18

USING BIOTECH: SOCIAL AND ENVIRONMENTAL IMPACTS

Should I use a product of biotechnology? You have a decision to make. But, be careful. Get good information. You are probably already using many biotech products of one kind or the other. It is hard to know. Most likely you are thinking about products from GMOs (genetically modified organisms). Get the facts before rushing to a decision!

New areas of biotechnology are developing very rapidly. Some of them have big impacts on people. Other areas of biotech may have big impacts on the environment. Biotech applications are becoming more diverse and are expanding each day. Many GMOs are now a part of human life.

Clearly, the products of biotech efforts are not limited to the laboratory. They are having direct contact with the environment. The same is true with health care, food manufacturing, and other areas.

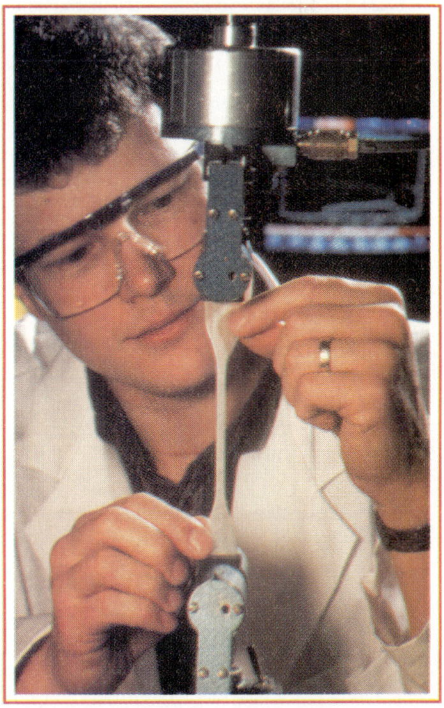

18-1. **Dried starch and other materials are being added in making experimental plastics to promote microbe degrading. (Courtesy, Agricultural Research Service, USDA)**

OBJECTIVES

This chapter provides information on the social and environmental impacts of biotechnology. It has the following objectives:

1. Name and classify innovations.
2. List and explain factors to consider with new technology.
3. Describe the meaning and importance of the social impacts of biotechnology.
4. Describe environmental impacts of biotechnology.

TERMS

abiotic factor
biotic factor
decision-making
innovation
patent
primary innovation
refuge
secondary innovation
selective social impact
social impact

BIOTECHNOLOGY AS INNOVATION

An ***innovation*** is something that is new. It has not been widely used before. Innovation includes new products as well as new processes. Biotechnology includes both things and methods; some of these build on each other.

TYPES OF INNOVATIONS

There are of two types of innovations: primary and secondary. A ***primary innovation*** is one that makes a major advance in a new area. Many modifications and improvements are possible. Primary innovations make it possible to generate other innovations. An example is the ability to create organisms with new genetic information. Once we have created a transgenic organism, we can go about using the procedures to create a number of such organisms in different species.

A ***secondary innovation*** is one that builds on a primary innovation. Having a variety of genetically altered plants is a secondary innovation. Altering the first plant of a species is a primary innovation. Methods have been created to enhance the process, such as the development of the gene gun. Many secondary innovations are possible.

SOURCES OF INNOVATION

Many, but not all, innovations are the products of science and research activity. Some are from the biological sciences; others are from the physical

18-2. The ability to use electronic imaging to view DNA was an innovation in biotechnology. (Courtesy, USDA-ARS Plant Stress and Water Conservation Lab, Lubbock, Texas)

18-3. The computer chip promoted innovation in science.

sciences. Specialists in these areas often work together to make innovation possible, especially biotechnology.

Innovations may be from:

- efforts of highly trained lab researchers
- technicians striving to solve practical problems
- agricultural producers experimenting with ways to improve their efficiency
- businesses seeking new entrepreneurial opportunities
- people who have experienced major losses and are seeking solutions
- consumers who want something better
- government-funded efforts to solve problems
- other possible sources

Gaining recognition for an innovation requires that the innovator make known his or her findings or developments. In some cases, individuals will seek patents. A ***patent*** is a government-granted right for an individual or business to produce and sell a product for a number of years. The holder of the patent makes money from an innovation if it is successful. However, an infrastructure must usually be in place to produce and market the innovation.

Innovations lead to new technology. People must often make decisions about the technology they are going to use.

CONSIDERING TECHNOLOGY

Some people rush to be the first to use new things that are available. Other people are slow to use new things and may never do so. Using new technology requires careful attention to assure that a good decision has been made. The first to use a new product are innovators. But, innovators often must pay higher costs and help refine the new technology. Innovators typically prosper if they have made a good decision to try new technology.

A good example is the use of bST on dairy farms. The producers who used it first learned from their mistakes. They learned the importance of good cow management. The need for more feed and nutrients as well as a time for the cow to restore her body system came about as producers learned how to use the technology.

DECISION-MAKING

Decisions are not easy to make. ***Decision-making*** is choosing among the available alternatives. A decision can be no better than the quality of the information that is used in making it. Good information is always needed.

In making decisions:

- Get all the information you can on the technology. Read reports, attend conferences or field days, and talk with experts.

18-4. Making decisions requires careful consideration of the best possible information.

- List the advantages and disadvantages of the technology. Compare them and interpret the meaning in terms of your situation.
- Select the alternative that is best for you. Make your choice based on good information with a research base. Do not make decisions based on emotion.
- Try out major decisions on a small-scale basis. A good example is the crop producer who plants only a portion of all cropland to a GMO. Always carefully assess your experiences.

WHAT TO CONSIDER

Adopting new technology requires good information and decision-making. A lack of good information results in poor choices.

Overall, here are some questions to answer when considering the use of new technology:

- What testing has the new technology received? How reliable is the testing? What is the reputation of the individual or lab who carried out the testing?

Career Profile

ENVIRONMENTAL CHEMIST

An environmental chemist promotes environmental quality. They identify pollutants in the air, water, and land. Their work includes measuring the level of pollution. Identifying a pollutant involves extracting the pollutant and studying the traits of the polluting substances. Chromatography and mass spectrometry may be used to gain information on the chemical structure of the pollutant.

Most environmental chemists have college degrees in chemistry or biochemistry. Advanced positions may require a masters or doctors degree.

Jobs are with government environmental agencies, colleges and universities, research units, and associations with environmental interests. This photo shows two environmental chemists studying fungi that can convert toxic chemicals to harmless forms. (Courtesy, Agricultural Research Service, USDA)

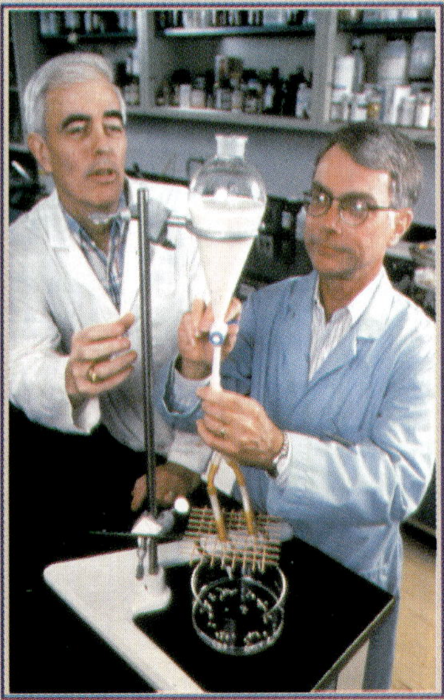

- What are the experiences of others with the technology?
- What is the cost of the technology in terms of the returns it might provide?
- What are my capabilities? Can I properly use the new technology?
- What is the nature of the biotechnology that is involved? What risks does it pose?

The answers to these questions will be helpful. In most cases, applications specific to the use of the technology should be considered.

Agriculture and Horticulture Producers

Making decisions about the use of new technology in agriculture and horticulture should consider several questions:

- What are the costs versus the returns? Will the new technology improve my profit?
- How does the new technology fit into my overall program? (For example, crop varieties with terminator genes may not work well for those who save seed. A "terminator gene" is one that allows the benefits of the genetic improvement to be apparent only the first year. Saving seed to plant later does not provide benefits.)
- What are the experiences of other producers with the product?
- What is the nature of the biotechnology that is involved with the product? Does it pose hazards to people, other organisms, or the environment?

18-5. Two methods of vaccinating cattle with a genetically engineered vaccine for shipping fever are being investigated. The methods are as a feed additive and with a hypodermic needle and syringe. (Courtesy, Agricultural Research Service, USDA)

- What legal or contractual regulations apply? Will I need to sign a contract and what are the provisions of the contract?
- How is the product received in the marketplace? Will I receive a penalty or a bonus for producing the product?

18-6. The effects of salinity on various rice varieties are being tested. (Courtesy, Agricultural Research Service, USDA)

Food Products

As consumers, all people are affected by food products involving biotechnology. Some people are involved as food handlers and preparers. Questions to answer include:

- Is the product safe? How has it been tested and proven safe?
- Does the product serve as a good source of nutrients?
- Can the product be readily stored? How is it stored?
- Will I feel comfortable preparing the food? Eating the food?
- Will the use of biotechnology promote consumption of the product?

Health Care

Biotechnology health care products and services can be viewed from the standpoint of the provider of health care as well as the user of health care. Some questions to answer are:

- What have been the experiences of other people who have had the product or service?
- What is the prognosis without the product? With the product? (This is an important question for a person with a disease that could possibly be cured with biotechnology.)
- What risk am I willing to take to use the product or service?
- Can I afford the procedure? Will my health care insurance cover most or all of the costs?

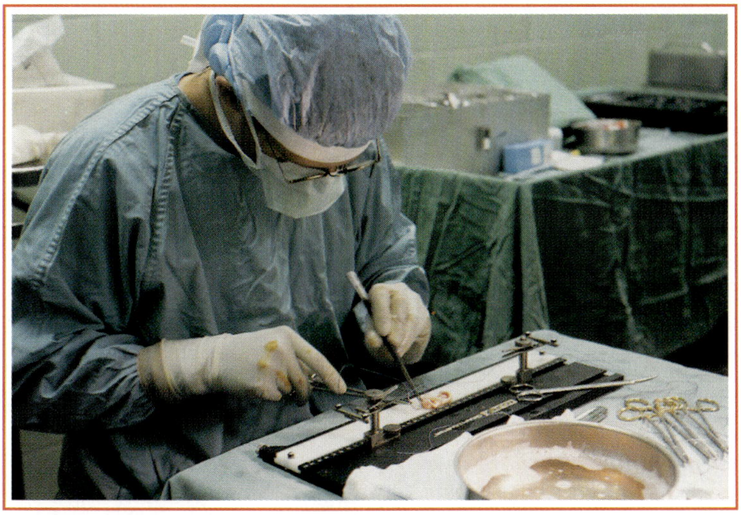

18-7. **A ligament is being prepared for surgery.**

SOCIAL IMPACT

Social impact is important. ***Social impact*** is the influence of biotechnology on the status of human beings. This shapes the society in which we live. People are sometimes surprised by how humans are affected by new technologies. Some are beneficial; others are not. Major changes result from some technology. Just think of how penicillin has influenced health care!

SELECTIVITY

Social impacts may be selective. A ***selective social impact*** is one that is beneficial for some people and not for others. The selectivity may be the consequence of geographic location, education, occupation, political decisions, or social status. For example, the income level of a person may determine whether or not they can use a new technology. Most things require money. Without money, we cannot use or benefit from certain technology.

18-8. **The ability of people to use biotechnology varies with their social status.**

Biotech Connection

NO-ROT FRUIT

We like fresh fruit. We do not like fruit that shows signs of rotting. Biotechnology can help solve rot problems.

Ripe fruits tend to rot rather quickly after they have been picked. Fungi often grow on ripe fruits. For example, bananas and peaches develop rot quickly after ripening if not properly stored. We have tried to avoid this problem by picking green fruits that are not ripe (and do not taste as good) or spraying the fruits with a chemical to prevent the rotting process. The peaches shown here were picked slightly before the ripe stage to prevent rot before they were marketed.

Thanks to biotechnology, we envision fruits that will protect themselves against rot without the need for chemical treatment. This would lead to good-tasting fruits that would not be covered with chemicals. When washing the fruit before eating it, no chemical will go in the water system.

Government policies may control the technologies that are available to people. What people can get is determined by a government agency or law. People in some countries, such as France, have access to medicine that is not available in other countries, such as the United States. The same is true with agriculture and other areas.

QUALITY OF LIFE

Biotechnology may benefit the quality of life of some people and not others. Overall, the use of biotechnology should improve the quality of life of all people.

Greater efficiency from biotechnology may displace some people. This is true in the developed as well as developing nations. An example is the dairy industry in the United States. Some people have feared the social consequences of using bST to promote milk production. bST increases per cow milk production and, hence, per farm productivity. Those not adopting the technology will be at a disadvantage. They will lose efficiency. Their farm may become unprofitable and cease operation. When this happens, how will the dairy producer earn a living?

The social consequences of using a new biotech product or procedure must always be considered. Good education is needed so informed decisions can be made.

THE ENVIRONMENT

The environment is the combination of biotic and abiotic factors that surround an individual or community. ***Biotic factors*** are the living organisms in the environment. This includes plants, animals, and microbes. The ***abiotic factors*** are the nonliving parts of an environment, such as soil, water, air, minerals, and others. Biotic and abiotic factors interact to shape the environment.

SUSTAINING THE ENVIRONMENT

Our quality of life depends on the environment and our actions affect the environment. The influence of humans can be felt every day on our immediate surroundings. It can also be seen on a larger scale with every living species. The quality of the air we breathe and the water we drink depends on hu-

18-9. Masses of people make big demands on the environment.

man activity. Our actions and their impacts on the environment also have an effect on plants and animals.

Many human activities have negative impacts on the environment. An example is the negative impact of the dams built on rivers in the northwestern United States on the salmon population. Salmon swim upstream to reach specific locations where they lay their eggs. Because of the dams, the salmon cannot reach those locations and do not reproduce. Some salmon species are endangered and might become extinct. Steps have been taken to solve the problem and assure a salmon population.

Big cities are faced with pollution problems. On some days, the air quality deteriorates and everyone suffers. Although sometimes unnoticed in healthy people, polluted air can be harmful to young children and people with respiratory problems.

Depletions of the ozone layer and the presence of ozone holes in the North and South Poles are consequences of human activity. Some types of propellant gases as well as exhaust from cars, factories, lawn mowers, and other sources react with the ozone layer and destroy it. The ozone layer is important to humans. It protects us from radiation coming from the Sun. Steps are being taken to reduce ozone depletion and other damage to the atmosphere.

Changes in the environment require organisms to adapt. If not, they will not continue to live. With biotechnology, we can take steps to sustain the environment.

BIOTECHNOLOGY AND THE ENVIRONMENT

Biotechnology is or will be used to produce new crops, chemicals, plastics, fibers, sources of energy, drugs, vaccines, and other products. The products of biotechnology are diverse and will have a major impact on our lives. Several genetically engineered crops are being produced. Many more new products are being tested in experimental fields and could become the prominent crops of the future. A few negative and positive impacts of biotechnology are introduced here. You will need to assess the likely consequences of some of them.

Negative Impacts

Negative impacts relate to the potential risks of biotechnology. Some people feel that risks exist when GMOs, whether animals, plants, bacteria, or other microorganisms, are released into the environment. The GMOs will come into contact with the land, water, and air that all living organisms share.

Herbicide Resistance. Crop plants engineered to carry genes giving them herbicide resistance are now used. The most common GMO crops are soybeans and corn. Using these varieties allows spraying fields with herbicides. The weeds are killed without crop injury. Two potential risks to the environment associated with herbicide-tolerant crops are: reducing genetic diversity and developing "super weeds."

The herbicides to which these plants are resistant are effective in killing unwanted plants. Some people feel that their use could result in weed "wipe-out." Although too many weeds are bad in crops, they are essential for environmental balance. They provide diversity, habitat, and food for other organisms.

Another potential problem linked to herbicide-resistant crops is the possibility that the resistance gene will be passed from crops to wild relatives. This could happen through natural pollination. The new plants would then produce offspring that can reproduce and expand throughout the environment. These plants could become "super weeds" impossible to kill because of resistance to herbicides.

Scientists have shown that herbicide tolerance could spread rapidly from transgenic canola to wild relatives. The result is an herbicide-resistant weed. The offspring of these "super weeds" would lose their tolerance to herbicide.

18-10. These potato plants have been engineered to be resistant to the Colorado Potato Beetle. (Courtesy, Ashton Hi-Tech Seed Company, Ashton, Idaho)

Some long-term damage to the environment is not very likely in terms of herbicide resistance gene transfer.

Insect Resistance. Crops have been engineered to resist specific insects or diseases. For example, Bt maize (corn) and cotton have a gene that produces a toxin that kills certain worms. It is likely that the targeted pests will develop resistance to the Bt varieties of crops. Further, some beneficial insects may be damaged by the Bt plants.

A concern is that the pest-resistance gene might be transferred to weeds. If the main control of weed expansion is the pest or virus the weeds are now immune to, nothing can stop these weeds from spreading and possibly overrunning areas covered with other types of plants. The biodiversity of that area could be affected.

Introducing pesticide genes into plants might disrupt the ecological balance between pests and beneficial insects. Crop pests include caterpillars, bollworms, and aphids. Ladybugs are beneficial insects because they feed on crop pests. Ladybugs help control the population levels of other insects. This is a natural way of reducing pests. Some genetically engineered potatoes produce a toxin to control aphids. A study was done on the effects of these potatoes on ladybugs. Ladybugs that were fed aphids exposed to the toxin-producing potatoes had shorter lives and produced more unhatched eggs than the ladybugs that were fed aphids from non-genetically engineered potatoes.

Insect resistance to insecticide is another area of consideration with GMO plants. Chemical and biological insecticides have evolved over the years. There has been a constant need for new insecticides with different mechanisms of action because insect pests build resistance to the insecticides used. For example, before Bt was used to genetically engineer plants, it was used in insecticides. When first used, most insecticides are very effective and destroyed pests. After prolonged exposure to the insecticide, the pests learn to live with it and find ways to protect themselves against it.

Insecticides are sprayed on fields when insects pose a threat. If few insects are present, a producer would not spray. Some insecticides degrade rapidly after use. Degrading does not occur with Bt plants. Bt is present in every genetically engineered Bt plant. It is present as long as the plant is alive. Some people view a field planted with Bt plants as a field being continuously sprayed with Bt insecticides. Constant exposure to Bt could push insect pests to evolve more rapidly than with insecticides.

The only way to minimize insect resistance is to provide a refuge. A ***refuge*** is an area of the field planted with the same non-GMO crop as the field but where no insecticide is used. This refuge allows insects without resistance to live. These insects can mate with resistant insects and, hopefully, provide offspring sensitive to the insecticide.

Positive Impacts

Positive impacts of biotechnology relate to agriculture, food, medicine, and other areas.

Biotechnology influences how plants and animals are and will be grown. It also influences the way chemicals, plastics, and pharmaceuticals will be produced. Because it is a new approach, biotechnology can help in protecting and enhancing the environment.

Reduce Pesticide Use. Crops are being genetically engineered to be herbicide resistant or to produce their own pesticide. Some pesticides can be detrimental to the environment. When sprayed on fields, poisonous chemicals can leach into the water system and harm animals and humans who drink this water. With herbicide-resistant crops, the need for herbicides can be reduced. These crops allow the use of very effective herbicides that have to be applied only a few times. There is no use for insecticide when dealing with plants that can produce their own. These plants will release insecticide when

18-11. **Biotechnology can reduce the use of pesticides in crop production. Aerial application and air pollution would be reduced.**

attacked by an insect. Old-style chemical insecticides will become obsolete and unnecessary with such crops.

Waste Disposal. Biotechnology provides alternative methods for making chemicals and pharmaceuticals. Traditional methods use organic solvents that may be harmful to the people who handle them and to the environment. Solvent vapors are released in the air and contribute to poor air quality. Because these solvents are toxic, it is necessary to destroy them after use. The only way to dispose of them is with burning. Vapors produced by burning can be toxic. Biotechnology provides safer ways based on the use of aqueous solvents, which are easy and safe to dispose of.

Engine exhausts are a major cause of air pollution. Burned fuel generates toxic gases that contribute to many environmental problems: poor air quality, depletion of the ozone layer, and the greenhouse effect. Ethanol is a safe alternative to gasoline. Biotechnology can be used to produce ethanol from agricultural crops and wastes. Biotechnology can enhance the environment by providing a non-polluting fuel and by using agricultural waste.

Human Activity. Human activities have produced areas contaminated with oil, toxic chemicals, and nuclear wastes. Contaminated areas are hazards for humans and for wildlife. Cleaning up these hazardous areas is a major concern and in many cases there are no known methods to do it. With biotechnology, microbes can be engineered to break down oil or other toxic materials. GMOs are being developed to break down plastic materials that accumulate in landfills.

18-12. **Special metal-scavenging plants can be grown on soil polluted with heavy metals, such as lead and mercury. These plants, known as hyperaccumulators, have a massive root system. The plant tops are cut like hay and hauled away thus removing heavy metals that have been absorbed. (Courtesy, Agricultural Research Service, USDA)**

REVIEWING

MAIN IDEAS

Biotechnology is viewed as innovation. Each major area is an innovation. Some are primary innovations; others are secondary.

People must often make choices that involve biotechnology. They need good information so they can make informed choices. Emotional choices usually result in poor decisions.

Social impacts are also important. These influence the status of human life and relationships. Some social impacts are selective. Most biotechnology influences the quality of life of people.

The most popular use of biotechnology has been in agriculture. Some agricultural biotechnology has a large influence on the environment. The effect on the environment is a compromise between "good" and "bad."

GMOs can harm the environment by generating species of pests resistant to some of the most effective means to destroy them. These "super weeds" or "super bugs" might overrun other species and lessen the biodiversity of an area.

Biotechnology can be seen as an environment-friendly alternative to traditional methods. The volume of pesticides used will decrease with the implementation of genetically modified crops. The amount of toxic materials will also decrease. Biotechnology will allow the replacement of toxic solvents by water-based systems. Biodegradation, the use of microorganisms to break down a product, can greatly contribute to the enhancement of the environment.

QUESTIONS

Answer the following questions using correct spelling and complete sentences.

1. Distinguish between the two types of innovations.
2. What are the sources of innovations?
3. Why is good information needed in making decisions?
4. Why should decisions be made based on information and not emotion?
5. Why is social impact important?
6. How are innovations selective in terms of society?
7. Distinguish between biotic and abiotic factors in the environment.
8. How does quality of life relate to sustaining the environment?

9. Give an example of a possible negative impact of biotechnology on the environment. Explain it.

10. Give an example of a possible positive impact of biotechnology on the environment. Explain it.

EVALUATING

Match the term with the correct definition. Write the letter of the term in the blank that is provided.

a. innovation
b. patent
c. decision-making
d. social impact
e. abiotic factors
f. biotic factors
g. refuge

_____ 1. The living plants and animals in an environment.

_____ 2. Government-granted protection for the developer of a product or process.

_____ 3. Something that is new and has not been used before.

_____ 4. Choosing between alternative courses of action.

_____ 5. An area of a field planted to traditional crops or not sprayed with pesticides.

_____ 6. The influence of biotechnology on the status of humans.

_____ 7. The nonliving parts of an environment.

EXPLORING

1. Use the World Wide Web, school library, and newspapers to find examples of recent impacts of human activities on the environment. Prepare a report on your findings.

2. Go around your town and surroundings and make a list of the environmental problems you encounter. Link them to a specific human activity. Prepare an oral report for your class on your findings.

3. Visit a local farm that uses refuges during the crop-growing season. Compare the insect population between refuges and fields. What do you see? Prepare a written report on your observations.

19

ISSUES AND ETHICS

Biotechnology has a wide array of applications. These may change our lives and the way we see science. Should we make the changes that are possible?

Because biotechnology often involves working with living organisms, issues and ethical questions may arise. Some people believe it is unsafe and unethical to change the genetics of an organism using biotechnology. Others believe that people have the ethical obligation to find ways to improve the quality of life for all people. What do you think?

Science will have a big role. People need a good understanding of science to identify and deal with the issues and ethics that are involved. This chapter is to help you in assessing issues and ethical considerations.

19-1. A pig born using biotech processes of sperm sorting and *in vitro* fertilization is normal in all regards. (Courtesy, Agricultural Research Service, USDA)

OBJECTIVES

This chapter provides information on human and societal considerations in the use of biotechnology processes. It has the following objectives:

1. Identify issues and explain how issues are studied.
2. Explain the interaction of ethics and biotechnology.
3. Identify three major areas of biotechnology applications that create debates on ethical issues.

TERMS

bioethics
ELSI
ethics
eugenics
germ cells
issue
religion
somatic cells

ISSUES AND BIOTECHNOLOGY

Controversy sometimes surrounds biotechnology. People have differing opinions. We need to be sure that we go beyond the emotional level of an issue and get the facts.

IDENTIFYING AN ISSUE

An ***issue*** is a subject or problem that has more than one point of view. They are sometimes stated as questions. The questions can be answered or the problems solved in more than one way. Selecting the best answer is not easy.

Some people become activists for particular issues. This takes considerable time and effort. A person must be sure they are willing to make the commitment and sacrifices to aggressively pursue important issues.

Areas where issues in biotechnology are likely include:

19-2. Studying bees disrupts their normal life routine. Some people feel that bees should not be disturbed; others think it is okay. What do you think? (Courtesy, Agricultural Research Service, USDA)

- changing the genetic material of an organism using artificial means
- planting genetically modified crops
- raising genetically modified animals
- consuming foods from genetically modified plants and animals
- using substances that alter or enhance the normal processes of plants and animals
- maintaining genetic diversity in wild plants and animals
- taking drugs produced through genetic modification
- injecting foreign genes into the human body to promote good health

The two most controversial areas in biotechnology are genetic engineering and cloning of organisms. With cloning, we produce a genetically identical copy of a parent organism. A clone is grown from a single somatic cell of its parent. ***Somatic cells*** are the cells that compose tissues, organs, and parts of the body, other than ***germ cells,*** which make up the reproductive system in organisms.

19-3. **A scientist is studying a differential DNA profile to gather information about a grass species. (Courtesy, USDA-ARS Plant Stress and Water Conservation Lab, Lubbock, Texas)**

ACTION PLANS

People who are impacted by an issue or problem should have an opportunity to participate in solving it. Several steps can be followed in handling issues:

- Properly identify the problem or issue. Sometimes people only deal with symptoms or evidence. The facts are needed. Avoid emotion and hype. Promotion and flamboyant approaches by people who have a vested interest in only one solution can make it difficult to properly identify the problem.
- Good information is needed to properly deal with an issue. Regular reading of newspapers, news magazines, and science-oriented materials will help keep people up-to-date on information surrounding issues. Be sure the information is balanced. Some associations have materials that reflect the bias or mission of the group. These should be carefully assessed for scientific accuracy. Policy making involvement in national, state, and local government may be needed. Go to a meeting of the local school board, city council, or other government body. Contact members of the state legislature or U.S. Congress.
- Identify possible solutions to the issue. Make a list of the different ways an issue or problem can be handled. Review the list to identify the best approach. In some cases, combinations of approaches may work best. Also assess which solutions are most likely to be acceptable to the people who are involved.
- Identify the consequences of using possible solutions. Courses of action often have consequences. Some solutions have major impacts on people, such as forcing them to move or get a new occupation. Some alternatives also involve considerable costs. Taxes may be raised or property owners may be forced to spend money that they do not have.
- Focus on the one best solution. The best solution is often a compromise. The one chosen may not be the perfect solution, but it is the one that has the greatest opportunity for success.
- Implement the solution. Solving a problem or an issue requires action. Identifying the best solution is important, but action must follow.

ETHICS AND BIOTECHNOLOGY

Ethics is a set of principles that guide human actions. Some people refer to ethics as the rules of appropriate human conduct. Ethics deals with moral principles and values. Our ethics focus on the rightness or wrongness of particular actions. These are the values that guide our behavior. They shape how we respond to issues associated with biotechnology.

It is difficult to set specific ethical rules. This is because ethics relates to personal judgement and how a person thinks in terms of morals, culture, and/or religion.

ETHICS IN SCIENCE

People who do research in biotechnology are particularly under scrutiny. Their work requires careful adherence to science ethics.

Bioethics is the area of ethics that deals with the life sciences. It often goes beyond ethics and includes political and social areas. Many people involved in biotechnology regularly study or consider bioethics. It is recognized as an important area in the acceptance of biotechnology. For example, the Human Genome Project has allocated 3 percent of its funds to bioethics and associated issues.

Ethics in science can be divided into three areas: scientific misconduct, questionable research practices, and other misconduct. These three categories apply to biotechnology as well. However, because biotechnology deals with genes in living organisms, the ethical implications go even further.

19-4. Students and their instructor in veterinary medicine discuss large animal care. This helps develop appropriate ethical standards in new veterinarians. (Courtesy, College of Veterinary Medicine, Texas A&M University)

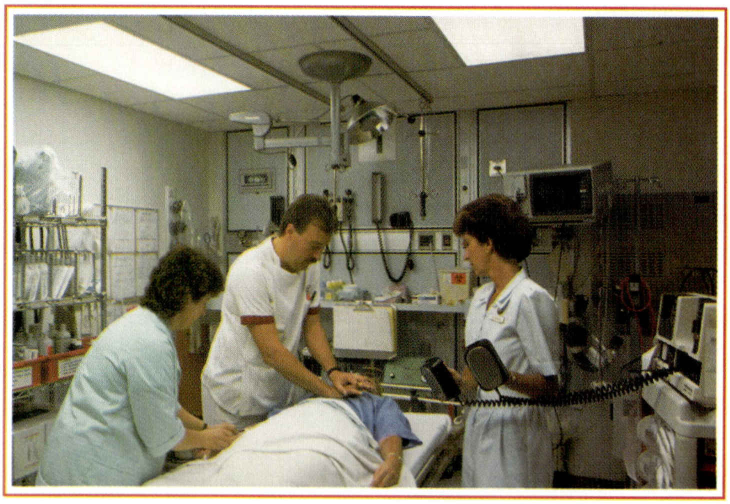

19-5. Human health care providers often deal with life or death matters.

Using the wrong protocol and making inaccurate observations violate science ethics. Purposefully reporting data that are false or fabricated is beyond ethics and is a criminal activity.

Students can begin developing appropriate ethics in science when carrying out their science experiments and projects.

RELIGION AND BIOTECHNOLOGY

A ***religion*** is a system of beliefs and practices that an individual chooses to follow. These may focus on one supreme God or Deity. Some people are members of organized religious groups; others practice religion in their own individual way. Some people do not practice religion at all. To others, religion is very important. There are many different views around the world about religion.

Religion influences biotechnology and its advances in two different ways. First, religions set moral rules that will influence the way a person perceives what is good or bad. Biotechnology, as well as any other scientific achievements, will be judged according to these religious beliefs. Second, many religions have in common the fact that a single entity, God, created life. Biotechnology scientists are often depicted as "playing God." They bring up the issue of creating life in an unnatural way, through cloning or genetic manipulation, which contradicts many religious beliefs.

Many religious practices have been around for thousands of years. Some tend to retain beliefs that were appropriate long ago. They fail to accommodate the findings of science. Others may reject science outright. Change occurs slowly. Education is a key part of how religion embraces biotechnology.

ETHICS AND THE FUTURE OF BIOTECHNOLOGY

Biotechnology developments are limited by advancements in science and by ethical rules. Government policies have major impacts, particularly in human biotechnology. Funding from tax dollars often supports basic research. Without it, research is hampered.

Some of the progress and applications of biotechnology are considered morally unacceptable. No government funding will be available for these. An example is the June 1997, statement by U.S. President William J. Clinton. He stated, "Banning human cloning reflects our humanity. At its worst, [this new technology] could lead to misguided and malevolent attempts to select certain traits, even to create certain kinds of children - to make our children objects rather than cherished individuals." He, as well as other world leaders, has proposed a ban on human cloning.

Genetic engineering allows the insertion of new genes from any source into organisms. One could imagine transgenic plants having genes from a rat or worm. Right or wrong, most people would refuse to eat such a crop. Already, European consumers request that transgenic crop and food resulting from transgenic crops be labeled as such. In fact, actions of the European nations are having big impacts on some U.S. agricultural producers. Who is right? The answer is based on our values.

19-6. Restoring wildlife (such as this falcon) to health and releasing it back to the wild is usually considered a "right" thing to do. What do you think?

SPECIFIC ETHICAL ISSUES

Several areas of biotechnology covered in this book are surrounded with major ethical debates. Three of those areas are covered here from an ethical perspective.

THE HUMAN GENOME PROJECT

The Human Genome Project involves the mapping of all the genes in humans. This project has the following goals:

- identify all genes in human DNA (80,000 to 100,000)
- determine the sequences of the chemical base pairs
- maintain databases of the information
- develop methods of data analysis
- study ethical issues associated with human genome information and uses

One outcome of the project is the possibility that diseases will be detected before they occur. The disease could then be treated or avoided. The advantages linked to the Human Genome Project therefore seem tremendous. There are, however, ethical issues intimately associated with the Human Genome Project. Scientists managing the project have addressed these concerns by creating The Ethical, Legal and Social Implications Program (ELSI). ***ELSI*** provides specific guidelines for those involved with the Human Genome Project.

Knowing that a person will develop a specific disease later during her or his life could lead potential employers to choose not to hire this person. In-

Biotech Connection

ELSI

ELSI is the acronym for Ethical, Legal, and Social Issues Program of the Human Genome Project. ELSI was instated by the U.S. Department of Energy (DOE) and the National Institute of Health (NIH). It is the biggest bioethics program in the world.

ELSI focuses on four areas:

- Privacy and confidentiality of genetic information.
- Safe and effective introduction of genetic information in the clinical setting.
- Fairness in the use of genetic information.
- Professional and public education.

ELSI takes into account ethnic, cultural, social, and psychological influences. It is meant to protect the qualities unique to human beings and their well-being.

surance companies might decide not to give medical insurance to people that have a high chance for developing a disease. The common answer to this problem is that laws could prevent this from happening.

A person who knows she or he will contract a specific disease can take actions against it. The person can take medicine, undergo preventive surgery, or change their lifestyle to avoid becoming sick. There are, however, other diseases that cannot be cured or avoided. Living with the knowledge that we will develop such a fatal disease in the near future might be morally very difficult and could cause psychological disorders. But, perhaps by knowing the genes that contribute to the disease, a cure can be found. Do you want to go through your life knowing that you are likely to develop a disease?

What if genetic information becomes available only to people who can afford it? Poor people could be left with a disease to fight and little means with which to do it.

Career Profile

BIOETHICIST

A bioethicist studies ethics as related to living organisms. Many focus on humans. They deal with the ethics of human organ transplants, genetic engineering, care of the terminally ill, and other areas. Bioethicists may deal with human reproduction areas, such as superovulation, *in vitro* fertilization, embryo transfer, and use of unwanted embryos.

Bioethicists need good backgrounds in science, philosophy, sociology, and other areas that help interpret science. College degrees may be in philosophy, biological science, or a related area. Most have advanced degrees. Some have degrees in theology.

Jobs are limited to large medical and science labs, including those of government agencies. They may be in association with philosophy departments in universities. Many bioethicists are part-time and work in areas such as philosophy and theology.

One area a bioethicist might address is the issue of gender preselection. This shows a high-speed sperm sorter being used to separate X and Y sperm. The gender of offspring can be predicted with 90 percent accuracy using this method. (Courtesy, Agricultural Research Service, USDA)

CLONING

Cloning animals, such as sheep or cattle, is seen as a way to increase production or gain products more efficiently. The purpose might be to have animals capable of producing pharmaceuticals, products in their milk, or other things. Cattle producing an abundance of quality beef would be interesting to clone.

Cloning might have advantages in testing pharmaceutical compounds. New drugs are sometimes tested on animals to determine the activity and assess the risk. This is typically done on laboratory animals that are chosen because they have characteristics very similar to humans. Each animal's response to the drug is determined. This depends on the animal and its genetic material. Testing drugs on clones would remove some interference.

One fear linked to cloning animals is that it can adversely affect the diversity in species.

The main ethical issue associated with cloning is the possibility of cloning human beings. Besides the moral and religious aspects, there is the fear of eugenics. ***Eugenics*** is the science of altering the hereditary qualities of a species or breed. The primary goal is to improve the organism. There is a general consensus among biotechnology scientists that human beings should not be cloned.

19-7. A growing transgenic cotton plant will soon be removed from the jar along with the medium and grown in a greenhouse. (Courtesy, USDA-ARS Plant Stress and Water Conservation Lab, Lubbock, Texas)

Much debate focuses on the cloning of human embryos. An embryo is an animal in the early stages of growth and differentiation. The study of embryos could lead to a better understanding of diseases, such as cancer and Parkinson's. Embryos could be cloned for many purposes. Some cultures and religions see embryos as human beings; others consider life to begin only after birth. Cloning embryos could, therefore, be perceived as cloning human beings.

GENETIC ENGINEERING

Genetic engineering is introducing genes from any organism or microorganism into another organism. The barrier between species does not represent a problem with this technique. People could argue against genetic engineering on a moral or religious base.

Crops genetically modified to carry herbicide or pest resistance are the products of long and costly years of development. Companies that develop these plants apply for a patent. A patent grants an inventor the exclusive right to make, use, or sell the invention. Once a patent has been delivered for a specific genetically engineered crop, only the company that owns the patent can sell or license the crop. Anyone wanting to use the crop has to pay the company. Companies might genetically engineer crops so the seed are sterile. This means that the seed could not be saved for the next growing season. The seed would have to be bought again the following year. In a biotechnology world, only those who could afford the modified crop seed would be able to grow the crop. This is a major issue as it relates to poor nations that might need a genetically modified crop for their survival, but cannot afford it.

REVIEWING

MAIN IDEAS

Numerous issues are found in biotechnology. Almost everyone has an opinion on some aspect of biotechnology. People need to get the facts and make informed, non-emotional decisions.

Ethics deals with moral principals and values. Ethical issues in biotechnology revolve around the modification of living organisms. Religion and morals determine how people view and relate to science and biotechnology.

A special area of ethics relating to the life sciences is bioethics. Bioethics often goes beyond ethics into the political and social areas.

Cloning is a controversial topic for many people. Ethical issues have contributed to the banning of human cloning by most world governments including the U.S. government. Genetic engineering of organisms has also led to ethical issues. Many governments in Europe require that genetically modified products be clearly labeled. The products of biotechnology research must be considered with the moral views of the people.

QUESTIONS

Answer the following questions using correct spelling and complete sentences.

1. What is an issue? Identify one area with issues in biotechnology.
2. What steps should be involved in addressing issues?

3. What forms the basis for ethics?
4. Why is bioethics important?
5. How do religious beliefs influence how an individual views biotechnology?
6. What is ELSI?
7. How can genetic information help people avoid disease?
8. How does cloning human embryos relate to cloning humans?

EVALUATING

Match the term with the correct definition. Write the letter of the term in the blank that is provided.

a. eugenics
b. ethics
c. issue
d. bioethics
e. germ cell
f. somatic cell

_____ 1. A problem or question that has more than one side or answer.
_____ 2. Cells that make up reproductive systems.
_____ 3. Cells that make up the tissues of an organism.
_____ 4. The area of ethics that deals with life sciences.
_____ 5. A set of principles that guides human actions.
_____ 6. The science of altering the hereditary material of a species or breed.

EXPLORING

1. Take a look at the Humane Genome Project World Wide Web site and more specifically the part dedicated to ethics: www.ornl.gov/hgmis/. Prepare a report on your findings.

2. Form teams in the classroom with the goal of debating the use of cloning, genetic engineering, or other area of biotechnology. One team will advocate the practice and the other team will oppose it. Get the facts before you debate. Use the library, Internet, or other sources to gather information that can be used to defend each position.

3. Make a study of "when life begins." Use reliable sources, such as books, science magazines, and the Internet. Interview qualified scientists. Prepare a written report on your findings.

20

REGULATIONS IN BIOTECHNOLOGY

Do we want regulations? Most areas of our lives involve conforming to various regulations. Our schools have schedules, dress codes, and other rules that help us have orderly education. Our society has many regulations–speed limits, tax assessments, and housing zones, to name a few.

Sometimes we like regulations; other times we don't. Most regulations are in place to protect us. They provide us with freedom and restrict our freedom. We know what we can and cannot do. We like those controls that protect us from outside threats. We may not like those that require us to conform. Nevertheless, we follow all of them to demonstrate our care for others, ethics toward our society, and patriotism to our nation.

Biotechnology has regulations as well. These are intended to allow us to safely use the benefits of biotechnology. Sometimes, we think they slow down the rate of progress. Overall, they help us in developing and using biotech products and services.

20-1. Regulations are intended to protect people and allow them to enjoy their environment. This shows outdoor fun at the Great Sand Dunes National Park. (Courtesy, Gary Halverson, U.S. Fish and Wildlife Service)

OBJECTIVES

This chapter introduces the importance and nature of regulations and how they control biotechnology. It has the following objectives:

1. Explain the role and importance of regulations in biotechnology.
2. Identify government agencies and the areas that are regulated.
3. Explain the role of organizations and identify one biotech organization.
4. Describe the interaction of regulations with ethical issues in biotechnology.

TERMS

APHIS
confidential business information
confidentiality
EPA
FDA
permit
regulated organism
regulation
USDA

NATURE AND PURPOSES OF REGULATIONS

A ***regulation*** is a law, principle, or other directive. Regulations usually apply to a specific activity or area of endeavor. Some kinds of products are included. Regulations shape how we go about our work as well as daily living activities. They influence the way agriculture, medicine, and other areas are carried out.

NATURE OF REGULATIONS

Regulations apply to the production, sale, and use of biotechnology products. Most refer to genetically engineered organisms. However, regulations apply to most activities that involve commerce or the sale of products.

Before GMOs are made available, certain standards must be met. These standards are to assure that products are safe. They involve careful testing and documentation of testing procedures. Review by government officials is needed to assure that no harm would be done if the products are available.

When GMOs are released, the products must be labeled and used according to instructions. In some cases, users must have prescribed training and sign contracts indicating that the products will be properly used.

20-2. Ladybugs are beneficial insects. They destroy harmful insects. What if the plant they are on had been genetically engineered with an insecticide? Regulations are intended to protect and maintain biological diversity. (Courtesy, Ron Singer, U.S. Fish and Wildlife Service)

20-3. **Some organisms have been developed to grow in space. People who work with these organisms need to follow proper protocol.**

Agricultural producers must segregate GMO crops from other crops. For example, harvested GMO soybeans must be stored separately from soybeans that are not from GMO varieties. This is done to assure that GMO soybeans do not get used where their use is forbidden. It also allows producers to receive appropriate payments for their crops, such as a bonus for the crop that is in greatest demand.

PURPOSE OF REGULATIONS

In biotechnology, regulations are designed to protect people, other living organisms, and the environment. Most people try to do the right thing but some people do not. They fail to respect the rights of others. Regulations help those who are inclined not to do the right thing to follow acceptable procedures.

Regulations help commerce occur in an orderly manner. International trade is an example. Some nations refuse to accept GMO grain, cotton, and animal products. The harvested crops must be kept separate to assure that they can be exported to nations where restrictions exist.

The overall purpose of regulations is to protect consumers. They also have the effect of protecting producers who do not use GMOs. Economic and political purposes are also served.

AGENCIES AND RESPONSIBILITIES

Agencies that regulate biotechnology are at the federal, state, and local levels. State and local level regulations vary, but must not violate those of the federal government. In addition, international regulations may also be placed on some kinds of activities, especially those involving commerce.

Career Profile

PLANT PHYSIOLOGIST

A plant physiologist studies the processes in living plant materials. The work may also include substances to alter plant growth, such as herbicides or nutrients. This photo shows a plant physiologist using emission spectrometry to determine the amount of zinc and iron in a liquid from acid-digesting bean seeds. (Courtesy, Agricultural Research Service, USDA)

Plant physiologists need college degrees in physiology, botany, agronomy, or a closely related area. Most have masters and doctors degrees. Begin preparation in high school with courses in science, horticulture, agriculture, and related areas. Practical experience working with plants is beneficial. Good communication skills are useful.

Jobs for most plant physiologists are with research stations of universities and government agencies, private companies involved in research, and regulatory agencies.

WHO MAKES REGULATIONS

Regulations are made by the three branches of the federal government: legislative, executive, and judicial.

Legislative regulations are rooted in the laws enacted by the Congress of the United States. Once passed by Congress, the president must sign approving the law. In some cases, laws can be enacted without the president's signature. Agencies may issue guidelines or interpretations of the laws. Most of the regulations on biotechnology in the United States originated as legislation.

The executive branch of government can issue executive orders that influence biotechnology. Only the president has the authority to issue such orders. These have the same effect as laws passed by Congress.

Decisions made by courts are handled as laws. A decision may set a precedent and influence future cases. Once a court case has been decided, an appeal can be made until all appeals are exhausted with the U.S. Supreme Court.

RESPONSIBLE AGENCIES

Three agencies in the federal government have primary responsibilities for administering laws related to biotechnology. These are the U.S. Department of Agriculture (***USDA***), the Environmental Protection Agency (***EPA***), and the Food and Drug Administration (***FDA***).

Products are regulated according to their intended use. Some are regulated by only one agency. Others are regulated by two and some by three.

Biotech Connection

CRYOPRESERVATION

Cryopreservation is using very low temperatures to preserve living materials. Cells, tissues, organs, and other living materials may be kept indefinitely. This includes embryos that have resulted from superovulation and *in vitro* methods. This photo shows a sow with pigs from cryopreservation and transferred embryos. (Courtesy, Agricultural Research Service, USDA)

Cryopreservation methods include freezers, dry ice, and liquid nitrogen. Liquid nitrogen is the coldest, with a temperature of –196°C. Freezers and dry ice are both approximately –80°C. At these very low temperatures, living substances remain stable and inactive. The life of such materials can be quite long.

What regulations are needed with cryopreservation? Think about how unused embryos should be disposed of. Formulate a regulation that you think would be appropriate. Discuss it with your class members.

The USDA

The USDA focuses on biotechnology from the standpoint of products being safe to grow. This includes whether products are safe to use on cultured crops or animals. Within the USDA, the Animal and Plant Health Inspection Service (***APHIS***) has the major responsibility. APHIS is responsible for plant materials and veterinary biologics. The Food Safety Inspection Service (FSIS) is responsible for the safe use of genetically engineered livestock, poultry, and products derived from them. Other agencies of the USDA also are involved depending on the product and how it is marketed.

New crops may be initially regulated. After a time, the regulation may be lifted. For example, the restriction on virus resistant squash was lifted in 1996. This allowed the developer of the GMO to pursue widespread seed marketing.

The EPA

The EPA is an agency of the United States government that is independent of any department. Its role is to assure a quality environment. The EPA sets and enforces standards to protect the environment. All biotechnology uses are assessed from the standpoint of safety for the environment.

The EPA regulates genetically modified microbes, pesticides, and microorganisms that could pose threats to the environment.

The FDA

The FDA is an agency in the U.S. Department of Health and Human Services. It is responsible for all laws that are designed to assure the safety of food and drugs. The responsibilities include cosmetics and a wide range of health care products. The FDA sets standards for labels and enforces truthfulness in labeling.

The FDA regulates food, feed, food additives, veterinary drugs, human drugs, and medical devices. Regulations have been developed by the FDA on food products derived from genetically altered plants. All such products must receive approval before being marketed.

GETTING A NEW PRODUCT APPROVED

Government agencies have detailed procedures to assure that all regulated products are okay for use. A large amount of testing is needed. Field tri-

20-4. Getting a new product approved may require discussion with permitting agencies.

20-5. Cotton with lepidopteran (moth and butterfly) insect resistance developed by the Calgene company is no longer regulated by the USDA. (Courtesy, Texas Department of Agriculture)

als must be carried out in an acceptable and appropriate manner. Application forms that include reports of all testing are needed to gain approval.

In some cases, the use of biotech products is granted only for experimental purposes. Widespread and general use is not allowed for experimental products. Restrictions may later be lifted and allow widespread use.

The USDA requires that two copies of applications be submitted to APHIS for a 120-day review period prior to its intended introduction. Application forms are available from the Biotechnology Permit Unit of APHIS at the USDA office in Riverdale, Maryland.

Permits are needed to introduce products that are regulated, such as a new GMO crop. A ***permit*** is written approval granted by the appropriate government agency. The permit follows an appropriate application process. Each permit relates to one particular product.

A ***regulated organism*** is one that has been genetically engineered, or one that is a vector agent for a plant pest, or contains plant pest components. This includes donor as well as recipient organisms. Permits are needed to transport regulated organisms.

A permit applicant must comply with the regulations of all government agencies. For example, a regulated organism might be subject to rules of the USDA and the EPA or the FDA.

After a time and with no negative consequences, a product may be removed from the regulated list.

THE FEDERAL REGISTER

The *Federal Register* is the official document of the U.S. government for informing the public about the regulation of biotech products. In published form, this document contains massive amounts of information on rules and regulations in every aspect of government. Much of the information is available on the Internet, such as through the APHIS web site: www.aphis.usda.gov/biotechnology/

ORGANIZATIONS

Organizations of various types are found in biotechnology. These serve a range of roles. Subgroups of medical, education, agricultural, and environmental organizations often address biotechnology areas. For example, the National Science Teachers Association (NSTA) provides workshops and information for its members in biotechnology. However, biotechnology is not the major focus of the NSTA.

Organizations provide guidelines and opportunities for communication among biotechnologists. They do not have the authority, as do the government agencies, to regulate the use of biotechnology products. Organizations that are formally organized may be known as associations.

One major organization and a subsidiary of it are briefly introduced here.

BIO

The Biotechnology Industry Organization (BIO) is the largest trade organization. It provides leadership and guidance to help its members advance the overall industry. Its membership includes biotechnology companies, research units, centers, and colleges and universities involved with biotechnology. Major worldwide biotech companies are members of BIO.

One area of leadership by BIO is bioethics. BIO has a committee that is charged to define, articulate, and communicate information relative to biotechnology issues. Among its positions are those of a moratorium on human cloning and confidentiality of identifiable medical information. Bioethics will continue as an area of high interest.

BIO promotes informed citizens. It has programs to promote the study of biotechnology by students. For example, it sponsors an essay contest each year. The contest is intended to have students use biotechnology to solve problems. Information on the essay contest is available from BIO, 1625 K Street, N.W., Suite 1100, Washington, D.C. 20006-1604.

CBC

The Council of Biotechnology Centers (CBC) is an offshoot of BIO. Its membership is made up of research and development centers of universities, government agencies, and others. The CBC is intended to provide dialogue among individuals with shared interests in biotechnology. Some of their efforts focus on the social issues associated with the use of biotechnology.

ETHICS AND REGULATIONS

Ethics and regulations sometimes conflict. What is good for people and what the regulations allow do not always agree. Of course, it is good to have regulations to protect people. Regulations need to be changed when they interfere with people gaining the benefits from biotechnology.

INTERNATIONAL PERSPECTIVE

Nations benefit differently from biotechnology. Some nations do research and provide people with the benefits. Some nations do not have the research and rely on other nations for their innovations.

Nations are often classified as developing or developed. A developing nation is one that has low levels of economic activity. Its people often have a low standard of living and quality of life. A developed nation is one that is highly industrial. It has large amounts of technology. Its people benefit from the innovations that are made. This illustrates differences in access to biotechnology. Question: Is it ethical for the people of some nations to have innovations and others not to have them?

Sometimes, developed nations do not readily share among themselves. Innovations in one nation may not be accepted in another. Some of the differences are related to political and economic competitiveness. Medicine is a common example. Drugs available in some countries may not be available in another until additional testing has been done. Question: Is it ethical for tested and approved innovations in one nation not to be available in another nation?

20-6. The relocation of gray wolves has involved considerable debate. Actions were to assure biodiversity and balance in wildlife. (Courtesy, U.S. Fish and Wildlife Service)

FREE ENTERPRISE

Companies engaged in biotechnology development are in business to make a profit. Sometimes, the benefit of people and the welfare of a company may conflict. Question: Is it appropriate for a company to have a proven innovation that is not available to people? An example could be medicine to control a dreaded disease.

The profit incentive motivates companies to do research and develop new products. Without a profit, there is little or no incentive to take the large financial risks associated with developing new products. For example, Americans are particularly vulnerable to increased costs for drugs. Drugs bought in the United States cost more than when bought in adjoining countries. Question: Is it ethical for companies to charge Americans more for drugs than people are charged in other nations? (Remember, the United States is wealthier than most countries. Its citizens may be better qualified to pay higher costs. But, not all citizens may have the income to afford the needed drugs.)

CONFIDENTIALITY

Confidentiality is keeping information secret or private. The information is not made available to the public. Those who are involved with the information have an ethical responsibility to assure confidentiality. It includes several areas. These relate to businesses and human information.

Applications to government agencies to gain approval to transport or release a regulated organism allow the applicant to indicate information that is ***confidential business information*** (CBI). This information is protected and not released to insure that the investment of the company is protected. The information is sometimes known as trade secrets. The applicant designates which information is CBI. If it were released, a company that had not made an investment in researching the product could take the information and possibly produce the product without an investment. Information about the financial position of a company is also treated as CBI if the applicant so designates.

20-7. Some information is confidential and not to be released by agencies, organizations, or individuals.

Humans are often involved in biotechnology. They may be the subjects or beneficiaries of biotechnology products. Information that would allow the identification of individuals should be safeguarded. This includes genetic traits that may be used against an individual. An example is the predisposition to a particular disease. If not kept confidential, biotechnology information might be used to discriminate against individuals.

REVIEWING

MAIN IDEAS

Regulations provide uniformity and safety in biotechnology. They are intended to protect people, other living things, and their environment.

Regulations are made by the legislative, executive, and judicial branches of government. Three government agencies have responsibilities in biotechnology: U.S. Department of Agriculture, the Environmental Protection Agency, and the Food and Drug Administration. Products are regulated according to their intended use. Agency responsibilities sometimes overlap, as with biotech crops that are used for food. Regulated products must go through a permitting process before being used.

Organizations have useful but nonregulatory roles in biotechnology. The Biotechnology Industrial Organization is the largest and, likely, the most important.

Regulations sometimes clash with ethical issues. Most of the disparity is between nations and people in their access to biotechnology.

QUESTIONS

Answer the following questions using correct spelling and complete sentences.

1. What activities do biotechnology regulations include?
2. What is the major purpose of regulations in biotechnology?
3. What three government agencies have responsibility for regulating biotechnology in the United States? Briefly describe the roles of each.
4. What is a regulated organism?
5. What steps are needed to get a regulated product approved?
6. Name an organization in biotechnology. Briefly describe its role.
7. How do ethics and regulations sometimes conflict?
8. Why is confidentiality needed? What should be covered by confidentiality?

EVALUATING

Match the term with the correct definition. Write the letter of the term in the blank that is provided.

a. confidential business information
b. regulation
c. permit
d. regulated organism
e. confidentiality
f. APHIS

_____ 1. An agency of the U.S. Department of Agriculture that has major responsibility for agricultural biotechnology.

_____ 2. Keeping information secret or private.

_____ 3. Information on a permit application that is a trade secret.

_____ 4. A law or directive that applies to a specific activity or product.

_____ 5. Approval granted by a government agency.

_____ 6. A genetically engineered organism, vector, and similar forms.

EXPLORING

1. Get the details of the essay contest from the Biotechnology Industry Organization. Write and submit an essay following guidelines provided by the organization. (BIO, 1625 K Street, N.W., Suite 1100, Washington, D.C. 20006-1604.)

2. Contact a government agency that regulates biotechnology. Get details on the permit process and how a biotech organism is removed from the regulated list. Prepare and give an oral report on your findings.

3. Invite a biotechnologist to serve as a resource person in class and discuss the roles of regulations. Prepare a written report on the presentation.

APPENDIXES

Appendix A
Common Weight Conversions and Equivalents Used in Biotechnology

Customary System		
1 tablespoon (tbs)	=	3 teaspoons (tsp) or 0.5 fluid ounce (fl. oz)
1 cup	=	8 fl. oz
1 pint (pt)	=	2 cups or 16 fl. oz
1 quart (qt)	=	2 pt or 4 cups or 32 fl. oz
1 gallon (gal)	=	4 qt
1 bushel (bu)	=	8 gal or 32 qt
Metric System		
1 milliliter (ml)	=	1 cubic centimeter (cc) or 1 gram (g)
1 centiliter (cl)	=	10 ml
1 deciliter (dl)	=	10 cl or 100 ml
1 liter (L)	=	10 dl or 1,000 ml or 1 kilogram (kg)
1 dekaliter (dal)	=	10 L
1 hectoliter (hl)	=	10 dal or 100 L
1 kiloliter (kl)	=	100 hl or 1,000 L or 1,000 kg
Equivalents		
1 ounce	=	28.3495 grams
1 pound	=	453.59 grams or 0.45359 kilograms
1 kilogram	=	35.27 ounces or 2.2046 pounds
1 metric ton	=	2.204.6 pounds
1 hectare	=	2.47 acres
1 liter	=	1.057 quarts
1 liter	=	0.2642 gallons
1 gallon	=	3.79 L
1 meter (m)	=	39.37 inches
1 acre	=	0.40468 hectares
1 inch	=	2.54 centimeters (cm)
1 centimeter (cm)	=	0.4 inch

Appendix B
Common Volume Measures Used in Biotechnology

Customary Fluid Measures				
4 gills (gi)	=	1 pint (pt)	=	28.88 cubic inches
2 pints (pts)	=	1 quart (qt)	=	57.75 cubic inches
4 quarts (qts)	=	1 gallon (gal)	=	231 cubic inches
31.5 gallons (gals)	=	1 barrel (bbl)	=	4.21 cubic feet
2 barrels (bbls)	=	1 hogshead (hhd)	=	8.42 cubic feet
7.5 gallons	=	1 cubic foot	=	1,728 cubic inches
Customary Dry Measures				
2 pints	=	1 quart	=	67.2 cubic inches
8 quarts	=	1 peck (pk)	=	537.6 cubic inches
4 pecks (pks)	=	1 bushel (bu)	=	2,150.4 cubic inches or 1.244 cubic feet

Appendix C
Weight Measures Commonly Used in Biotechnology

Customary Weights		
Avoirdupois Weight:		
437.5 grains (gr)	=	1 ounce (oz)
16 ounces	=	1 pound (lb)
100 pounds	=	1 hundredweight (cwt)
2,000 pounds	=	1 ton (T)
2,240 pounds	=	1 long ton
Troy Weight:		
24 grains	=	1 pennyweight (pwt)
20 pennyweight	=	1 ounce
12 ounces	=	1 pound
Apothecaries Weight:		
20 grains	=	1 scruple (sc)
3 scruples	=	1 dram (dr)
8 drams	=	1 ounce
12 ounces	=	1 pound
Metric Weights		
1 centigram (cg)	=	10 milligrams (mg)
1 decigram (dg)	=	10 centigrams and 100 mg
1 gram (g)	=	10 dg, 100 cg, and 1,000 mg
1 kilogram (kg)	=	1,000 grams
1 metric ton (t)	=	1,000 kg

GLOSSARY

abiotic factors–the nonliving parts of an environment, such as soil, water, air, and minerals.

abstract–a brief (no longer than a page) description of the information contained in a research report.

aerobe–a microbe that requires free oxygen for growth.

agar–a gelatin-like laboratory growth medium that is prepared using hot water.

allele–a different or alternative form of a gene; source of genetic variation.

anaerobe–a microbe that does not require free oxygen for growth.

ancient biotechnology–a period of biotechnology that begins with the early history of human civilization and includes important developments in agriculture and food production.

anestrous–the absence of cycling in the female reproductive cycle.

animal biotechnology–the use of biotechnology with animals.

animal well-being–caring for animals so their needs are met and the animals do not suffer.

antibiotic–a drug used to combat bacterial infections.

antibody–a protein molecule produced as defense against disease organisms.

APHIS–the Animal and Plant Health Inspection Service in the U.S. Department of Agriculture; responsible for plant materials and veterinary biologics.

appendix–a separate supplemental section at the end of a report.

applied research–an investigation or research that is designed to directly solve an immediate problem.

artificial insemination–the transfer of collected semen to a recipient female.

artificial selection–the process of choosing individuals from each generation that possess a superior quality.

asepsis–something without germs; sterile.

aseptic environment–an area free of microbes.

aseptic technique–taking action to have an environment that is free of germs.

Asilomar Conferences–the name for a series of conferences that represent the first efforts by scientists to develop guidelines for biotechnology.

atom–the smallest unit in which an element can exist.

atomic number–the number of protons in the atom.

autoclave–a steam sterilizer that uses pressure to exceed normal water and steam temperatures.

autosome–a chromosome that is not a sex chromosome.

bacteria–single-celled prokaryotes with a single chromosome and no cellular organelles; singular form is bacterium.

bacteria culture–growing bacteria under laboratory conditions and often using a medium, such as agar.

bactericide–a chemical or physical agent that kills some kinds of bacteria.

bacteriostat–an agent that prevents the formation of bacteria.

ballistic gene transfer–the use of tiny DNA-coated projectiles as carriers to transport DNA through the walls of intended recipient cells.

basic research–inquiry that adds to the body knowledge about a specific issue.

bibliography–a list in a report of all of the resources used in the literature review and in the study.

binomial name–the scientific name of an organism; consists of the genus and species.

biochemical process–the series of reactions occurring within an organism that enables the production of specific compounds or energy.

biocompatibility–the process of implanting devices that must be mechanically fit for the purpose and not rejected by the body.

biodegradation–the natural process by which microbes break down a product to environmentally friendly or useful compounds.

bioethics–the area of ethics that deals with the life sciences; it is especially important in certain areas of biotechnology.

biogeochemical processes–the processes by which microbes transfer nutrients from living forms to physical forms and back to living forms.

biohazard–a risk potential from working with or developing organisms.

biological seeding–used to assure the presence of the appropriate microbes by placing the required bacteria with the waste.

biology–often known as life science; the study of living things.

biopharmaceutical–a drug or vaccine that is developed through biotechnology.

biopharming–the production of pharmaceuticals in cultured organisms.

bioremediation–a method that uses biotechnology and microbes to reclaim a contaminated site.

biotechnology–using scientific processes to get new organisms or new products from organisms.

biotic factors–the living organisms in the environment including plants, animals, and microbes.

biotransformation–a process that promotes the production of chemicals by microbes.

bovine somatotropin (bST)–a hormone secreted by the pituitary gland that regulates milk secretion by the mammary glands of female mammals.

callus–a mass of undifferentiated plant cells.

carrion–the flesh of a dead animal.

catalysis–the alteration of the rate at which chemical reactions occur.

catalyst–a substance that alters the rate of a chemical reaction without being changed itself.

cell–a unit of life that has a definite structure.

cell membrane–a membrane that acts as a gatekeeper, controlling what enters and leaves a cell and separates individual cells from each other.

cell specialization–the difference in cells; may be specialized and form tissues, organs, and organ systems.

cell theory–a theory that states that all living things are made of one or more cells, that cells live and are the basic building blocks of life, and that cells reproduce with new cells coming from old cells.

cell wall–a tough outer covering or membrane that provides structure and protects the cell from pressure caused by the movement of water; found in plant cells.

chemical energy–energy present in chemical bonds that have formed molecules.

Chi-square–an inferential statistic that compares observed frequencies with expected frequencies.

chloroplast–a green body in a plant cell that contains chlorophyll.

cilia–short threadlike projections from protozoa that move back and forth and help the organism with locomotion.

circulation–the movement of substances within an organism.

classical biotechnology–the period of biotechnology development that follows ancient biotechnology when widespread use was made of the methods from ancient biotechnology, especially fermentation.

cloning–the process of producing new organisms by separating cells or tissues of existing organisms; the new organism has identical genetic material to its parent.

competent cell–one that is capable of taking up DNA.

composting–promoting the biological decomposition of organic materials.

conclusion–an inference about the findings of research or experimentation.

confidential business information(CBI)–information on an application that is protected and not released to insure that the investment of the company is secure.

confidentiality–keeping information secret or private and not available to the public.

conjugation–the contact of bacteria that involves the exchange of DNA with a mating tube.

constant–a quality that is being controlled or kept the same.

containment–taking steps to control the accidental release of potentially harmful organisms or their products.

control–in experimentation, it is the environment or condition that is not changed.

copulation–the sexual uniting of animals so the male can deposit (ejaculate) semen near the eggs in the reproductive tract of the female.

crop–a plant that is grown and harvested for food and other products.

crop residue–the leaves, stems, and other plant materials that remain on a field after a crop is harvested.

crossbreeding–breeding individuals of different varieties or species to have offspring with desired qualities.

cultivar–a cultivated variety of a crop species distinguished by minor differences, such as length of time required to reach maturity.

culture–the ways to go about life that individuals have developed and transmitted from one generation to the next.

cytoplasm–a jelly-like material that surrounds the organelles within a cell.

data–the items of information collected in an experiment; singular form is datum.

decision making–choosing among the alternatives that are available.

defense–an oral presentation to a professional conference, corporation, review board, or peer group on a thesis or dissertation.

descriptive statistics–summary statistics where numbers define the data from an experiment.

designer drug–a drug that is developed as a three-dimensional model on a computer before it is tried out in a lab.

development–the creation of new products or methods based on the findings of experimentation and field testing.

developmental stages–the combination of biological and cultural advancement of human beings from birth to death.

diagnostics–the identification of a human health problem or other condition.

dialysis machine–a kind of prosthesis that performs functions normal kidneys would do in removing wastes from the blood.

diffusion–the movement of molecules, especially gases and liquids, from areas of high concentration to areas of low concentration.

diploid–organisms that have two chromosomes that control a single trait.

directional hypothesis–a hypothesis that indicates a definite relationship between the problem and the action.

disease–an abnormal condition of the body.

disinfection–removing potential disease-causing microbes.

display board–a graphic report on a research project.

dissertation–the final project of a candidate for a doctoral degree.

diversity–a variety of different forms and types exist.

DNA (deoxyribonucleic acid)–the molecule in a chromosome that furnishes the genes with information for their development.

DNA isolation–the process of extracting and separating DNA from all other cell materials.

DNA profiling–identifying an organism based on regions of DNA that vary greatly from one organism to another.

DNA sequencing–the process of determining the order of the nucleotides in DNA fragments.

domestication–the adaption of organisms so they can be cultured.

dominant–the trait that can be observed in an organism if the alleles are different.

donor cell–the cell that provided the DNA fragment for the recipient cell.

ectotherm–an organism whose body temperature adjusts to its environment, such as snakes, fish, and lizards.

electron microscope–a device used to observe very small things; uses a beam of electrons passing through an object to visualize and magnify.

electroporation–the use of an electric shock to momentarily open or disrupt cell walls so genetic information can enter the cell.

element–any substance that cannot be broken down into simpler material by ordinary means; the fundamental units of a chemical substance.

elimination–the process by which an organism rids wastes and harmful substances from its body.

ELSI–Ethical, Legal, and Social Implications Program; provides specific guidelines for those involved with the Human Genome Project.

embryo stage–the first stage of development during gestation.

embryo transfer–the harvesting of fertilized ova from a donor and implanting them into a recipient.

endoplasmic reticulum–a layered, membranous organelle that may be found surrounding the nucleus or as a separate organelle within the cytoplasm.

endotherm–an organism that maintains a constant body temperature, such as dogs, mice, and chimpanzees.

energy–the capacity or ability to do work; forms of energy include chemical energy, heat, sound, light, and electricity.

energy transfer–the exchange of energy by organisms at different levels in the environment.

environment–our surroundings.

environmental biotechnology–the study of biotechnological processes that promote a good environment.

environmental quality–the general condition of the natural resources and other factors where we go about life.

enzyme–a kind of protein that acts as a catalyst in chemical reactions needed for the living condition to occur.

EPA–the Environmental Protection Agency; an agency of the U.S. government that is independent of any department.

equilibrium–a system that is balanced, or equal.

estrous cycle–the time between estrus.

estrus–one of the periods in the cycle when a female is fertile and receptive to a male.

ethics–a set of principles that guide human actions.

eugenics–the science of altering the hereditary qualities of a species or breed.

eukaryote (also spelled eucaryote)–an organism made up of cells that have a nucleus.

euthanasia–causing death without pain or stress.

experimentation–the process of conducting tests that establish the validity of a hypothesis.

explants–small pieces of plant material that are removed from the parent plant.

F_1–the first filial generation of offspring.

F_2–the second filial generation of offspring.

FDA–the Food and Drug Administration; an agency in the U.S. Department of Health and Human Services.

fermentation–a process in which yeast enzymes chemically change compounds into alcohol.

fermenter–a specially designed chamber that promotes the fermentation process.

fertilization–the union of the sperm and egg.

fetal stage–the later stage of development during gestation.

fiber–slender threadlike material used to make clothing, paper, and shelter.

field plot–a small area of land that is used to test questions or hypotheses.

figure–a graph, line drawing, photograph, or other illustration in the text to provide additional information.

fission–prokaryotic cells reproduce by the process of the cell growing larger and making a copy of its DNA (nucleoid).

flagellum–a long, moving thread on a protozoan.

flaming–exposing the tips or surfaces of instruments to a flame to destroy any microbes.

food–the solid and liquid material humans and other organisms consume for nutrients.

food biology–the study of the characteristics of food products obtained from living organisms.

food biotechnology–the use of biological processes in producing and processing food.

food-borne illness–a disease that people get from the food they eat.

food chain–the feeding sequence in which organisms get their food.

food chemistry–the study of the substances that make up foods.

food physics–the study of the physical characteristics of food.

food preservation–using processes that prevent or slow spoilage by heating or cooling foods to keep microorganisms from growing.

food processing–the preparation of food materials for consumption.

food quality–the ability of a food to meet nutritional needs.

food science–the study of the substances humans eat.

food web–a graphical way of showing how food chains are interconnected.

fungi–complex, eukaryotic organisms (meaning their cells have a nucleus).

fungicide–a chemical compound used to control fungi, especially fungal diseases.

gamete-a sex cell capable of uniting with another sex cell to produce a new individual.

gel electrophoresis–a process using an electrical field in agar to separate DNA and RNA molecules based on molecule size.

gender preselection–choosing the gender of potential offspring.

gene–a segment of a chromosome that contains the heredity traits of an organism.

gene expression–the process of making DNA information available to a cell.

gene therapy–changing the genetic makeup of a human being.

gene transfer–moving a gene from one organism to another.

genetic engineering–changing the genetic information in a cell.

genetics–the study of the heredity of living organisms.

genotype–the genetic makeup of an organism that is not observable.

germ–a microorganism that can cause problems with biotechnology, food keeping, and other areas; some cause disease.

germ cells–special cells that comprise the reproductive systems in organisms.

germ-line gene therapy–changes the cells that produce sperm or ova in an individual.

gestation–the period of pregnancy in mammals; length varies by species.

gland–a specialized organ or tissue that produces secretions.

GMO–a genetically modified organism; the product of genetic engineering.

GMO-free food–a food product in which no transgenic materials were used in its manufacture.

Golgi bodies(or Golgi apparatus)–cell parts that appear similar to the endoplasmic reticulum; adds modifications to unfinished proteins and makes lysosomes.

gram stain–a common test used to identify bacteria.

growth–the process by which the cellular components and structures of an organism increase in size.

hazard–a danger; potential risk for injury or harm.

hazardous waste–a substance, tissue, or organism that is potentially dangerous to human health or the environment.

health–the state of the human body or other organism as related to disease and vigor.

healthcare–promoting good health; includes the services, information, equipment, and other conditions for good health.

herbicide–a chemical used to control weeds.

heredity–the study of how traits are passed from parents to offspring.

heterozygous–if the alleles for a trait are different.

homozygous–if both alleles for a trait are the same.

human development–the progressive maturation of an individual from birth until death.

human genome–all of the genes that comprise the genetic makeup of human beings.

Human Genome Project (HGP)–an international effort involving scientists, educators, and students in locating and identifying every human gene.

human medicine–the area of medicine that deals with human health.

hybridization–the process of breeding two individuals from distinctly different varieties; also known as cross-pollenation.

hydrocarbon–any substance that contains only carbon and hydrogen.

hypothesis–an "educated guess" or a statement that gives a solution to a problem or anticipated outcome of an experiment.

incineration–using high temperatures to destroy or burn waste materials.

industrial biotechnology–the mass production of substances or organisms using a factory approach.

inferential statistics–methods using numbers that allow a scientist to expand the data in the experiment to predict what would happen in a larger environment.

innovation–products or processes that are new or have not been widely used before.

insecticide–a chemical used to control insects.

invasive procedure–entering the body of an animal with a needle, catheter, or other medical instrument.

in vitro–a Latin term meaning "in glass."

in vitro fertilization–the process of collecting ova from a donor female and fertilizing them, outside the body, with collected semen usually in a glass test tube.

issue–a subject or problem that has more than one point of view.

kinetic energy–the energy a body possesses because it is in motion.

laboratory animal–an animal used for laboratory or research purposes.

lactation–the secretion of milk by the mammary glands.

laminar flow hood–a small work environment in a device that helps provide aseptic conditions.

landfill–an excavated area of land for the permanent disposal of wastes.

leachate–liquids produced and released by the decomposition or oxidation process in landfills and by precipitation on a landfill.

life expectancy–the number of years that an individual may be expected to live.

life process–an essential function for an organism to be in the living condition.

life science–the study of living things.

ligation–uniting or attaching two DNA fragments.

light microscope–the most commonly used microscope in the scientific community.

line breeding–breeding offsprings (successive generations) among themselves.

lipid–an organic compound of a fatty acid that is soluble in organic solvents, but not in water.

literature review–the step in the scientific method of gathering information to get a good understanding of the topic and to learn what has already been done; the use of books, journals, and similar materials.

livestock–species of domesticated animals found on farms and ranches.

living condition–the ability of the protoplasm in the cells of a living organism to carry out complex chemical processes.

lysosome–a small, round organelle that stores enzymes that break down food into chemical compounds.

marker gene–a gene that identifies which organisms have been successfully transformed; evidence of transformation; any genetic element that produces variation in expression of a trait.

material–a physical object or chemical that has been produced by humans.

material safety data sheet (MSDS)–a summary of information about the safe use of a chemical.

mathematics–the science of numbers.

matter–anything that has mass and takes up space.

mean–the sum of the scores divided by the number of scores; the average.

median–the value in a distribution of scores that represents the point at which half the scores are higher and half the scores are lower.

medicine–includes the art and science of promoting health among living animal organisms; any substance used to treat conditions that impair human health.

medium–a substance that provides the needed environment, including nutrients, for the growth of the bacteria; plural form is media.

meiosis–division through which sex cells or gametes are formed.

microbial biotechnology–using microbes for biotechnology research and to produce materials.

microbiology–the study of microbes (microorganisms).

microdissection–a method of obtaining pure cells from diseased sections of tissues.

microinjection–the injection of materials directly into cells using a small glass pipette.

microorganism–a living organism or microbe that is not normally seen with the naked eye.

microscope–any instrument that uses lenses to enable the observation of very small objects.

mitochondria–the "energy factories" of a cell that change food into molecules that can be used for energy.

mitosis–the division of a eukaryotic cell into two identical "daughter" cells.

mode–the most frequently occurring score in a set of scores.

models for animal research–the use of living or nonliving animals for biotechnology work; four models are available.

modern biotechnology–the period of biotechnology that involves the manipulation of genetic material within organisms based on genetics and the use of microscopy, biochemical methods, and related sciences and technologies.

molecule–the smallest group of atoms that acts together to form a stable, independent substance.

molecular biotechnology–changing the genetic make-up of organisms.

multidisciplinary–involving many disciplines or branches of learning.

multiple hypotheses–more than one hypothesis for experimentation.

mutation–a genetic variation that occurs naturally and without prediction that it will occur.

negligence–failing to take a reasonable degree of care.

nonhuman primate–an animal similar to humans with thumb and forefinger opposition, but is not a human.

nucleus–a relatively large structure within a cell; directs cell activity.

null hypothesis–a premise that no relationship exists between a problem and the action of an experiment.

nutraceutical–a food that has benefits other than nutrition.

nutrient–a substance in food that supports life processes.

objective–sources that deal only with facts; avoid emotion.

oral report–a verbal or spoken presentation made to inform others about an experiment.

organ–a collection of tissues that work together to perform a function.

organelles–any subcellular, membrane-enclosed structure in a cell with specific functions.

organismic biotechnology–working with complete, intact organisms or their cells.

organ transplant–replacing bad or damaged organs in a recipient with organs from a donor.

organ system–a collection of organs that work together to perform a specific function for an organism.

osmosis–a type of diffusion; the movement of water from areas of high concentration to areas of low concentration.

ovum–egg; the sex cell from the female parent; plural is ova.

parasite–an organism that lives in or on another organism (host) and gains sustenance from its host.

parent plant–the plant that is the source of explants.

parturition–the birthing process.

patent–a government-granted right for an individual or business to exclusively produce and sell a product for a number of years.

penicillin–one of the first effective antibiotics discovered by Alexander Fleming in 1929.

per capita consumption–the average amount of food or other products consumed by each individual in a year.

permit–written approval granted by the appropriate government agency related to a specific product.

personal protective equipment(PPE)–devices that people wear to protect themselves from injury.

pharmaceutical–a chemical substance (drug) used in the prevention, treatment, cure, or diagnosis of a disease.

pharmacology–study of the preparation, use, and effect of drugs.

phenotype–the outward or physical appearance of an organism.

photosynthesis–the process of food manufacture by plants.

physical science–study of the nonliving things around us.

plant biotechnology–the use of biotechnology to improve plants and the products that are produced from plants.

plant breeding–the improvement of plants by selectively breeding chosen individuals.

plantlet–a plant with tiny leaves, stems, and roots that have not developed into the normal-size parts of the parent plant.

plant tissue culture–a method used to clone new plants.

plasmid–a piece of free-floating DNA in the cytoplasm of bacteria.

polymer–the product of the chemical union of two molecules to the same compound to produce another compound.

polymerase chain reaction–a procedure that uses controlled temperatures and the enzyme, *taq* polymerase, to replicate pieces of DNA.

potential energy–the capacity for doing work that a body possesses because of its position.

poultry–domesticated fowl produced for food and other products.

primary innovation–an innovation that makes a major advance in a new area.

primate–an animal with thumb and forefinger opposition.

prokaryote(also spelled procaryote)–a single-celled organism in the Kingdom Monera that lacks a nucleus.

pronucleus–the haploid(single) nucleus of the sperm or ovum that have united in fertilization to form a zygote.

prosthetics–the study of the use of artificial parts as substitutes for human body parts that no longer function; includes the study and use of mechanical devices to replace or supplement natural human parts.

protein–a substance consisting of amino acids; also, an important food nutrient.

protocol–the plan or procedure for carrying out a scientific research procedure.

prototype–a research model that has been developed for testing.

puberty–the time at which an animal is capable of reproduction.

Punnett square–a simple graphic approach to manipulate and visualize genotypes.

quality of life–having a good environment that meets all of an individual's basic needs.

recessive–traits that cannot be observed; a gene whose trait is masked by a dominant trait.

recipient cell–the cell that is receiving the "new" DNA.

recombinant DNA process–technique in which genetic material is moved from one chromosome to another.

recycling–reusing the materials in a product or the product itself.

refuge–a place that offers protection; an area of a field planted with the same crop as the field, but where no insecticide or other conditions that threaten the well-being of an organism are used.

regeneration–the regrowing of a lost body appendage.

regulated organism–one that has been genetically engineered or that is a vector agent for a plant pest or contains plant pest components.

regulation–a law, principle, or other directive that apply to a specific activity or area of endeavor.

religion–a system of beliefs and practices that an individual chooses to follow; often concerned with the origin of the universe.

repair–the replacement of cells in a living organism.

replication–running an experiment more than once.

reproduction–the process by which offspring are produced.

research–the use of systematic methods to answer questions based on a problem that needs to be solved.

research log–a written record of the processes, actions, and observations made in a research project.

research proposal–a document that describes the first three steps before any experimentation.

research protocol–how the research or experimentation is designed and will be carried out.

research report–a complete written description of the activities during each step of the scientific process.

respiration–an oxidation process in the living cells of an organism that releases the chemical energy of foods.

restriction endonuclease–an enzyme produced by bacteria that is used in recombinant DNA.

ribosome–an extremely small, round organelle where protein synthesis takes place.

risk–the chance of a loss or injury.

safety–using practices that prevent injury or loss.

sanitization–reducing the level of microbe contamination to an acceptable level.

science–knowledge about the world we live in that involves making observations, measurements, and experiments to set up general truths or laws.

scientific method–a series of steps used in solving problems to insure that conclusions about scientific issues are drawn from complete and accurate information.

secondary innovation–an innovation that builds on a primary innovation.

secretion–the process by which the glands of organisms produce and release substances.

selection–choosing parents of the next generation.

selective social impact–one that is beneficial for some people and not for others.

semen–the fluid produced by the male reproductive organs that contains sperm.

sensation–awareness by an organism of its environment.

sexual reproduction–the union of a sperm and an egg to ultimately produce a new individual.

shelf life–the length of time that food or other perishable products can be stored.

social impact–the influence of biotechnology on the status of human beings.

social science–study of the interactions of people as individuals and groups.

solid waste–garbage, refuse, sludge, and other discarded material.

solid waste management–collection and disposal of solid waste materials.

somatic cell gene therapy–an approach in gene manipulation used with cells in the body not associated with sperm or ova production.

somatic cells–the cells that compose tissues, organs, and parts of the body; do not produce sperm or ova.

species diversity–variation in the genetic makeup within and between species.

sperm–the sex cells from the male parent.

spoilage–a condition that makes food unsafe to eat.

spore–a thick-walled dormant form of a fungus.

statistics–analysis of numerical data to provide useful information.

stereomicroscope–a magnifying instrument that provides an image that is three dimensional.

sterilant–an agent that kills microbes.

sterilization–the destruction of all microbes.

strain–a group within a species that has some distinct characteristics.

subjective–research based on general observations or opinions.

substrate–a substance on which a specific enzyme reacts.

symbiosis–the close relationship between two different types of organisms for their mutual benefit.

synthetic biology–creating lifelike characteristics through the use of chemicals.

table–a list organized into rows and columns in a research report that gives numerical information about the experimentation.

taxol–a substance made from the bark of a Pacific yew tree that holds promise in treating human cancer.

taxonomy–the classification of living things.

technical journal–a publication written by specialists in a given field for their peers.

technical writing–a direct and factual report of information gathered by the scientific method.

thallus–an aggregate of algal filaments in plantlike form.

therapeutant–a product used to maintain health or treat disease.

thesis–the final project for a candidate for a master's degree.

tissue–a group of cells that is alike in structure and activity.

tissue engineering–the process of causing new, healthy tissues to grow.

totipotent–the ability of an organism to grow from a single cell.

transcription–the process of making RNA that is complimentary to a strand of DNA.

transformation–the process of introducing free DNA into bacteria.

transformed cell–a cell with new DNA; the basis for an organism being transgenic or genetically altered.

transgene integration–an expression by the animal of the desired trait.

transgenic–an organism whose genetic material has been altered.

translation–the process of a cell beginning to build a protein (amino acid).

USDA–the U.S. Department of Agriculture.

vaccine–a substance introduced into the body to prevent disease.

vacuole–a large organelle without a specific shape that functions for storage of any possible particles.

variable–the "thing" in an experiment that is changed.

variability–differences of characteristics between species.

vector–a carrier for DNA during the recombinant DNA process.

vesicle–a tiny, rounded structure with cell-like traits.

veterinary medicine–a branch of medicine that deals with the health of animals other than humans.

virus–a very small particle composed of a core of nucleic acid surrounded by a protein coat.

visualization–an approach used by individuals in giving public speeches to help them more effectively prepare the report by imagining themselves speaking in as much detail as possible.

visual report–a presentation that relies primarily on photographs, real things, and other devices to communicate information.

wastewater–used water often containing suspended or dissolved materials.

weed–a plant that is growing where it is not wanted.

xenograft–the grafting an organ or a tissue from one individual in a species to another individual in a different species.

yeast–a group of single-celled fungi of the phylum Ascomycota that reproduce by fission or budding.

yogurt–a product made by fermentation from the milk of various animals, including cows, goats, and buffaloes.

zygote–the one-cell embryo formed by the union of a sperm and ovum.

BIBLIOGRAPHY

Alcamo, I.E. and Elson, L.M. *The Microbiology Coloring Book*. New York: Harper Collins College Publishing, 1996.

Alexander, M. *Biodegradation and Bioremediation*, 2nd Edition. San Diego, California: Academic Press, 1999.

Bains, W. *Biotechnology from A to Z*. Oxford, Great Britain: Oxford University Press, 1998.

Baker, M. and Mikesell, R.E. *Animal Science Biology & Technology*. Danville, Illinois: Interstate Publishers, Inc., 1996.

Barnum, S. *Biotechnology: An Introduction*. Belmont, CA: Wadsworth Publishing Company, 1998.

Biondo, R.J. and Lee, J.S. *Introduction to Plant and Soil Science and Technology*. Danville, Illinois: Interstate Publishers, Inc., 1997.

Biondo, R.J. and Noland, D.A. *Floriculture: From Greenhouse Production to Floral Design*. Danville, Illinois: Interstate Publishers, Inc., 2000.

Black, J.G. *Microbiology*, 2nd. Englewood Cliffs, New Jersey: Prentice-Hall, Inc., 1993.

Bolsover, S.R., Hyams, J.S., Jones, S., Shephard, E.A., and White, H.A. *From Genes to Cells*. New York: John Wiley & Sons, Inc., 1997.

Chrispeels, M. J. and Sadava, D. E. *Plants, Genes, and Agriculture*. Boston: Jones and Bartlett Publishers, 1994.

Collin, C.H., Lyne, P. and Grance, J.M. *Microbiological Methods*, 6th Edition. Oxford, United Kingdom: Butterworth and Co. Ltd. 1989.

Darling, D.C. and Morgan, S.J. *Animal Cells Culture and Media*. New York: John Wiley & Sons, Inc., 1994.

Gartland, K.M.A. and Davey, M.R. *Agrobacterium Protocols*. Totowa, New Jersey: Humana Press, 1995.

Glick, B.R. and Pasternak, J.J. *Molecular Biotechnology*, 2nd Edition. Washington, DC: American Society for Microbiology, 1998.

Grace, E.S. *Biotechnology Unzipped*. Washington, DC: Joseph Henry Press, 1997.

Hanlin, R.T. and Ulloa, M. *Atlas of Introductory Mycology*. Winston-Salem, North Carolina: N.C. Hunter Textbooks, Inc., 1988.

Harrington, R.B. *Animal Breeding: An Introduction*. Danville, Illinois: Interstate Publishers, Inc., 1995.

Herman, H.A., Mitchell, J.R., and Doak, G.A. *The Artificial Insemination and Embryo Transfer of Dairy and Beef Cattle*, 8th Edition. Danville, Illinois: Interstate Publishers, Inc., 1994.

Jackson, N.S., Greer, W.J., and Baker, J.K. *Animal Health*, 3rd Edition. Danville, Illinois: Interstate Publishers, Inc., 2000.

Janick, J. and Simon, J.E., ed. *New Crops*. New York: John Wiley & Sons, Inc., 1993.

Jones, G., Robertson, A., Forbes, J. and Hollier, G. *The Harper Collins Dictionary of Environmental Science*. New York: Harper Collins Publishers, 1992.

Krenzer, H. and Massey, A. *Recombinant DNA and Biotechnology*. Washington, DC: American Society for Microbiology, 1996.

Krimsky, S. and Wrubel, R. *Agricultural Biotechnology and the Environment*. Urbana, Illinois: University of Illinois Press, 1996.

Krimsley, V.S. *Introductory Chemistry*, 2nd Edition. Pacific Grove, California: Brooks/Cole Publishing Company, 1995.

Kyte, L. *Plants from Test Tubes*. Portland, Oregon: Timber Press, 1987.

Lee, J.S. *Natural Resources and Environmental Technology*. Danville, Illinois: Interstate Publishers, Inc., 2000.

Lee, J.S., Hutter, J., Rudd, R., Westrom, L., Mohr, C.E., and Pollock, J. *Introduction to Livestock and Companion Animals*, 2nd Edition. Danville, Illinois: Interstate Publishers, Inc., 2000.

McGee, G. *The Perfect Baby*. Lanham, Maryland: Rowman & Littlefield Publishers, Inc., 1997.

Micklos, D.A. and Freyer, G.A. *DNA Science*. Burlington, North Carolina: Carolina Biological Supply Company, 1990.

Morgan, E.M., Chelewski, R.E., Lee, J.S., and Wilson E. AgriScience Explorations, 2nd Edition. Danville, Illinois: Interstate Publishers, Inc., 2000.

Munn, M. and Ehrman, P. *High School Human Genome Program: Classroom Protocols for DNA Sequencing and Ethics*. Seattle: University of Washington, n.d.

Osborne, E.W. *Biological Science Applications in Agriculture*. Danville, Illinois: Interstate Publishers, Inc., 1994.

Ptashne, M. *A Genetic Switch*, 2nd Edition. Cambridge, Massachusetts: Blackwell Scientific Publications, Inc., 1992.

Rainis, K.G. and Nassis, G. *Biotechnology Projects for Young Scientists*. New York: Franklin Watts, 1998.

Sanderson, J. B. *Biological Microtechnique*. Oxford, United Kingdom: BIOS Scientific Publishers Ltd., 1994.

Saxon, J.H., Jr. *Physics: An Incremental Development*. Norman, Oklahoma: Saxon Publishers, Inc., 1993.

Schroeder, C.B., Seagle, E.D., Felton, L.M., Ruter, J.M., Kelley, W.T., and Krewer, G. *Introduction to Horticulture*, 3rd Edition. Danville, Illinois: Interstate Publishers, Inc., 2000.

Smith, C.W. *Crop Production: Evolution, History, and Technology*. New York: John Wiley & Sons, Inc., 1995.

Smith, J. E. *Biotechnology*, 3rd Edition. Cambridge, United Kingdom: Cambridge University Press, 1997.

Snyder, C.H. *The Extraordinary Chemistry of Ordinary Things*. New York: John Wiley & Sons, Inc., 1992.

Steinberg, M. L. and Cosloy, E. H. *The Facts on File Dictionary of Biotechnology and Genetic Engineering*. New York: Facts on File, Inc., 1994.

Terborgh, J. *Diversity and the Tropical Rain Forest*. New York: W. H. Freeman and Company, 1992.

Turner, P.C., McLennan, A.G., Bates, A.D., and White, M.R.H. *Notes in Molecular Biology*. Oxford, United Kingdom: BIOS Scientific Publishers Ltd., 1997.

Wells, D. *Biotechnology: Inventors and Inventions*. New York: Benchmark Books, 1996.

Winter, P. C., Hickey, G. I., and Fletcher, H. L. *Notes in Genetics*. Oxford, United Kingdom: BIOS Scientific Publishers Ltd., 1998.

Witt, S.C. *Biotechnology, Microbes and the Environment*. San Francisco, California: Center for Science Information, 1990.

_____. *The Handy Science Answer Book*, 2nd Edition. Detroit, Michigan: Visible Ink Press, 1997.

_____. *Short Protocols in Molecular Biology*, 2nd Edition. New York: John Wiley & Sons, Inc., 1992.

_____. *HIP Biology*. Tucson, Arizona: Center for Image Processing, 1996.

INDEX

A

abiotic factors, 379
aerobe, 245
aflatoxin, 242
agar, 256-257
agriculture, 224, 259, 375-376
agriscience, 3
***Agrobacterium* spp., 65**
  transformation, 233-234, 269
algae, 249-250
allele, 200
anaerobe, 245
ancient biotechnology, 18-23
anesthesia, 171
animal
  laboratory, 286-287
  research models, 285-286
  well-being, 69-70
antibiotics, 24-25
antibody, 47
APHIS, 405-407
apron, 60
artificial insemination, 294-296
aseptic technique, 155-173
Asilomar Conferences, 70-71
atom, 131
autoclave, 165

B

***Bacillus thuringinesis*, 42**
bacteria, 162, 231
  culture, 256-258
  food, 339
  identification, 253-256
ballistic gene transfer, 234, 270
bibliography, 112-113
binomial name, 135-136
biodegradation, 260, 354-355
bioethics, 391
biogeochemical processes, 362-364
biological seeding, 359
biopharming, 7
bioremediation, 353-356
biotechnology,
  animal, 9, 43-45, 283-305
  definition, 3
  environmental, 8, 347-366, 380-384
  food, 9
  history, 17-33
  human, 307-324
  molecular, 5
  multidisplinary, 4
  organismic, 4
  plant, 9, 41-43, 263-281
  stages of development, 19
Biotechnology Industrial Organization (BIO), 407-408
biotic factors, 379
biotransformation, 258
bleach, 164, 169
blood
  banks, 322
  transfusions, 322
Borlaug, Norman E., 30
breeding
  animal, 292-297
  plant, 256-268
***Brucella abortus*, 239**
bST, 11, 300, 378-379

C

classical biotechnology, 19, 23-25
career profiles,
  A.I. technician
  biochemist, 38

bioechicist, 395
environmental biotechnologist, 8
environmental chemist, 374
food scientist, 334
genetic counselor, 203
genetic engineer, 228
human biologist, 314
laboratory technician, 160
marketing specialist, 190
microbiologist, 248
molecular mycologist, 360
plant breeder, 276
plant geneticist, 81
plant physiologist, 403
plant technician, 130
poultry producer, 24
research veterinarian, 62
technical writer, 100

carpal tunnel syndrome, 323
carrion, 354
Carver, George Washington, 331
catalysis, 341
cells, 145-146
donor, 231
germ, 389
processes, 183-184
recipient, 231
reproduction, 189-190
somatic, 389
structure, 175-183
centrifuge, 55
cheese, 22
chemistry, 128, 130-133
cholesterol, 344
chromosomes
DNA construct, 230-231
chi-square, 207
circulation, 140-141
classification, scientific, 133-136
cloning, 84, 396
vectors, 230-231
composting, 245, 364-366
confidential business information (CBI), 410
Congress, U.S., 3, 403
conjugation, 232
containment, 70-71
Council of Biotechnology Centers (CBC), 408
Crick, Francis, 29-30
crops, 41-42, 265-268, 396
residue, 350-351
cryopreservation, 404
CT scanner, 321
cultivar, 265

D

data, 86
DDT, 259
defense, 114
development, 32
diagnostics, 323
dialysis machine, 322
diffusion, 183-184
disease, 38
disinfection, 159
display board, 117-119
diversity, 244
species, 352-353
DNA, 148-149, 209-218
human, 314-317
isolation, 211-213
microbe, 243
profiling, 218, 268
recombinant, 223, 228-233
domestication, 20-21
drench shower, 66
dress, appropriate, 168

E

ear plugs, 62-63
ectotherm, 71
electroporation, 231
elimination, 144
ELSI, 393-395
embryo transfer, 298-299
endonucleases, 228-229
endotherm, 71
energy, 133
transfer, 351-352
environment, 52-53, 260, 369-384
quality, 349
Environmental Protection Agency (EPA), 404-405
enzymes, 138, 151, 338, 341-344
estrous cycle, 291-292, 294-295
ethanol alcohol, 164
ethics, 391
eukaryote, 178, 243
euthanasia, 69
experimentation, 79, 89-90
eye wash, 66

F

Federal Register, 407
feral animals, 23
fermentation, 22-24
fiber, 40
field plot, 31
flaming, 167-168
Fleming, Alexander, 29, 244
food, 21-22 40, 139-140, 279, 376
biology, 332-333
biotechnology, 327-344
borne illness, 340-341
chemistry, 330-331
consumption, 333-334
physics, 331-332
preservation, 22
processing, 335
quality, 335-3337
science, 329
spoilage, 335-336
Food and Drug Administration (FDA), 404-405
Food Guide Pyramid, 139
forensic science, 208
Franklin, Rosalind Elsie, 29
free enterprise, 409
fungi, 162-163, 250-251
fungicide, 278

G

gamete, 203-204

gel electrophoresis, 216-218, 317-320
gender preselection, 297
gene, 200
dominant, 201
expression, 200-201
marker, 232
promoter, 230
recessive, 201
transfer, 268-272
gene therapy, 49, 320-321
genetic engineering, 221-236, 396-397
animal, 300-304
definition, 6
plants, 268-272
genetically modified organism (GMO), 224, 369, 402
genetics, 25, 197-218
Mendelian, 199-208
molecular, 208-211
genotype, 202
geothermal pools, 244
gestation, 290-291
gloves, 63-64, 170
goggles, 61-62
gram staining, 254-256
growth, 137-139
guayule, 5
Gutenberg, Johannes, 99

H

hazard, 57-59
microorganism, 172
hazardous waste, 67-68
health, 38, 225-226, 376
herbicide, 278-279
heredity, 25
homozygous alleles, 201
hormones, 11, 225-226, 233, 299-300, 304
human
biotechnology, 309-324
development, 311-313
needs, 10-13
waste, 347
Human Genome Project (HGP), 314-320, 324, 394
hybridization, 199-200, 267-268
hyperaccumulators, 384
hypothesis, 84-86

I

incineration, 358
industrial biotechnology, 70
innovation, 371-372
insecticide, 277
insulin, 227
invasive procedure, 170-171
***in vitro*, 232**
fertilization, 396-297
isolation, DNA, 211-213
issue, 389-390

J

Janssen, Zacharias, 27
journals, 113

K

King, Mary-Claire, 30

L

lactation, 291
laminar flow hood, 165-166
landfill, 357-358
Leeuwenhoek, Anton Van, 27, 241
life expectancy, 39
life processes, 136-145
ligation, 231-232
lipids, 151-152
literature review, 82
livestock, 44-45
log, research, 102-103

M

material safety data sheet (MSDS), 68-69
materials, 50-53
matter, 131
media, plant, 276-277
medicine,
biotechnology, 46-59
genetic engineering, 223, 226
human, 6
veterinary, 7
meiosis, 192-194, 203
Mendel, Gregor Johann, 27-28, 199-200
microbiology, 241-243
food, 336-341
microdissection, 324
microinjection, 233-234, 302
microorganisms, 155, 157, 161-164
in bioremediation, 355
in biotechnology, 233, 241-260
microscope, 185-189
electron, 188
light, 186-187
stereoscopic, 187-188
Miescher, Johan Friedrich, 28
mildew, 241
milk fever, 6
mitosis, 190-192
movement, 143-144
mutation, 274

N

nuclear magnetic resonance, 19
nutraceutical, 43

O

Office of Technology Assessment, 3, 125
organ systems, 147-148
organ transplant, 49

ovum, 288-290

P

parturition, 291
patent, 372
peanuts, 334
penicillin, 242
permit, 406-407
pharmaceuticals, 48-49
  genetic engineering, 226-227
  production, 280
phenotype, 202-204
photosynthesis, 184
plant breeding, 265-268
plasmid, 231
polymerase chain reaction, 213-216
ppe, 59-64
primates, 287
prokaryote, 177-178, 246
prosthetics, 322
protein, 149-151
  synthesis, 210
protocol, 64-66, 87-88, 168
prototype, 32
protozoa, 247-249
pST, 300
puberty, 290
Punnett square, 204-207

Q

quality of life, 37, 378-379

R

recombinant DNA, 26
recycling, 361
regulated organism, 407
regulations, 401-402
religion, 392
replication, 88
reports, 93-94, 97-123
  oral, 100-101, 114-117
  visual, 101, 117-120
  written, 99-100, 103-113
reproduction, 288
  development, 290-292
research, 31, 75-95
  applied, 78
  basic, 77-78
  log, 102-103
respiration, 142
risk, 57
Ruska, Ernst, 28-29

S

safety, 12, 55-73
Saguaro, 137
sanitation, 159
science, 127-130
scientific method, 79-80
secretion, 142-143
selection, 201, 266-267
sensation, 143
shelf life, 336-337
slide, microscope, 181, 256-259
social impact, 377-378
sperm, 194, 203, 288-289, 295-296
  parts, 297
  sorting, 297-298, 387
statistics, 90-92
  descriptive, 91
  inferential, 91-92
sterilization, 158-159
stoma, 145
streaking, 257
Sutton, Thomas Hunt, 28
synthetic biology, 13-14

T

taxol, 280
taxonomy, 133-136
technology, 373-377
thesis, 100
tissue, 146-147
tissue culture, 234-235, 273-277
tissue engineering, 50
tomato, 13
totipotent, 233
transcription, 210-211
transfer area, 169
transformation, 231-235
translation, 211
transgene integration, 236
transgenic, 6
  animals, 301-304
  organisms, 227-232

V

vaccine, 47-48, 321
variability, genetic, 25
vectors, 230-231
vesicle, 14
virus, 163-164
visualization, 117

W

waste
  disposal, 383
  hazardous, 360
  solid, 356-359
  water, 360-364
Watson, James, 29-30
Wilmut, Ian, 30
writing, 103-113
wolf relocation, 409

Y

yeast, 22, 163, 175, 258, 338

Z

zygote, 303